REMEDIATION ENGINEERING

Design Concepts

Suthan S. Suthersan

LEWIS PUBLISHERS

Boca Raton   New York   London   Tokyo

Library of Congress Cataloging-in-Publication Data

Suthersan, Suthan. S.
   Remediation engineering : design concepts / Suthan S. Suthersan.
       p.   cm.
   Includes bibliographical references and index.
   ISBN 1-56670-137-6
   1. Soil remediation. 2. Groundwater--Purification. I. Title.
TD878.S88 1996
628.5′2—dc20
                                                     96-33432
                                                         CIP

© 1997 by CRC Press, Inc.
Lewis Publishers is an imprint of CRC Press

No claim to original U.S. Government works
International Standard Book Number 1-56670-137-6
Library of Congress Card Number 96-33432
Printed in the United States of America   2 3 4 5 6 7 8 9 0
Printed on acid-free paper

Foreword

Remediation of contaminated soil and groundwater at hazardous waste sites has been in full swing for about 10 years. During this relatively short period, the environmental industry has seen tremendous changes in both strategy and technique. In the beginning, it was thought that success could only be achieved if you moved and treated as much soil and water as possible, often at great expense and additional risk to the environment. Little, if any, thought was given to the simple processes of nature that could be harnessed to accomplish the job at a fraction of the cost and no added risk. Who would have thought that feeding molasses to naturally occurring microorganisms could result in the destruction of technology-defying chlorinated hydrocarbons? Yet, the application and manipulation of such processes, achieved through a highly specialized blend of science and engineering, represents the future of remediation technology at hazardous waste sites around the world.

This book is truly a "first of its kind." It focuses on the key innovative technologies that are beginning to dominate the remediation field, most of them *in-situ*, and most of them relying on natural processes. Each method is described in detail, and the reader is guided through its evaluation design and implementation. The material, amply illustrated, qualifies both as a textbook for the engineer or scientist and a guide for the technician already in the field applying the remedy.

Only someone with the author's professional background and commitment to seeking creative solutions to tough problems could have put together this book. For the past 15 years, Suthan Suthersan has been a pioneer in the development and application of new methods for the solution of soil and groundwater contamination problems. Starting out in the hydrocarbon field, where regulatory flexibility allowed for on-site testing of new approaches, Suthan was one of the first to apply *in-situ* methods such as air sparging, which have become some of the most widely used technologies today. He holds several patents and continues to develop new ideas that should increase the efficiency and reduce costs of soil and groundwater cleanup.

I enjoyed this book. It takes the mystery out of terms like *hydro-fracturing* and *phytore-mediation*. It is a valuable contribution to the field both here and abroad.

David W. Miller
Plainview, New York

The Author

Suthan S. Suthersan, PhD, PE, is Vice President and Director of Remediation Services at Geraghty & Miller, Inc., an international environmental and infrastructure services company. He is considered one of the country's leading and pioneering experts in the field of environmental remediation. His primary responsibilities include development and application of innovative *in situ* remediation technologies, implementation of technology transfer programs, and provision of technical oversight on projects across the entire country and internationally. His technology development efforts have been rewarded with four patents awarded and many pending.

Dr. Suthersan has a PhD in Environmental Engineering from the University of Toronto, a Masters in Environmental Engineering from the Asian Institute of Technology (AIT) and a BS in Civil Engineering from the University of Sri Lanka. He is also a registered Professional Engineer in several states. In addition to his consulting experience, Dr. Suthersan has taught courses at Northeastern University, University of Wisconsin-Madison, and University of Toronto.

Dr. Suthersan's primary strength lies in developing the most cost-effective site-specific solutions utilizing cutting edge techniques. He has developed a national reputation for persuasion of the regulatory community to accept the most innovative remediation techniques by unraveling the concepts associated with any new application. He has also co-authored another book and has published many papers in various aspects of environmental remediation.

Dr. Suthersan is a member of the American Society of Civil Engineers, Water Environment Federation, and various committees in many professional societies.

Preface

Remediation engineering as a discipline has evolved only in the last few years and continues to evolve even today. There is an increased level of awareness regarding the efficiencies and limitations of the applicable technologies. Pump and treat systems, the primary remediation technique during the early days, have been found to be ineffective due to better understanding of contaminant fate and transport mechanisms. Many new and innovative *in situ* technologies have been introduced recently to develop faster and more cost-effective solutions for the responsible parties. As a result, engineers and scientists practicing remediation engineering have to continuously learn the development and application of new concepts and techniques.

This book is intended as a text for engineers and nonengineers to understand the applications of various conventional and innovative remediation technologies. In this industry, a technology considered to be innovative will become "conventional" in a much shorter time frame than in many other industries as a result of the need and urgency to develop cost-effective solutions. The individual chapters in this book attempt to provide a basic understanding of the various technologies and a detailed discussion of the design concepts. Description of the detailed design steps of individual technologies would require more than a single text: each chapter could easily be expanded into an entire book. To that end, I have attempted to present a level of discussion that provides a working knowledge of the basic principles, applications, advantages, and limitations of a wide spectrum of remediation technologies. I have also provided case studies to describe the application of some of the technologies.

Even as this manuscript goes to press, some new technologies may be in the process of experimentation and development to be ready for wider applications. This book covers

- Soil Vapor Extraction (currently considered to be a conventional technology)
- *In Situ* Air Sparging (in transition from innovative to conventional)
- *In Situ* Bioremediation (conventional and innovative applications)
- Vacuum-Enhanced Recovery (in transition from innovative to conventional)
- *In Situ* Reactive Walls (Innovative at present)
- *In Situ* Reactive Zones (Innovative at present)
- Hydraulic and Pneumatic Fracturing (Innovative at present)
- Phytoremediation (Innovative at present)
- Pump and Treat Systems (Conventional)
- Stabilization Technologies (Conventional)

In addition, a chapter on basic principles of contaminant characteristics and partitioning has been provided.

I have written the book to reach a wide audience: remediation design engineers, scientists, regulatory specialists, graduate students in environmental engineering, and people from the industry who have general responsibility for site cleanups. I have tried to provide detailed information on basic principles in various chapters for interested readers. Readers who are not interested in basic principles can skip these passages and get the general knowledge that they need.

Acknowledgments

First of all, I want to thank my assistant, Amy Weinert, for all the typing, which at times seemed unending, and for her patience in keeping me organized during the writing of this book. Arul Ayyaswami and David Share deserve special mention for the help provided in writing Chapter 6 (Vacuum-Enhanced Recovery) and Appendix B (Flow Devices). I am greatly indebted to Dave Schafer for being my guru and guide on various theoretical aspects of subsurface fluid flow. Special thanks go to Pete Palmer, Mike Maierle, Tom Crossman, Eileen Hazard, Matt Mulhall, Arul Ayyaswami, Don Kidd, Angie Gershman, and Doug Newton for providing valuable insight by graciously reviewing the various chapters. Angie Gershman's help with the equations is well appreciated.

As the saying goes, "a figure speaks a thousand words," I hope the figures in this book enhance understanding and easy reading. I would like to thank Bill Cicio, Eileen Schumacher, Ron Padula, Rufus Faulk, and Steve Gozner for drafting all the figures in the book. Bill Cicio deserves special mention for his input and creative help with the excellent design of the cover on this book.

I have to thank the management of my employer, Geraghty & Miller, Inc., and especially Fred Troise for all the support given to me in writing this book. The opportunities I had to develop and experiment with various innovative technologies at Geraghty & Miller were immense. I have a special debt to those engineers and project managers who helped me to implement many innovative and challenging remediation projects.

To my wife Sumathy, daughter Shauna-Anjali and son Nealon Aaron, whose love is manifested in so many ways, but particularly in their unstinting patience, understanding, encouragement, and support.

To my loving parents for their encouragement and motivation.

Table of Contents

1

REMEDIATION ENGINEERING

1.1 INTRODUCTION

Remediation engineering as a discipline has evolved only in the last few years. Some of the traditional engineering disciplines such as civil engineering, mechanical engineering, and electrical engineering have been taught and practiced in an organized fashion during the last few centuries. There is a wealth of knowledge available for the practicing engineers in these disciplines. Some of the younger subdisciplines such as structural, geotechnical, transportation, and water resources engineering within the major area of civil engineering have also taken firmer roots within the organized world of engineering. These subdisciplines have benefitted from an enormous amount of research and developmental efforts in academic institutions in the U.S. and around the world. Environmental engineering, probably one of the youngest subdisciplines in engineering, is still evolving with respect to society's expectations and demands for a cleaner environment. Remediation engineering is an even younger subdiscipline of environmental engineering.

Beginning in the late 1960s and gathering momentum ever since, the whole picture concerning our environment has changed. Prior to the 1960s, society's demands on environmental engineers were limited to the provision of clean drinking water and disposal of domestic wastes. Hence, this discipline was aptly called public health or sanitary engineering. With time, environmental engineers started to focus on activities related to solid waste, water quality, and air quality.

The application of highly sensitive analytical techniques to environmental analysis has provided society with disturbing information. As a result, significant changes in requirements for environmental protection occurred in the late 1960s and early 1970s. The late 1970s and early 1980s have seen an emerging scientific and public awareness of the potential for detrimental health effects due to the accumulation of hazardous compounds in the various environmental media such as soil, groundwater, surface water, and air. The occurrence and fate of trace levels of organic and inorganic compounds in the environment and the passing of new regulations to address these concerns spawned the need for a new group of specialists known as "remediation engineers."

What is remediation engineering? It could be simply defined as the next phase in the evolution of environmental engineering. More precisely, it could be defined as *the development and implementation of strategies to clean up (remediate) the environment by removing the hazardous contamination disposed in properties since the beginning of the industrial revolution.*

Scientists and engineers practicing remediation engineering have to learn the nuances of investigative techniques, data collection, and treatment technologies. This education includes a new understanding of the physical and chemical behavior of the contaminants, the geologic and hydrogeologic impacts on the fate and transport of these contaminants, the human and environmental risks associated with contamination, and the selection of appropriate technologies to provide maximum mass transfer and destruction of the contaminants. Hence, remediation engineering is a multidisciplinary field in the truest sense, requiring knowledge of civil, chemical, mechanical, and electrical engineering, geology, hydrogeology, chemistry, physics, microbiology, biology, toxicology, geochemistry, statistics, data management, etc.

1.2   PRACTICE OF REMEDIATION ENGINEERING

In the traditional world of engineering practice, engineers are familiar with the *design–bid–build* process. In this process, the responsibilities of the owner, the engineer providing design services, and the contractor providing the construction services are well defined and understood. The design and construction of a dam, a multi-storied building, or a highway has to take into consideration that these structures have to be built to last a lifetime. The structural strength and stability of these structures are very important in addition to meeting the performance objectives.

When we design remediation systems, these systems are designed as "temporary" systems expected to last only until the cleanup standards are achieved. The primary design objective is to meet the treatment process efficiencies that would meet the cleanup standards, albeit at trace quantity levels, and thus ensure that the contaminated sites are restored to meet the minimum requirements of the public. Another major objective is to provide adequate health and safety for the workers who install, operate, and maintain these systems. Hence, there are strong advocates among the practitioners of remediation engineering for the *design–build* process. This process is also referred to as the *turnkey* process. Proponents of this process value it as a method of delivering completed projects faster and cheaper than the traditional design–bid–build process.

The practice of remediation engineering itself has evolved from the mid 1970s and continues to evolve even today. There is an increased level of awareness regarding the efficiencies and limitations of the applicable technologies. As a result, there is an increased level of effort to experiment with and develop new and innovative technologies. As the industry rapidly matures, a greater emphasis is placed on providing on-site remedies involving *in situ* technologies. The use of technology, risk assessment, and statistical concepts as a combination to get the "best answer" for site cleanup is becoming widespread.

The last decade has seen a significant evolution of remediation technologies from the early containment techniques to today's very aggressive site closure techniques (Appendix A). Pump and treat systems, the primary remediation technique during the early days, have been found to be ineffective due to better understanding of contaminant fate and transport mechanisms. Many new and innovative *in situ* technologies have been introduced to develop faster and more cost-effective solutions for the responsible parties. Incorporation of natural, intrinsic transformations of the contaminants in the subsurface is taking firmer root today.

Development and implementation of innovative technologies requires a significant level of interaction between the design team and the construction team due to many truly unanswered questions related to any new technology. The conventional design–bid–build relationship does not promote value-added integration and experimentation, which is required in remediation engineering today to fine-tune the innovative concepts into conventional techniques.

Experience and empirical knowledge with various technologies is still the key in designing remediation systems incorporating on-site and *in situ* technologies. Hence the control exercised by a single entity providing both consulting and contracting services will be mutually beneficial and crucial for the success of the remediation project.

2 CONTAMINANT CHARACTERISTICS AND PARTITIONING

2.1 INTRODUCTION

For many years, there was a general lack of concern for the environment and a widespread but unfounded assumption that the subsurface environment would adsorb or degrade almost unlimited amounts of chemical contaminants. Historically, prevailing popular views held that the passage of water through soil exerted a purifying effect and that wastes dumped into the ground somehow were cleansed from the system. As improved techniques of analytical chemistry revealed the extent of contamination in soil and groundwater, public concern about subsurface contamination greatly increased. Once transport mechanisms were understood, it became clear that contaminants introduced at or near the surface could find their way into underlying aquifers and to receptors that might be used for drinking or recreational purposes.

Two basic elements affecting the transport and fate of contaminants in the subsurface are properties of the subsurface materials or the subsurface environment and physicochemical and biological properties of the contaminants. Nonreactive (conservative) chemicals will move through the subsurface environment with the groundwater (hydrodynamic processes) and will not be affected by abiotic (nonbiological) or biotic (biological) processes that may be active in the subsurface. Conversely, contaminants that have the potential to be reactive (nonconservative) will not be affected during groundwater transport if the subsurface environment is not conducive to the reactions that affect the contaminant (e.g., a contaminant that is susceptible to aerobic degradation but is in an anaerobic subsurface environment). Thus, for interactions between the subsurface environment and the contaminant to occur, it is necessary that both the contaminant property and the subsurface environment be conducive to these interactions. The goal of this chapter is to provide an overview of subsurface and contaminant properties that may affect the behavior and partitioning of the contaminants in the subsurface.

The general categories of processes affecting subsurface behavior and partitioning of contaminants are hydrodynamic processes, abiotic processes, and biotic processes.[1] *Hydrodynamic processes* affect contaminant transport by impacting the flow of groundwater in the subsurface. Examples of hydrodynamic processes are advection, dispersion, and preferential flow. *Abiotic processes* affect contaminant transport by causing interactions between the contaminant and the stationary subsurface material (e.g., adsorption, volatilization, and ion exchange) or by affecting the form of the contaminant (e.g., hydrolysis, redox reactions). *Biotic processes* can affect contaminant transport by degrading the contaminant (e.g., organic contaminants) or by immobilizing the dissolved contaminant (e.g., dissolved heavy metals) or by utilizing the contaminant in the metabolic process (e.g., nutrients, and nitrate during denitrification). Examples of biotic processes are aerobic, anoxic, and anaerobic biodegradation.

The saturated zone is the region within which chemical pollution is generally of most concern, because the saturated zone is a source of drinking water. Before contaminants can reach the saturated zone they must first move through the unsaturated zone (vadose zone).[1] While vadose zone contamination itself is of concern, the behavior and partitioning of chemicals in the vadose zone are also of interest because they affect the transport of chemicals to the saturated zone.

The variety of contaminants that can be released to the subsurface and cause an adverse impact includes organic compounds, inorganic compounds, and elements. Each contaminant released either as an individual constituent or as a mixture has its own distinct set of physicochemical characteristics that govern its behavior in the environment. When releases of organic compounds take place, the contaminants may exist in the subsurface as four distinct phases: (1) mobile free product or nonaqueous phase liquid (NAPL), (2) adsorbed phase, (3) dissolved phase, and (4) vapor phase. The distribution of contaminants into these different phases, while a result of dynamic transport, is ultimately a function of their physical and chemical properties and the hydrogeologic and geochemical characteristics of the subsurface formation. Table 2.1 represents the estimated phase distribution of a 30,000-gal gasoline spill in a medium sand aquifer, with the depth to the water table being 15 ft.[2]

Table 2.1 Phase Distribution of a 30,000-Gallon Gasoline Spill

Phase	Contaminated spatial volume (yd³)	Percent of total	Contaminated gasoline volume (gal)	Percent of total
Free Product	7,100	1.0	18,500	62
Adsorbed (soil)	250,000	20.0	10,000	33
Dissolved (water)	960,000	79.0	333	1–5
Vapor	Not quantified	—	—	<1

2.2 CONTAMINANT CHARACTERISTICS

It is essential to have a fundamental understanding of the classes of contaminants and their properties to understand their behavior. Almost all the contaminants of concern can be divided into organic and inorganic constituents. By definition, compounds that contain organic carbon are considered organic and those that contain no organic carbon are inorganic.

2.2.1 Organic Contaminants

There are thousands of both naturally occurring and anthropogenic organic compounds. More than 1600 organic compounds have been identified to be present in both natural and polluted environments.[3] Those that are typically of interest to subsurface contamination are often associated with refined petroleum products, or are associated with combustion of fossil fuels. Also of environmental concern are chlorinated and nonchlorinated solvents and degreasers, and organic compounds used as raw materials in various manufacturing processes.

Historically, the term "NAPL" was coined during studies of a hazardous waste landfill in Niagara Falls, New York.[4] Separate phase "dense nonaqueous phase liquids" such as chlorinated solvents are known as DNAPLs. In contrast, separate phase "light nonaqueous phase liquids" such as petroleum products are known as LNAPLs.

Refined petroleum products are generally complex mixtures of a variety of organic compounds (predominantly hydrocarbons) with minor fractions of organic and inorganic additives that fall into a number of chemical classes. The mixtures are, for the most part, less dense than water (and, therefore, float on top of the groundwater table), sparingly soluble in water, relatively volatile (have high vapor pressures), and vary from product to product and manufacturer to manufacturer. For example, gasoline, kerosene, diesel fuel, and waste oil have different physical and chemical characteristics and, thus, behave differently in the environment. Not only are the products different, but the same refined product type (i.e., gasoline) varies as a function of the crude source from which it was manufactured, the facility in which it was refined, the geographic area in which it is to be used, and the individual manufacturer's formulation (i.e., composition and additives). Because of these differences, the behavior of refined petroleum products in the environment is variable, as is the partitioning pattern.

Polynuclear aromatic hydrocarbons (PAHs) are complex fused aromatic ring compounds that can be naturally occurring or of anthropogenic origin. They are generally high molecular weight, readily adsorbed, sparingly soluble, low volatility compounds that originate from incomplete combustion of fossil fuels, wood, and a variety of industrial processes (i.e., coal gasification, petroleum refining, coking). PAHs having fused ring numbers from two (naphthalene) to seven (coronnene) have been observed.

In order to enter into a discussion of chemical characteristics, it is useful to have a fundamental understanding of the organic chemistry and nomenclature of the compounds that comprise the contaminants of interest. The following is an overview of relevant aspects of organic chemistry.

Some organic compounds contain only hydrogen and carbon and are, therefore, known as hydrocarbons. On the basis of structure, hydrocarbons can be divided into two primary classes: aliphatic and aromatic. Aliphatic compounds can be further divided into subcategories which include alkanes, alkenes, alkynes, and other cyclic analogs.

Alkanes, which are also called paraffins, are hydrocarbons in which the carbon atoms are joined by single covalent bonds. Alkane molecules can be straight (*n*-alkanes), branched, or cyclic. They also can exhibit a significant amount of isomerism. Alkane nomenclature always has –ane as a base and a prefix that denotes the number of carbon atoms. As noted above, the names methane, ethane, propane, butane, and pentane are used to denote alkanes with one, two, three, four, and five carbon atoms, respectively. Except for the first four, the name is simply derived from the Greek (or Latin) prefix (i.e., *pent*ane for five, *hex*ane for six, *hept*ane for seven carbon atoms, and so on). The linear structure, however, is not a requirement for alkane compounds.

As an introduction to chemical nomenclature, one learns that it is useful to have names for certain groups of atoms that compose only part of a molecule and yet appear many times as a unit. These units are called functional units, or more frequently, functional groups. For example, CH_3Cl or *methyl* chloride and C_2H_5Br, *ethyl* bromide. These functional groups are named simply by dropping –ane from the name of the corresponding alkane and replacing it with –yl. These functional groups are known collectively as alkyl groups. Alkyl groups are further classified according to whether they are primary, secondary, or tertiary. An alkyl group is referred to as *primary* if the carbon at the point of attachment is bonded to only one other carbon, as *secondary* if bonded to two other carbons, and as *tertiary* if bonded to three other carbons.

The principal source of alkanes is petroleum and natural gas. Decay and millions of years of geological stresses have transformed the complicated organic compounds that once made up living plants or animals into a mixture of alkanes ranging in size from 1 carbon to 30 or 40 carbons.

Alkenes, also called olefins, differ from alkanes in that they contain less hydrogen, carbon for carbon, than the alkanes. The loss of hydrogen atoms causes the carbon atoms to share a pair of electrons to maintain the required octet of electrons around the carbon. This sharing of electrons is referred to as a double bond. It is this double bond that distinguishes alkenes from other chemical classes. The carbon–carbon double bond is the distinguishing feature of the alkene structure. Unlike alkanes, alkene nomenclature can be confusing. This confusion results from the existence of common and formal (International Union of Pure and Applied Chemistry — IUPAC) nomenclature. The common names use –ylene as a root preceded by a prefix that denotes the number of carbon atoms. IUPAC nomenclature uses –ene as a root and the same prefix conventions.

Alkenes are obtained in industrial quantities by the cracking of petroleum. The lower molecular weight alkenes can be obtained in pure form by fractional distillation and are thus available as feed stocks for a large number of more complex aliphatic compounds. Higher molecular weight alkenes, which cannot be separated from the complicated cracking, remain as integral components of gasoline.

The *alkynes* contain even fewer hydrogen atoms per carbon than the alkenes. The loss of additional hydrogen atoms causes the carbon atoms to share three pairs of electrons, which means that they share a triple bond. The carbon–carbon triple bond is the distinguishing feature of the alkyne structure. For higher alkynes, the nomenclature (IUPAC) is similar to that of alkenes, except that the ending –yne replaces –ene. The parent structure is the longest continuous chain that contains the triple bond, and the positions both of substituents and of the triple bond are indicated by numbers. The triple bond is given the number of the first triply bonded carbon encountered, starting from the end of the chain nearest the triple bond.

The interest in many oxygen-containing species is increasing due to the presence of various compounds in the subsurface environment. Alcohols, aldehydes, ketones, carboxylic acids, phthalates, and esters are some of the oxygen-containing compounds of environmental concern.

Other organic compounds such as nitrogen-containing amines and nitrosoamines and sulfur-containing alkyl thiols and mercaptans, to name a few, have also been detected in the subsurface environment.

Unlike the aliphatic compounds which have open chain and cyclic configurations and undergo addition and free-radical substitution, aromatic compounds are ring compounds that have a tendency to undergo heterocyclic substitution. Aromatic compounds include benzene and compounds that resemble benzene in chemical behavior. Aromatic properties are those properties of benzene that distinguish it from aliphatic hydrocarbons.

Structurally, aromatic compounds have a ring configuration. For many benzene derivatives, one simply prefixes the name of the substituent group to the word –benzene, for example, chlorobenzene, or nitrobenzene. Other derivatives have special names that may bear no resemblance to the name of the attached substituent group. For example, methylbenzene is always known as toluene, aminobenzene as aniline, hydroxybenzene as phenol, and so on.

If several groups are attached to the benzene ring, not only is the functional group named, but their relative positions are also identified. The three possible isomers of a di-substituted benzene are differentiated by the use of the names ortho–, meta–, and para–. If the two functional groups are different, and neither group gives a special name to the molecule, the two groups are simply named successively and precede –benzene, for example, chloronitrobenzene. If one of the two groups gives a special name to the molecule, then the compound is named as a derivative of that special compound as, for example, nitrotoluene, chlorophenol, etc.

If more than two groups are attached to the benzene ring, numbers are used to indicate their relative positions. If all of the groups are the same, each is given a number, the sequence being the one that is given the lowest combination of numbers; if the groups are different, then the last-named group is understood to be in position 1 and other numbers conform to that, as for example, in 3-bromo-5-chloronitrobenzene. If one of the groups that gives a special name is present, then the compound is named as having the special group in position 1; thus for 2,6-dinitrotoluene the methyl group is considered to be at the 1-position.

With this general overview as a foundation, the discussion will shift more to contaminant characteristics and behavior in the environment.

Gasoline is a complex mixture of relatively volatile hydrophobic hydrocarbons from a number of chemical classes and families. The bulk gasoline mixture and the compounds that comprise it have their own unique physicochemical characteristics. These characteristics determine the behavior of gasoline and its constituents in the environment. The hydrocarbons in gasoline include aliphatics (alkanes, cycloalkanes, alkenes), aromatics, and additives and are dominated by compounds with 4 to 12 carbon atoms.

Diesel fuel is composed primarily of straight-chain unbranched hydrocarbons. While this product class does contain some aromatic hydrocarbons (benzene, toluene, ethylbenzene, and xylenes), these are minor constituents and their percent composition can be used to distinguish them from gasoline and other refined petroleum products. Perhaps the most significant differences between gasoline and diesel fuel are chain length, carbon number, molecular weight, and vapor pressure. Diesel fuels contain higher percent composition of longer chain, heavier, less volatile hydrocarbons. In automotive diesel fuel (no. 2), the carbon number typically ranges from 8 to 27. Other diesel fuels ranging from no. 4 (railroad) to no. 6 (Bunker, C) have progressively longer chains, higher molecular weights, lower vapor pressures, and are less soluble in water.

Motor oil and waste oil are also hydrocarbon mixtures and are dominated by carbon chains with 20 or more carbon atoms. These typically have very high molecular weights and are less dense than water, relatively insoluble in water, and minimally mobile in the environment. Motor oil typically contains little or no mono-aromatic hydrocarbons. However, waste oil typically has a highly variable composition. It contains aromatic hydrocarbons as a result of engine blow-by and polynuclear aromatic hydrocarbons (PAH) as a result of exposure to high temperatures in the internal combustion engine. Waste oil can also contain a variety of other constituents including gasoline, solvents, and antifreeze, as a result of activities associated with collection.

Chlorinated solvents and degreasers are generally released to the environment in a more or less pure form as opposed to a complex mixture. The contaminants are typically chlorinated methanes, ethanes, and ethenes. They have been used extensively since the late 1940s and are typically minimally soluble in aqueous systems, have variable vapor pressures, and may be more dense than water, and thus sink through an aquifer (DNAPLs) and may accumulate on a confining unit. Although these compounds are typically released in a more or less pure form as opposed to complex mixtures, they often contain minor amounts of other constituents as a result of degradation and the manufacturing process impurities.

A very wide range of industries use chlorinated solvents in large quantities. The primary chlorinated solvents are: 1,2-dichloroethane (1,2-DCA), 1,1,1-trichloroethane (1,1,1,-TCA), carbon tetrachloride (CTET), methylene chloride (DCM), chloroform (TCM), tetrachloroethylene (PCE), trichloroethylene (TCE), chlorobenzene, and dichlorobenzene.

Many other organic compounds besides the chlorinated solvents are also classifiable as DNAPLs. These compounds include the chlorinated benzenes, the polychlorinated biphenyls (PCBs), some pesticides, coal tar, and creosote. Some of these other DNAPL chemicals are much less soluble than the chlorinated solvents.

Other compounds such as acetone, methyl ethyl ketone (MEK), and methyl iso-butyl ketone (MIBK) are also used as solvents and cause environmental concern when leaked into the subsurface. In contrast to chlorinated solvents, these compounds are extremely soluble in water and travel faster with the moving groundwater than most compounds discussed in this section. A simplified classification tree of the potential environmental contaminants is shown in Figure 2.1.

2.2.1.1   Lump Parameters

During investigations of subsurface contamination, various "lump parameters" are used to quantify the extent of contamination. These lump parameters include total petroleum hydrocarbons (TPH), total organic carbon (TOC), total dissolved solids (TDS), biological oxygen demand (BOD), and chemical oxygen demand (COD), to name a few.

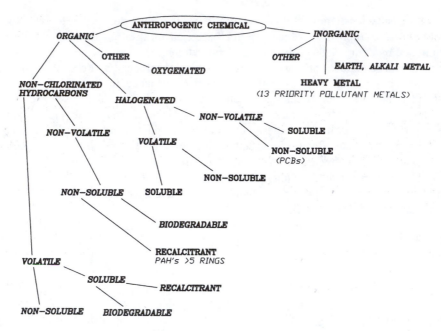

Figure 2.1 Contaminants classification tree.

Considerable attention should be paid to evaluate the significance and impact of the individual compounds that constitute the above "lump parameters."

2.2.1.1.1 Total Petroleum Hydrocarbons (TPH)

Hydrocarbons comprise a remarkably wide array of compounds, from the simplest hydrocarbon, methane, composed of one carbon atom surrounded by four hydrogen atoms (CH_4), to much longer, heavier molecules such as tetraoctane, composed of 40 carbon atoms and 82 hydrogen atoms ($C_{40} H_{82}$). Petroleum products such as gasoline and diesel fuel(s) differ by the range of hydrocarbons that comprise them. The major classes of petroleum hydrocarbons form a continuum from gasoline that is mainly short-chain hydrocarbons to middle-range fuels such as diesel fuel, fuel oil, jet fuels, and kerosene to lubricants and waxes, which are primarily long-chain heavier hydrocarbons. Properties of these hydrocarbon classes vary, and this is an important analytical concern. Important defining characteristics of these products include distillation range, boiling points, flash point, vapor pressure, and API gravity.

2.2.1.1.2 Total Organic Carbon (TOC)

All natural waters contain some carbonaceous material. Even rainwater has been shown to include very low levels of organic carbon. Some of this organic material is in the form of "truly dissolved" soluble compounds, whereas other fractions consist of colloidal or particulate organic matter. It is important to differentiate between the naturally occurring organic fraction in the samples as opposed to the organic compound levels present as a result of the contamination caused. Determination of the natural organic carbon content in uncontaminated areas will give an indication of the contribution of the contaminants toward the total organic carbon value in the contaminated areas.

A mass balance performed on the TOC measured with the target compounds analyzed may raise some questions regarding the presence of any unexpected compounds not targeted for analysis.

2.2.1.1.3   Total Dissolved Solids (TDS)

The amount of dissolved solids present in groundwater may give an indication of the presence of various dissolved compounds, specifically, inorganic species such as metallic ions, SO_4^{--}, Cl^-, and NO_3^-. TDS analysis can be used as an initial screening tool; however, any unexpected values compared to background levels should be supplemented with analyses for target species.

2.2.1.1.4   Biological Oxygen Demand (BOD)

The most widely used parameter of organic pollution applied to both wastewater and surface water is the biological oxygen demand (BOD). With the increased application of biological treatment techniques for remediation, this parameter is being used widely during the investigation of remediation projects. This determination involves the measurement of the dissolved oxygen used by microorganisms in the biochemical oxidation of organic compounds. A shortened test performed for 5 days measures BOD_5 and a longer-term test measures $BOD_{ultimate}$.

2.2.1.1.5   Chemical Oxygen Demand (COD)

The chemical oxidation test is used to measure the content of organic and inorganic compounds that consumes oxygen during chemical oxidation. For municipal wastewater, BOD is approximately about 65% of COD. This analysis is used to estimate the total oxygen demand of contaminated groundwater.

2.2.2   Metal Contaminants

Metals are natural constituents in soils. In addition to the natural constituents, metals enter the soil via beneficial agricultural additives such as lime, fertilizer, manure, herbicides, fungicides, and irrigation waters.

Metal-containing waste materials that may impact soil and groundwater pollution include municipal solid wastes, sewage sludge, storm water run-off, dredged materials, industrial by-products, wastes from mining and smelting operations, filter residues from wastewater treatment and atmospheric emission control, ashes and slags from burning of coal and oil, and sewage sludge and hazardous wastes from municipal refuse. Some of the primary industrial activities that cause metal contamination are metal plating, tanning, and mining.

The common metals causing environmental concern when present in the subsurface environment are chromium (Cr), cadmium (Cd), zinc (Zn), lead (Pb), mercury (Hg), arsenic (As), nickel (Ni), copper (Cu), and silver (Ag).

2.2.3   Contaminant Properties

As noted previously, the behavior of contaminants and contaminant mixtures in the environment is influenced to a great extent by the physicochemical properties of the contaminants and mixtures. These properties govern the partitioning, transport, and fate of the contaminants either individually or in bulk mixtures.

2.2.3.1   Solubility

Whether a contaminant, organic or inorganic, "likes" or "dislikes" being surrounded by water molecules is one of the key factors determining its environmental behavior and impact. A direct measurement of how much a compound likes to be present as a solute in water is given by its aqueous solubility. *Aqueous solubility* is commonly defined as the maximum amount of the chemical per unit volume in the aqueous phase when the solution is in equilibrium with the pure compound in its actual aggregation state (gas, liquid, solid) at a specified temperature and pressure (e.g., 25°C, 1 atm).[5] Above this concentration, two phases will exist in the solvent–solute system.

Solubility controls the amount of solute that can partition into the aqueous environmental compartment and thus be transported by it. Solubilization of compounds in an aqueous environment is an equilibrium process. At the outset, we should discriminate between the thermodynamic quantity of equilibrium and the phenomenon of the steady state condition. *Equilibrium* implies that the system has reached its ultimate desired state where, for example, forward and reverse processes are occurring at equal rates. *Steady state* implies merely that the situation is unchanging with time. Dissolved contaminant levels of benzene at 10 mg/l within a gasoline plume may be at steady state, but it is not at equilibrium with the surrounding environment, which may be higher than 10 mg/l. A steady state condition such as this is dictated by the rates of solubilization of benzene into the groundwater balanced by the rates of consumption (via volatilization, adsorption, and natural biodegradation).

For pure compounds, the aqueous solubility is a direct reflection of molecular structure and electrochemical characteristics. As such, it is often considered a fundamental physical constant. Environmental contaminants have aqueous solubilities ranging over several orders of magnitude: from highly soluble compounds (infinitely soluble or completely miscible, e.g., acetone and methanol) to levels of saturation that are so low that the concentration can scarcely be detected (e.g., benzo(a)pyrene).

For organic mixtures such as gasoline, solubility is a function of the mole fraction of each individual constituent in the mixture.[6-8] The concentration of an individual compound in water at equilibrium with a hydrocarbon mixture can be expressed according to the following equation:

$$C_i^* = C_i^o X_i \gamma_i \tag{2.1}$$

where C_i^* = equilibrium solute concentration for component i in the mixture
C_i^o = equilibrium solute concentration for component i as pure compound
X_i = mole fraction of compound i in the hydrocarbon mixture
γ_i = activity coefficient of compound i in the hydrocarbon mixture.

When we imagine organic compounds mixed in water, we recognize that organic molecules and water molecules differ from one another in two primary ways: (1) they have very dissimilar shapes and sizes, and (2) in general, organic molecules are much less polar than water since they are chiefly constructed from atoms having comparable electronegativities resulting in evenly spaced electronic distribution. These dissimilarities between solute and solvent result in various enthalpic and entropic contributions. These contributions are lumped into one parameter, the activity coefficient of the compound in water, γ_w. The aqueous solubility of a solute is sometimes defined as equal to the reciprocal of its aqueous activity coefficient.[7]

$$\log C_w = -\log \gamma_w \tag{2.2}$$

where C_w = aqueous solubility of compound
γ_w = activity coefficient of the compound in water.

Another aspect of solution composition that can affect the solubility (or aqueous activity coefficient) of organic chemicals involves the presence of other organic compounds in the water. Codissolved organic compounds may influence the overall solubility of the solute of interest to us. Three general cases appear to describe the various observations reported.[5] (1) When the other organic compound(s) are present in relatively large quantities (more than 10% by volume), these compounds act as solvents themselves and dissolve the solute of interest approximately in proportion to their volume fraction in solution. (2) When the other organic compounds are present in somewhat smaller quantities, these molecules reduce the energy requirements for the solubilization of the solute of interest. This situation may be best referred to as *cosolvency*. An example of cosolvency effect is the increased solubilization of benzene within gasoline plumes in the presence of methyl tertiary butyl ether (MTBE) or tertiary butyl alcohol (TBA). (3) If the other organic compounds are present at low enough levels, there is a very low probability for the solubility of the solute of interest to be affected.

Solubility is generally expressed in terms of mass per unit volume (milligrams per liter (mg/l) or micrograms per liter (µg/l) or weight per weight (ppm, ppb)). Less common units are mole fraction and molal concentration (moles per kilogram of solvent). At low concentrations, all units are proportional to one another. At high concentrations this is not the case, and it becomes important to distinguish if the solubility is per volume of pure water or per volume of solution.

The solubilities of most organic compounds fall in the <1 to 200,000 mg/l range. Aqueous solubility data can be found in the literature of many scientific disciplines. Appendix C presents the solubility data of selected compounds of environmental concern.

2.2.3.2   Vapor Pressure

Vapor pressure, another important physicochemical property, is that parameter that can be used to estimate a compound's tendency to volatilize and partition into the gaseous phase. The vapor pressure is defined as the pressure exerted by the vapor of a compound at equilibrium with its pure condensed phase, be it liquid or solid, at a given temperature. Vapor pressure of a pure compound will be equal to 1 atm at its boiling point temperature.

Vapor pressures of organic compounds may differ by many orders of magnitude. These compound-to-compound variations arise from differences in molecular interactions. Vapor pressure is generally expressed in millimeters of mercury or in atmospheres. Constituents of gasoline (e.g., benzene, toluene, ethyl benzene, and xylenes) and chlorinated and nonchlorinated solvents (e.g., trichloroethylene, tetrachloroethylene, acetone) have relatively high vapor pressures and are, therefore, relatively volatile.

Vapor pressure is an important parameter in determining the behavior of contaminants in the subsurface environment. It is an indication of the tendency of a compound to volatilize from the adsorbed, liquid, and aqueous phases. Partitioning of a spilled compound adsorbed to the solid and/or present in the NAPL phase will be influenced by its vapor pressure. The physical parameters that have the greatest effect on vapor pressure are temperature and the nature of the compound itself (i.e., critical temperature, critical pressure, and heat of vaporization). For mixtures, the composition of the mixture also has a bearing on the vapor pressure according to the following expression:

$$P_i^* = X_i \gamma_i P_i^o \tag{2.3}$$

where P_i^* = equilibrium partial pressure of component i
X_i = mole fraction of compound i in the hydrocarbon mixture
γ_i = activity coefficient of compound i in hydrocarbon mixture
P_i^o = vapor pressure of pure compound i.

Appendix C presents vapor pressures of various environmental contaminants of concern.

2.2.3.3   Henry's Law Constant

Not only are solubility and vapor pressure important physicochemical properties in and of themselves in determining the partitioning of contaminant, but they also interact to control the air–water partitioning of volatile organic compounds. Henry's law constant, in other words air–water partition constant, quantifies the relative escaping tendency of a compound to exist as vapor molecules as opposed to being dissolved in water. Henry's law is essentially an expression of the partitioning equilibrium developed between a solute and water at dilute concentrations. Henry's law constant, K_H, is expressed as the ratio of a compound's abundance in the gas phase to that in the aqueous phase at equilibrium.

$$K_H = \frac{P_i}{C_w} \left(\text{atm} \cdot \text{m}^3/\text{mol} \right) \qquad (2.4)$$

where P_i = partial pressure of compound in the gas phase (atm)
            C_w = aqueous solution as a molar concentration (mol/m³).

The units of K_H are, of course, dependent on the choice of measures. If the abundance of a compound in air is expressed as moles per cubic meter of air (C_a), the so-called dimensionless Henry's law (K_H') constant can be obtained:

$$K_H = \frac{C_a}{C_w} \left(\text{mol} \cdot \text{m}_a^{-3} \cdot \text{mol}^{-1} \cdot \text{m}_w^3 \right) . \qquad (2.5)$$

K_H and K_H' may be related to one another by applying the ideal gas law for converting partial pressure in atmospheres to moles per cubic meters.

From this discussion, it is obvious that compounds with high vapor pressures and low solubility in water should partition appreciably from water to air. However, compounds with high vapor pressure and very high solubility, such as acetone, do not partition easily from water.

Henry's law constant is used extensively to evaluate the partitioning of compounds from soil moisture and groundwater into the soil gas. It is also used extensively in the design of air stripping systems.

Appendix C presents K_H values for various environmental contaminants of concern.

2.2.3.4   Density

The density of a compound is the ratio of its mass to its volume. The property varies not only with molecular weight but also with molecular interaction and structure; for instance, although primary and tertiary butyl alcohol have the same molecular weight, their respective densities at 20°C are 0.8098 and 0.7887 g/ml. In environmental remediation, the primary reason for knowing the density of a substance is to determine whether gases are heavier than air, or whether liquids will float or sink in groundwater.

Units of density express a mass to volume ratio. For liquids, the units are g/ml or mol/ml, and for solids, g/cm³. Liquid densities range from 0.6 to 2.9 g/ml.[6] Vapor densities are in units of g/l or mg/m³.

Various refined petroleum products range in density from 0.75 g/ml for automotive gasoline to 0.97 g/ml for no. 6 fuel oil. In contrast, most chlorinated solvents have densities that range from 1.1 g/ml for chlorobenzene to 1.63 g/ml for tetrachloroethylene (PCE). For a chemical to be regarded as a potential DNAPL in a given circumstance, it must have a fluid density greater than 1.01 g/cm³ and a solubility in water of less than 20,000 mg/l.[4]

2.2.3.5   Liquid Viscosity

The viscosity of a liquid is a measure of the forces that work against movement or flow when a shearing stress is applied. It has an important bearing on several problems relating to the movement and recovery of bulk quantities of liquid when spilled in the subsurface environment. For example, a knowledge of the viscosity is required in formulas relating to the pumpability of a liquid (NAPL), the rate of flow into the extraction well, or spreading of the NAPL.

Viscosity of refined petroleum products increases when the mixture has heavier compounds in the product. For example, the viscosity of automotive gasoline is significantly lower than no. 6 fuel oil. Similarly, in the DNAPL class of compounds, chlorinated solvents have lower viscosities in comparison to coal tar, creosote, and PCB aroclors.

Viscosity is commonly reported in units of centipoise (cP). Values of viscosity for organic liquids generally range from 0.3 to 20 cP.[6] The viscosities of mixed DNAPLs are generally in the range of 10 to 100 cP, which is significantly greater than either water or pure chlorinated solvents. Water has a viscosity of 1 cP at 20°C.

2.2.3.6   Interfacial Tension with Water

The interfacial tension between an organic liquid and water affects such processes as the formation of stable emulsions, the resistance to flow through capillaries, and the dispersion of droplets.[6] A measured or estimated value of the interfacial tension may be important when one attempts to determine the fate of a chemical of environmental concern or desires to remove a liquid from the aqueous environment.

When two immiscible or partially miscible liquids are brought into contact, the interface thus formed possesses free surface energy. This surface energy is numerically equal to the interfacial tension.

The low interfacial tension between a NAPL phase and water allows the NAPL to enter easily into small fractures and pore spaces, facilitating deep penetration into the subsurface. Low interfacial tension also contributes to the low retention capacities of soils for chlorinated solvents.[4]

2.3   HYDRODYNAMIC PROCESSES

Transport of dissolved contaminants in saturated porous media is controlled by several mechanisms: advection, mechanical (sometimes called kinematic) dispersion, and molecular diffusion. The combined effect of mechanical dispersion and molecular diffusion is known as hydrodynamic dispersion.

Advection is the bulk movement of solute at a velocity equal to the mean velocity of flow in the aquifer system. Convection may be caused by differences in the water density (natural convection), regional movement of the water (advection), and pumping forced convection.[9] The terms "advection" and "convection" are sometimes used interchangeably; however, the differences between the two should be noted.

Dispersion is the spreading of a solute as it moves through a porous media. This is the overall process that causes the zone of contaminated groundwater to occupy a greater volume than it would if the contaminant distribution was influenced only by advection. Dispersion can also be characterized at macroscopic and microscopic scales.

In the groundwater environment, mixing is normally not dominated by turbulence, as in most surface waters, because the flow of groundwater is much slower. Most mixing is due to the tortuous, winding paths that water must follow as it travels through porous media. Some parcels of water follow wide, direct routes, while others follow narrow paths that zigzag back and forth at substantial angles to the average direction of flow.

Groundwater dispersion is usually handled mathematically in the same way as turbulent diffusion and dispersion in surface water; Fick's first law is assumed to apply. In one dimension the dispersion coefficient, D, is often approximated by

$$D = \alpha \cdot \upsilon \qquad (2.6)$$

where D = mechanical dispersion coefficient
$\quad\quad\quad \alpha$ = dispersivity of the aquifer (approximately equal to the median grain diameter of the aquifer solids)
$\quad\quad\quad \upsilon$ = seepage velocity.

Fick's law states that the diffusion rate is proportional to the concentration gradient:

$$\frac{\partial C}{\partial t} = \nabla(D \cdot \nabla C) \qquad (2.7)$$

where D = diffusion coefficient
$\quad\quad\quad C$ = concentration of solute
$\quad\quad\quad t$ = time.

Aquifers tend to be heterogeneous at the macroscopic or field level (i.e., there exists significant variations in hydraulic conductivity with respect to position within an aquifer). These heterogeneities in hydraulic conductivity create the spatial variation in the velocity field that actually causes macroscopic dispersion or macrodispersion. For example, the presence of volumes of clay amid expanses of coarse gravel will cause portions of contaminated water to move either much more slowly or much more rapidly than average, resulting in mixing.

The microscopic dispersion is the result of two processes: mechanical dispersion and molecular diffusion. *Mechanical dispersion* is the result of velocity variations in the pore channels and the tortuous nature of flow in the porous medium. *Molecular diffusion* is the random movement of molecules in a fluid due to concentration gradients.

In two- or three-dimensional flow, dispersion occurs not only along the axis of flow but along axes perpendicular to the flow. Longitudinal dispersion, which occurs in the direction of seepage velocity, is normally larger than dispersion perpendicular to flow. In sand and gravel aquifers, almost without exception, dispersion is weak in the directions transverse to groundwater flow.[4] This results in a narrow, elongated plume in sand and gravel aquifers (Figure 2.2). However, it should be noted that the shape and size of a source zone has a strong influence on plume dimensions.

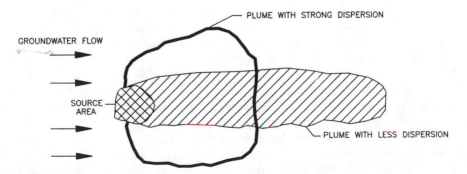

Figure 2.2 Influence of hydrodynamic dispersion on the dimensions of the plume.

In low permeability zones, molecular diffusion in the aqueous phase governs the migration of dissolved contaminants. Where DNAPL accumulations rest on unfractured clayey layers, diffusion of dissolved contaminants occurs into these layers at a very slow rate. The rate of penetration of the front is slow for chlorinated solvents—less than 1 m in the first decade—and decreases with time.[4] Therefore, molecular diffusion is generally not the cause of significant breakthrough of contaminants through aquitards.

The behavior of DNAPLs of chlorinated solvents as they penetrate the saturated zone has been studied by many researchers.[4] Two important conclusions have been drawn from these observations in the saturated zone: (1) subtle permeability contrasts can have a controlling influence on the movement of a DNAPL, and (2) typical groundwater flows do not have much effect on the flow behavior of a DNAPL. The latter point was clearly demonstrated under conditions of relatively high groundwater velocity (> 3 ft/day), and there was essentially no distortion of the penetrating solvent body.[4]

2.3.1 Groundwater Sampling

Means of studying differing composition of water in different parts of the saturated zone are not entirely adequate, mainly due to cost considerations. Wells commonly obtain water from a considerable thickness of saturated material and from several different strata. These components are mixed by the turbulent flow of water in the well before they become available for sampling. Groundwater monitoring wells specifically installed for obtaining water quality data and the sampling procedures should be designed to avoid some of these problems.

Although one rarely can be certain that a sample from a well represents exactly the composition of the bulk water in the vertical section at that point, it is usually a useful indication of the average composition of available water at that point. Areal variations in groundwater quality are evaluated by sampling wells distributed over the study area as appropriate to the amount of detail that is desired. Because rates of movement and mixing of groundwater are generally very slow, changes in composition of the water yielded by a well with respect to time can usually be monitored by seasonal sampling.

2.4 TRANSPORT IN THE UNSATURATED ZONE

Since many nonaqueous phase liquids (NAPLs) are immiscible with or sparingly soluble in water, when spilled, they begin to move downward through the unsaturated (vadose) zone as a separate phase, creating a system with three fluid phases: air, water, and NAPL, in addition to the soil phase itself. If no impermeable unit is encountered, the NAPL should be able to continue its migration, displacing air as it moves downward under the influence of gravity and capillary forces. As the NAPL passes downward through the porous medium, some of the NAPL coats the solid matrix, some dissolves in the soil moisture, some volatilizes, and some becomes trapped in the pore spaces in response to capillary forces. The NAPL is denied access to an interstitial pore until the capillary pressure exceeds the threshold value associated with the largest throat already in contact with the NAPL. The amount trapped is referred to as the residual saturation, S_r:

$$S_r = \frac{\text{volume of residual NAPL}}{\text{volume of pore space}}$$

The factors that affect the trapping can be quite complex, but in general, there is a tendency for the saturation of residual NAPL to increase as permeability of the porous medium decreases. Another way to express the trapping capability of the medium is to combine the residual saturation with the porosity (n) to produce the retention capacity, R, in liters per cubic meter, which can be expressed as

$$R = (S_r)(n)(1000) .$$

(2.6)

The retention capacity for oil in the vadose zone ranges from about 3 to 5 l/m^3 in highly permeable media to 30 to 50 l/m^3 in media with low permeability.[4]

Because of the ability of the porous medium to retain some of the infiltrating NAPL, as the fluid further infiltrates the unsaturated zone, the fraction of the spill that remains mobile is decreased. Depending on the size of the spill, as well as the depth to groundwater and the retention capacity of the soil matrix, it is possible for a spill to be completely immobilized before it reaches the water table. This does not, however, mean that dissolved phase groundwater contamination is prevented, since infiltrating precipitation can dissolve some of the immobilized NAPL and carry contaminants downward to the water table.

If the volume of the spill exceeds the retention capacity of the unsaturated zone soil, the NAPL will eventually reach the capillary fringe. The system now consists of just two immiscible phases. For the NAPL to continue its downward migration, it must be able to displace water from the pores. At this point, the density of the NAPL plays an important role in determining its subsequent movement (Figure 2.3). Since petroleum products, such as gasoline, are less dense than water, they will tend to spread on the surface of the water table in a "pancake" configuration. The groundwater contamination that ensues occurs as dissolution of the more soluble components in the mixture (i.e., benzene, toluene, and xylenes) takes place at the NAPL–groundwater interface. Dense nonaqueous phase liquids (DNAPL) will, however, continue to migrate through the saturated zone until either the volume required to sustain migration is inadequate due to solubilization or soil coating, or an impermeable unit is encountered. The capillary pressure at which the DNAPL becomes continuous in the macroscopic sense and is capable of flowing the material is known as the entry pressure.[4] Another characteristic capillary pressure that has found widespread use is the displacement pressure, which is defined as the minimum capillary pressure required to initiate invasion of a water-saturated porous medium by a DNAPL.

The spreading of NAPL on the water table surface is a result of capillary forces in the air–water capillary zone which prevent the contaminant mixture from entering the aquifer proper and the density of the NAPL. The presence of air and water in the porous medium reduces the migration velocity of the NAPL. The volume of the unsaturated porous medium affected by a limited quantity of NAPL is limited, because a threshold residual saturation must be exceeded before the NAPL can continue to migrate downward and spread laterally. Additionally, if the NAPL volume is sufficient to reach the water table, a minimum thickness must accumulate before lateral spreading can occur. It follows, then, that relatively large volumes of NAPL must be introduced into a permeable substratum before the presence of NAPL can be observed at some distance from the source. Nonaqueous phase liquid spreading on the water table will, for the most part, occur in the direction of the groundwater gradient. However, upon reaching the water table, the NAPL begins to "pancake" and thus spreads in all directions. In the case of a large release, NAPL may be observed at considerable distances upgradient.

As discussed previously, a significant vapor phase develops in the unsaturated zone following release of contaminants with high vapor pressures. Formation of the vapor phase is a function of volatilization from the parent NAPL, from the coated soil, and from the dissolved phase groundwater compartments. The behavior of the vapor plume is a result of interactions among and between these compartments.

The primary mechanisms of vapor transport in the subsurface are *gaseous diffusion* and *isothermal advection*. Following the release of a volatile NAPL, vapor phase transport of

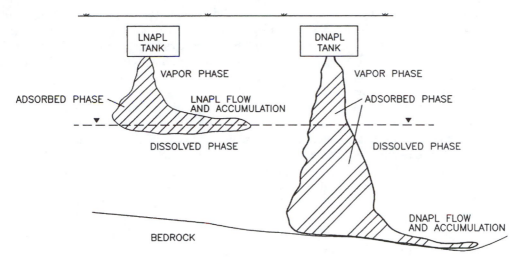

Figure 2.3 Behavior of LNAPL and DNAPL in the subsurface environment.

volatile constituents away from the source will occur, primarily through gaseous diffusion. As such, migration along a concentration gradient occurs. This migration, then, is independent of the groundwater hydraulic gradient and occurs in all directions in the absence of soil matrix heterogeneities and thermal gradients. Vapor phase migration will continue indefinitely into the soil atmosphere and outer atmosphere (assuming communication exists) until the concentration gradient is satisfied or an equilibrium is established. Partitioning mechanisms in the vadose zone are described in detail in Chapter 3.

From this discussion, it is apparent that as vapors migrate through the unsaturated zone, they interact with the soil matrix through which they are being transported, with the net result being a decrease in velocity. Partitioning in the unsaturated zone also serves to cause the soil and the three compartments (soil, soil air, and soil moisture) that comprise it to become a secondary contaminant source. As the contaminants in the unsaturated zone react with the vadose atmosphere and infiltrating precipitation, they are transported to the groundwater. As such, the vadose zone will continue to serve as a groundwater contaminant source until no contamination remains or the vadose zone contamination is remediated.

Not only does the soil residual in the vadose zone serve as a secondary source, but the contaminant vapor plume can also serve as a secondary source. Because vapor transport is a diffusion-dominated process and, therefore, occurs along a concentration gradient, uncontaminated groundwater can be a sink for contaminants in the vadose zone atmosphere via solubilization.

2.5 ABIOTIC PROCESSES

Contaminant properties and subsurface conditions that result in interactive processes often cause the rate of contaminant transport to differ from the rate of groundwater flow. Abiotic processes affect contaminant transport by causing interactions between the contaminant and the stationary subsurface material (e.g., adsorption, ion exchange) or by changing the form of the contaminant (e.g., hydrolysis, redox reactions) which may subsequently interact with the subsurface material.

2.5.1 Adsorption

The process in which chemicals become associated with solid phases is generally referred to as *adsorption*. This phase transfer process involves interaction between either vapor molecules or dissolved molecules with adjacent solid phases. Adsorption, absorption, and sorption are three terms that refer to similar phenomena.[1] Adsorption is defined as the accumulation occurring at an interface, absorption as the partitioning between two phases (accumulation from groundwater into organic carbon), and sorption as including both adsorption and absorption. Often, the terms adsorption and sorption are used interchangeably; adsorption will be the preferred term in this section.

In general, adsorption phenomena can be classified as either sorbent- or solvent-induced.[1] Sorbent-induced adsorption occurs when there is an attraction between the soil matrix (sorbent) and the contaminants (solute) and the contaminant accumulates at the surface due to the affinity of the surface for the contaminant (e.g., cation exchange at the mineral sites). Solvent-induced adsorption occurs when the contaminant is hydrophobic (nonpolar, water-disliking, and low solubility) and finds it energetically favorable to accumulate at an interface or partition into a nonpolar phase (e.g., partitioning onto the soil organic carbon) rather than remain in the water phase.

Polarity refers to the extent to which electrical charge is unevenly distributed within the molecule. Water itself is very polar, having an excess of negative charge, associated with the oxygen, balanced by an excess of positive charge, associated with the hydrogen atoms. Therefore, water readily dissolves polar (hydrophilic) compounds. In general, nonpolar (hydrophobic) compounds tend to associate more strongly with less polar phases, such as soil organic material or NAPLs. Polarity of a chemical is strongly correlated with K_{ow}, the octanol–water partition coefficient.

The way in which organic compounds, especially with low to moderate solubilities, partition themselves between soil and the aqueous phases is often described by the partitioning coefficient, K_d. The relationship between K_d, K_{ow}, and the fraction of organic content in the soils, f_{oc}, is described in detail in Chapter 3. It should be noted here that the fraction of organic content of the subsurface material is the dominant characteristic affecting partitioning via adsorption. Other factors that influence the partitioning coefficient are[6]

- temperature
- pH of soil and water
- particle size distribution and surface area of soil particles
- salinity of water
- concentration of dissolved organic matter in water
- suspended particulate matter in water
- nonequilibrium adsorption mechanisms or failure to reach equilibrium conditions
- solids to solution ratio

As a consequence of partitioning via adsorption, the transport of dissolved contaminants is slowed, or retarded. If the equilibrium between adsorbed and dissolved compounds occurs sufficiently rapidly, and if the concentration on the soil particles is proportional to aqueous concentration (i.e., a linear adsorption isotherm exists), the expressions for the transport of a compound can be readily modified by a retardation factor to describe the behavior.

The *retardation factor*, R, is defined as follows:

$$R = \frac{\text{dissolved (mobile) compound concentration} + \text{adsorbed compound concentration}}{\text{dissolved (mobile) compound concentration}}.$$

Therefore,

$$R = 1 + \frac{\text{adsorbed concentration}}{\text{dissolved concentration}}. \tag{2.7}$$

The rate at which the center of mass of an adsorbing compound in equilibrium with aquifer material moves through the aquifer is equal to the seepage velocity, υ, divided by R. As an example, if the retardation factor for a compound is 5, a plume of the dissolved compound will advance only one fifth as fast as a parcel of water. The impact of retardation on dispersion is negligible due to the long travel times involved.[10]

In equation (2.7), concentration is expressed as mass of compound per volume of total porous media. The volume of porous media, also termed aquifer volume, is defined to include both particle grains and pore water. Equation (2.7) can be rewritten in terms of the aqueous (dissolved) concentration, C_s, the adsorbed concentration, C_a, the water-filled porosity n, the partition coefficient K_d, and the bulk density of the aquifer material, ρ_b.

Equation (2.7) can be rewritten as

$$R = \frac{(C_s \cdot n) + (C_a \cdot \rho_b)}{C_s \cdot n} \ .$$

Since

$$K_d = \frac{C_a}{C_s}$$

$$R = 1 + K_d \cdot \frac{\rho_b}{n} \ . \qquad (2.8)$$

It must be noted that a retardation coefficient is strictly applicable only when a linear relationship exists between C_a and C_s. Mobility of various contaminants based on retardation can be described as shown in Table 2.2. It should be noted that the highly soluble compounds have a low retardation factor.

Table 2.2 Mobility Classes of Various Contaminants

R	1.4–2.0	3–6	9–20	40–100	>100
Mobility Compounds	Very mobile Phenols Alcohols Acetone	Mobile TCA, TCE Benzene	Intermediate Naphthalene	Low mobility Pyrene	Immobility PCBs Chlorinated dioxin

Adapted from Olsen, R. L. and Davis, A., Predicting the fate and transport of organic compounds in groundwater, Part 1, *J. Hazardous Mater. Control Res. Instit.*, May/June, 1990.

The influence of soil organic carbon content on the retardation factor of a few chlorinated compounds is shown in Table 2.3.

Table 2.3 Retardation Factor for Chlorinated Compounds

Compound	Organic carbon content (%)		
	0.01–0.02	0.001–0.01	<0.001
TCA	3.4–11	1.2–6	1–1.2
TCE	4–13	1.3–7	1–1.3
PCE	9–31	1.8–16	1–1.8

Adapted from Olsen, R. L. and Davis, A., Predicting the fate and transport of organic compounds in groundwater, Part 1, *J. Hazardous Mater. Control Res. Instit.*, May/June, 1990.

2.5.2 Ion Exchange

Ion exchange is a specific category of adsorption. Owing to the ubiquitous phenomenon of ionizable surface groups on wet particles, virtually every solid presents a charged surface to the aqueous solution. If this surface charge is of opposite sign to that exhibited by a contaminant's functional group, then there will be an electrostatic attraction between the solute in the bulk solution and the particle surface. Such solutes will accumulate in the thin film of water surrounding the particle as part of the population of charges in solution balancing the charges on the solid surface. Conversely, solutes with charges of like sign as the surface will be repulsed from the near-surface water.

The second interaction involves chemical bonding of the solute to the surface or to some component of the solid phase. This may also involve displacing an ion previously bound to the solid surface to achieve more efficient neutralization of charge deficiency of the adsorbent. For example, if sodium ions (monovalent) have accumulated at the interface and suddenly calcium ions (divalent) appear, the surface excess can be more efficiently neutralized by the calcium ions than by the sodium ions. As a result, the sodium ions will desorb and the calcium ions will adsorb.

Since clay minerals are frequently the dominant source of ion exchange in the subsurface, it is important to have a fundamental understanding of clay mineralogy and the source(s) of charge deficiencies in clay minerals. Clays are commonly layered aluminosilicates.[1] Two basic structural units include a tetrahedral structure about Si^{4+} and an octahedral structure about Al^{3+}. Clay minerals are made up of these basic structural units.

The surface charge of clays and many other mineral surfaces is a function of pH.[1] At low pH (excess hydrogen ions), the surface can exhibit a positive net charge, while at higher pH (excess hydroxyl ions), the surface can exhibit a negative charge. The nature of the surface, and adsorption, is thus impacted by the pH of the groundwater and/or of the contamination. At an intermediate pH the surface exhibits a neutral charge; this pH is classified as the point of zero charge (pzc). The pzc differs for the various mineral surfaces. At a pH above the pzc of the mineral, the surface exhibits a net negative charge, while at a pH below the pzc the surface exhibits a net positive charge. Thus, the surface charge exhibited by a mineral surface can be determined by evaluating the pH of the solution relative to the pzc of the mineral. At neutral pH clays will exhibit a negative surface charge.[1]

An important consideration in addressing the potential for ion exchange is to know the hierarchy of preferred ions on adsorption sites. It has already been stated that higher valency ions will displace lower valency ions. However, the question remains as to what sequence will occur for ions of the same valency (univalent). For univalent ions, it has been observed that smaller ions are preferentially exchanged for larger ions. The sequence in which ions are preferentially exchanged is referred to as the selectivity of the mineral surface for ions. However, this selectivity sequence can be negated by mass action effects at high concentrations of individual ions.

The level of adsorption for a subsurface media is greatly influenced by the surface area containing exchange sites. This parameter is typically discussed in terms of cation exchange capacity (CEC) of the medium and is expressed in units of milliequivalents of exchanging cations per 100 grams of solids (meq/100 g). Clay fractions are also high in specific surface area (SSA, surface area per unit mass of mineral, m^2/g) as compared to other size fractions. It is observed that, in general, increasing values of SSA correlate with increasing values of CEC.[1]

The cations most commonly thought of as participating in cation exchange processes are inorganic elements or compounds (e.g., Ca^{++}, NH_4^+). However, organic cations are equally susceptible to ion exchange.

Certain organic compounds can either gain or lose a proton (or hydroxyl) as a function of pH and thus go from a neutral form to an ionic form. This will greatly increase the solubility of the chemical in the groundwater (a polar solvent). The loss of a proton produces anionic compounds and the adsorption to clay minerals will be negligible. The gain of a proton (or loss of a hydroxyl) will result in a positive ion (cationic compounds), and the result will be the association with the cation exchange capacity of the clay minerals.[1] The overall impact on the mobility of the compound(s) will depend on the relative adsorption of the neutral and ionic forms of the compound.

Particulate natural organic matter may also contribute to the assemblage of charged sites of solids in the groundwater environment. This is mostly due to ionization reactions of carboxyl groups (–COOH), and at higher pH values, phenolic groups (aromatic ring –OH). The extent of charge buildup in the organic portion of natural particles will vary as a function of pH. Other solid phases common in nature, like carbonates, will also exhibit surface charging.

2.5.3 Hydrolysis

In general terms, *hydrolysis* is defined as a chemical transformation in which an organic molecule, RX, reacts with water, resulting in the formation of a new covalent bond with OH and cleavage of the covalent bond with X (the leaving group) in the original molecule. The net reaction is the displacement of X by OH^-:[12]

$$RX + H_2O \rightarrow ROH + H^+ + X^-.$$ (2.9)

Hydrolysis can occur when biodegradation cannot, and can result in byproducts susceptible to biodegradation. In this manner, hydrolysis can be a significant process affecting the fate and transport of various contaminants. Certain functional groups are potentially susceptible to hydrolysis:[11,12] amides, carbonates, epoxides, aliphatic and aromatic esters, alkyl and aryl halides, nitriles, ureas, and organo phosphorus esters. Organic functional groups relatively or completely inert to hydrolysis are:[6] alcohols, phenols, glycols, ethers, aldehydes, ketones, biphenyls, pesticides, etc.

When an organic compound undergoes hydrolysis, a nucleophile (an electron-rich nucleus seeker) attacks an electrophile (electron-deficient electron seeker such as carbon atom, phosphorus atom, etc.) and displaces a leaving group (chloride, etc.).

Hydrolysis may be biologically mediated or may occur independent of the biosystem.[1] For biotic hydrolysis, the microbial population and/or the specific enzyme activity will significantly impact the level of hydrolysis. Under abiotic conditions, environmental factors such as pH, temperature, dissolved organic matter, dissolved metal ions, etc., may impact the level of hydrolysis. Biotic and abiotic hydrolysis are not mutually exclusive processes; they may jointly contribute to the breakdown of a compound.[1]

The effect of pH can be attributed to either specific acid–base catalysis effects or to a change in the speciation of the compound. In addition to nucleophilic attack by H_2O (neutral hydrolysis), hydrolytic reactions are also sensitive to specific acid and specific base catalysis (i.e., catalysis by hydrogen ion H^+, and hydroxide ion OH^-, respectively). Accordingly, hydrolysis kinetics must take into account the potential for H_2O to dissociate. Even at pH 7.0, where the concentration of H^+ and OH^- is only 10^{-7} M, specific acid and specific base catalysis occurs because hydrogen ion and hydroxide ions provide an alternative mechanism for hydrolysis that is energetically more favorable.

The relationship between hydrolysis kinetics and pH is also dependent on the nature of the hydrolyzable functional group. For example, in the narrow pH range of 6 to 7, the neutral hydrolysis rate term for ethyl acetate will dominate. Only below pH 6 will the acid-catalyzed rate for ethyl acetate contribute to the overall hydrolysis rate. Likewise, base-catalyzed hydrolysis of ethyl acetate will not compete with neutral hydrolysis until well above pH 7.0. By contrast, the neutral hydrolysis of ethylene oxide and methylene chloride occurs over a much wider pH range (4 to 8). Either below or above this pH range, the rate terms for acid- or base-catalyzed hydrolysis will dominate.

Temperature effects on the rate of hydrolysis can be pronounced. For a 10°C change in temperature, the hydrolysis rate constant could change by a factor of 2.5 (increasing temperature corresponding to increasing rate constant).[1]

Metal ions, such as calcium, copper, magnesium, iron, cobalt, and nickel have been observed to catalyze the hydrolysis of certain compounds when the metal ions are present at high concentrations. However, when the metals are present at concentrations more common to natural groundwater, their impact on hydrolysis rates is insignificant. It has also been suggested that metal ions associated with soil medium (such as clay minerals) serve to catalyze the hydrolysis and thus are responsible for the increased hydrolysis rate, under certain circumstances.

2.5.4   Oxidation and Reduction Reactions

Oxidation is defined as a loss of electrons. Oxidizing agents gain electrons and are by definition electrophiles (liking for electrons). Oxidation can either be associated with the introduction of oxygen into a molecule or the conversion of a molecule to a higher oxidation state. For example:

$$R - CH_2 - H \rightarrow R - CH_2 - OH \rightarrow R - CHO . \qquad (2.10)$$

In equation (2.10), the first step in the sequence is the incorporation of oxygen by a formal "oxygen atom donor," whereas the second is formally a dehydrogenation, or oxidation of the carbon atom to a higher formal oxidation state.

Reduction, which is defined as a gain of electrons, occurs when there is a transfer of electrons from an electron donor "or reductant" to an electron acceptor "or oxidant." Examples of reduction reactions are

$$R - Cl + 2e^- + H^+ \rightarrow RH + X^- \qquad (2.11)$$

$$Fe(OH)_3 + 3H^+ + e^- \rightarrow Fe^{2+} + 3H_2O . \qquad (2.12)$$

It should be noted that in any electron transfer reaction, one of the reactants is oxidized while the other one is reduced. As such, it is only logical that we term such reactions *redox* reactions. Since our focus is on environmental contaminants, we speak of an oxidation reaction if the contaminant is oxidized, and of a reduction reaction if the contaminant is reduced.

Reducing environments abound in nature (subsurface waters and soils, aquatic sediments, wetlands, etc.). Typically, the subsurface is a closed or partially closed system (i.e., limiting the replenishment of oxygen). In this situation the available oxygen is readily depleted in contaminated zones, and reducing environments result. The realization that certain contaminants that are considered "persistent" under above-ground aerobic conditions may not be nearly as persistent in a reducing environment has generated a great deal of interest in the behavior of organic contaminants in reduced anaerobic environments. This interest has been intensified for the elucidation of reductive transformation pathways that may lead to *in situ* remediation technologies for the removal of contaminants from subsurface environments.

Reductive transformations are most conveniently categorized according to the type of functional group that is reduced. Reductive transformations known to occur in natural reducing environments include reductive dehalogenation, nitroaromatic reduction, aromatic azo reduction, sulfoxide reduction, *n*-nitrosoamine reduction, quinone reduction, and reductive dealkylation, to name a few. Although the reduction of these organic functional groups is performed routinely by synthetic chemists in the laboratory using strong reductants, the abiotic reduction of these functional groups in natural systems is not likely to occur.[12] The lack of reactivity observed for these classes of chemicals in natural reducing systems is probably due to a combination of unfavorable kinetic and thermodynamic factors. Microbially catalyzed reduction of some of these functional groups is described in Chapter 5.

Naturally occurring chemical elements that are most commonly impacted by abiotic reducing reactions are inorganic elements (for example, Fe and Mn) and inorganic metallic ions (for example, Cr^{6+}, As^{5+}).

In the subsurface environment, there are many sources of oxidants capable of facilitating abiotic oxidative reactions. These oxidants include molecular oxygen, ozone and related compounds, hydrogen peroxide, hydroxyl radical, peroxy radicals, clays, silicon oxides, aluminum oxides, iron oxides, and manganese oxides. Some of these oxidants (for example, iron and manganese oxides) are naturally occurring agents, while others (for example, hydroxyl radical and hydrogen peroxide) are formed as a result of photolysis of natural waters.

The ability of an oxidation–reduction (redox) reaction to occur in the natural environment is a function of the redox potential. Just as the concentration (activity) of free protons (hydrogen ions, H^+) can be utilized to assess the acid–base status of the subsurface environment, the activity of the free electron can be utilized to assess the oxidation–reduction (redox) potential. The redox potential is often defined in terms of the negative logarithm of the free electron activity:

$$pE = \log\left(e^-\right) \tag{2.13}$$

where e^- is free electron activity.

Large values of pE indicate low values of electron activity. Low values of electron activity indicate the existence of electron-poor (oxidized) environment. Small values of pE indicate high electron activity and thus correspond to electron-rich (reduced) environment.

2.5.5 Precipitation and Solubilization

Inorganic compounds can occur in many forms (species) depending on the environmental conditions (e.g., pH, pE, and presence of other organic and inorganic compounds). As the environmental conditions change, the speciation of the compounds may also change.

In groundwater, six categories have been suggested in which an element or compound may exist:[1] (1) free ions (Na^+, Fe^{2+}), (2) insoluble precipitates ($Cr(OH)_3$, PbS), (3) metal/ligand complexes (Cu-humate), (4) adsorbed species (arsenic adsorbed to ferric hydroxide surface, (5) species held on by ion exchange (Ca^{++} on clay surfaces), and (6) species that differ in oxidation states (Fe(II) vs. Fe(III), and Cr(III) vs. Cr(VI)). It is important to know the form of the element or compound, as this will greatly impact the fate, transport, reactivity, and toxicity of the compound. Knowledge of the total concentration of a contaminant is of little value when evaluating the environmental risk of the contaminant present.

Subsurface environmental conditions such as pH, redox potential, and temperature will influence the solubilization/precipitation of inorganic compounds. Weathering, which is a partial solubilization of certain elements within minerals, is an example of solubilization taking place in the subsurface environment. Manipulation of subsurface environments to achieve precipitation reactions for remediation of certain contaminants is discussed in Chapter 8.

2.6 BIOTIC PROCESSES

Another set of transformations that remove organic compounds from the environment is that group of reactions mediated by microorganisms. It should also be noted that inorganic compounds such as dissolved heavy metals can be immobilized in the subsurface environment by microbially induced reactions.

Biochemical transformations of organic compounds are especially important, because many reactions discussed in previous sections, although thermodynamically feasible, occur extremely slowly because of kinetic limitations. Microorganisms enable such reactions to proceed via two important approaches. The first approach involves the use of special proteins, called enzymes, that serve as catalysts. Enzymes can lower the activation energy of reactions, thereby speeding the transformations by many orders of magnitude. Second, microorganisms may invest energy to convert oxidative reagents into more reactive species. For example, molecular oxygen is converted to a more reactive species by a biochemical agent before it is used to oxidize some compounds

Microbial metabolic reactions can be broadly classified into aerobic, anaerobic, and anoxic modes. If biodegradation results in the formation of inorganic species (e.g., CO_2, H_2O, and mineral salts), it is referred to as mineralization. Complete mineralization typically involves oxidation using oxygen, but can also occur under anaerobic and anoxic conditions. Attention should also be paid to differentiation between degradation of compounds as a primary substrate in contrast to cometabolic degradation of certain compounds. Detailed description of these reactions involving various contaminants is provided in Chapter 5 and Chapter 8.

Microorganisms in the subsurface environment, especially in groundwater, are more likely to be attached to grains of soil matrix than to be moving with the water. It appears to be advantageous for microbes in the subsurface environment to remain in one place and harvest substrate (contaminants) and nutrients that are transported primarily by advection rather than move with the water and rely solely on dispersion.[12] If the metabolism of the compounds results in substantial energy yield and/or cell-building materials, then the microbial population will increase. As populations of attached microorganisms increase, the organisms start to accumulate on top of one another, gradually building up a layer of microbes with the help of attaching material (polysaccharides). This is referred to as the biofilm and can be very effective in degrading contaminants by diffusion into the biofilm.

2.7   SUMMARY

Various aspects of contaminant partitioning and the resulting fate and transport were discussed in previous sections of this chapter. It is important to understand the behavior of contaminants in the subsurface environment to properly design and successfully implement a remediation system. Conceptual models of contaminant partitioning in the subsurface are shown in Figures 2.4 and 2.5.

Remediation of contaminants present in the subsurface environment should be understood (from a chemistry point of view) as *perturbation of partitioning of the contaminants to the most favorable compartment and/or phase*. However, the efforts to perturb the system may have nonequilibrium limitations.[13] These nonequilibrium limitations have a significant impact on the *tailing effect* of contaminant concentration levels at various remediation sites.[14] This tailing effect has been observed even at sites where a significant amount of money and time have been spent implementing the most state-of-the-art remediation technologies.

The reasons for the tailing effect of contaminant concentration levels can be summarized as below:

- nonhomogeneous advective flow
- film limitations
- sorption effects
- transport effects
  - intraparticle diffusion
  - micropore diffusion
  - macropore diffusion

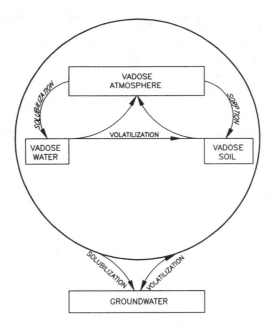

Figure 2.4 Unsaturated zone partitioning model.

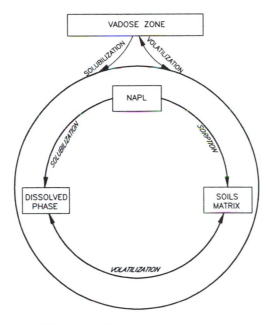

Figure 2.5 Saturated zone partitioning model.

REFERENCES

1. Knox, R. C., Sabatini, D. A., and Canter, L. W., *Subsurface Transport and Fate Processes*, Lewis Publishers, Boca Raton, FL, 1993.
2. Wilson, S. B. and Brown, R. A., In Situ Bioreclamation: A Cost Effective Technology to Remediate Subsurface Organic Contamination, *Ground Water Monitoring Rev.*, Winter, 1989.
3. Graedel, T. E., *Chemical Compounds in the Atmosphere*, Academic Press, New York, 1978.
4. Pankow, J. F. and Cherry, J. A., *Dense Chlorinated Solvents and other DNAPLs in Groundwater*, Waterloo Press, Portland, OR, 1996.

5. Schwarzenbach, R. P., Gschwend, P. N., and Imboden, D. M., *Environmental Organic Chemistry*, John Wiley & Sons, New York, 1993.

6. Lyman, W. J., Reehl, W. F., and Rosenblatt, D. H., *Handbook of Chemical Property Estimation Methods: Environmental Behavior of Organic Compounds*, McGraw-Hill, New York, 1982.

7. Yalkowsky, S. H. and Banerjee, S., *Aqueous Solubility: Methods of Estimation for Organic Compounds*, Marcel-Dekker, New York, 1992.

8. Burris, D. R. and MacIntyre, W. G., Water solubility behavior of binary hydrocarbon mixtures, *Environ. Toxicol. Chem.*, 4, 371, 1985.

9. Hemond, H. F. and Fechner, E. J., *Chemical Fate and Transport in the Environment*, Academic Press, San Diego, 1994.

10. Olsen, R. L. and Davis, A., Predicting the fate and transport of organic compounds in groundwater, Part 1, *J. Hazardous Mater. Control Res. Inst.*, May/June, 1990.

11. Lavson, R. A. and Weber, E. J., *Reaction Mechanisms in Environmental Organic Chemistry*, Lewis Publishers, Boca Raton, FL, 1994.

12. Mabey, W. R. and Mill, T., Critical review of hydrolysis of organic compounds in water under environmental conditions, *J. Phys. Chem. Ref. Data*, 7, 383, 1978.

13. Geraghty & Miller, Inc., (project summaries of various remediation projects), 1996.

3   SOIL VAPOR EXTRACTION

3.1 INTRODUCTION

Soil vapor extraction (SVE) is an accepted, recognized, and cost-effective technology for remediating soils contaminated with volatile and semivolatile organic compounds. This technology is known in the industry by various other names, such as soil venting and vacuum extraction. The process involves inducing airflow in the subsurface with an applied vacuum, and thus enhancing the *in situ* volatilization of contaminants. Depending on the depth of soil being remediated, extraction of air laden with contaminant vapors can be achieved with vertical extraction wells or horizontal extraction pipes.

The SVE process takes advantage of the volatility of the contaminants to allow mass transfer from adsorbed, dissolved, and free phases in the soil to the vapor phase, where it is removed under vacuum and treated above ground. In order for this process to be effective, the contaminants of concern must be volatile enough and have a low enough water solubility to be drawn into the soil gas for removal. These properties are usually expressed by the vapor pressure and the Henry's law constant of the compounds.

Bioventing is a process that uses a similar approach to soil vapor extraction in terms of system configuration, but has a different objective. The intent of bioventing is to induce airflow to provide oxygen to maximize the aerobic biodegradation of the compounds, in contrast to volatilization. This difference in approach renders less volatile compounds also to be treated by enhanced biodegradation.

A typical *in situ* soil vapor extraction system couples vapor extraction wells with blowers or vacuum pumps to remove contaminant vapors from zones permeable to airflow. The components of an *in situ* soil vapor extraction system are shown in Figure 3.1, and are usually readily available as off-the-shelf products. The choices available to treat the contaminants present in the effluent air are many, and will primarily depend on the type of contaminants and the mass loading rate.

Soil vapor extraction has many advantages that make this technology applicable to a broad spectrum of sites.

- SVE is an *in situ* technology that can be implemented with a minimum disturbance to site operations.
- SVE is very effective in removing the volatile contaminant mass present in the vadose zone.
- SVE has the potential for treating large volumes of soil at acceptable costs.
- The system can be mobilized and installed very quickly.
- SVE as a technology can be easily integrated with other technologies required for site cleanup.

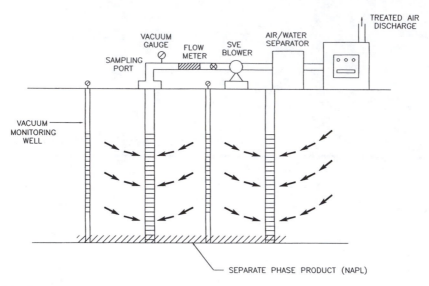

Figure 3.1 Schematic of typical components of an *in situ* soil vapor extraction system.

3.2 GOVERNING PHENOMENA

The basic phenomena that govern the performance of soil vapor extraction technology are very simple and easily described. The mechanisms that influence the success of SVE can be grouped into the following categories:

- Airflow characteristics in the vadose zone
- Mass transfer considerations that influence and limit the partitioning of the contaminants into the vapor phase and eventually into the soil gas

3.2.1 Airflow Characteristics

The basic principle of induced airflow during SVE is very simple. Airflow is induced in the subsurface by a pressure gradient applied through vertical wells or horizontal trenches. The negative pressure inside the extraction well, generated by a vacuum blower, causes soil air to move toward the well. Typical airflow patterns induced during SVE operation are shown in Figure 3.2. In order to design an SVE system, the subsurface airflow pathways and extraction flow rates must be properly defined.

The airflow field developed is dependent on many factors: the level of applied vacuum, available screen interval in the vadose zone, porosity, air permeability and its spatial variation, depth to groundwater, presence or absence of leakage from the ground surface, and subsurface conduits.

3.2.1.1 Mathematical Evaluation of Airflow

Three basic equations are considered in the description of airflow: the mass balance equation of soil air, the flow equation due to pressure gradients, and the equation of state.[1]

The mass balance of soil air is expressed by the equation of continuity.

$$\phi_a \frac{\partial \rho_a}{\partial t} = -\nabla \rho_a \cdot V \qquad (3.1)$$

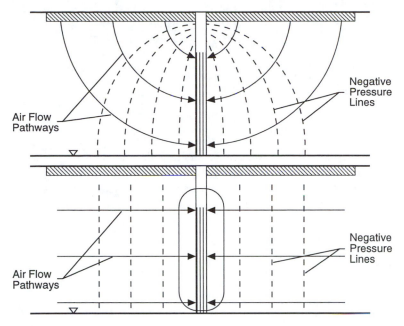

Figure 3.2 Typical airflow patterns and vacuum distribution during SVE operation.

where ϕ_a = air-filled porosity in soil
        ρ_a = the density of soil air
        V = velocity vector of airflow.

The airflow velocity due to the pressure gradient can be expressed by Darcy's law when the slip flow is negligible. In the case of airflow in sand and gravel, the slippage of air on the soil wall is negligible and Darcy's law for the flow in porous media can be applied.

$$V = \frac{K_a}{\mu_a} \cdot \nabla p$$
(3.2)

where K_a = the air permeability tensor
        μ_a = viscosity of air.

The density of air is a function of pressure and temperature. The relationship among these parameters is expressed by the equation of state. One of these equations of state is the ideal gas flow, which is simple and applicable for soil air at low pressure. The ideal gas law can be used because the operating pressure of the conventional SVE is close to ambient or lower. The ideal gas law for soil air is

$$\rho_a = \frac{p \cdot (MW)}{RT}$$
(3.3)

where (MW) = molecular weight of soil air
        R = ideal gas law constant
        T = the absolute temperature.

By combining the above equations, a general soil airflow equation can be obtained.

$$\phi_a \cdot \frac{\partial}{\partial t}\left\{\frac{p(MW)}{RT}\right\} = \nabla\left\{\frac{\rho_a K_a}{\mu_a} \cdot \nabla p\right\}. \tag{3.4}$$

This equation is nonlinear because the air density, ρ_a, in the right-hand side of the equation is a function of pressure.

3.2.1.2   Soil Air Permeability

Air permeability describes how easily vapors flow through the soil pore space. Since the airflow and the air permeability are linearly dependent, a higher air permeability will result in a higher flow rate at the same vacuum. There are numerous methods that could be employed for the determination of air permeability.[2] The methods are listed below.

- Correlation to soil's physical properties
- Laboratory measurement
- Field barometric fluctuation test
- Air injection test
- Pressure buildup test
- Pressure drawdown test

Correlation to soil's physical properties.   Soil air permeability is estimated from soil characteristics such as grain size distribution or hydraulic conductivity by assuming a linear correlation. This is a quick estimation method and provides only an order of magnitude estimate. This relationship is expressed by the following equation:

$$K_a = K_w\left\{\frac{\rho_a \cdot \mu_w}{\rho_w \cdot \mu_a}\right\} \tag{3.5}$$

where K_a = air permeability
K_w = hydraulic conductivity
ρ_a = density of air
ρ_w = density of water
μ_a = viscosity of air
μ_w = viscosity of water.

Example calculation:
At 68°F, for air and water, $\rho_a = 0.0012$, $\rho_w = 1$, $\mu_w = 10,000\ \mu P$, and $\mu_a = 183\ \mu P$. Therefore,

$$K_a = K_w \cdot \left\{\frac{0.0012}{1} \cdot \frac{10,000}{183}\right\}$$

$$= K_w\{0.0656\}.$$

If

$$K_w = 20\ \text{ft/day}$$

$$K_a = 20 \times 0.0656$$

$$= 1.31\ \text{ft/day}.$$

Laboratory method. Undisturbed soil samples are placed in pressure vessel (permeameter) and saturated with water. Air is injected to force fluid out of the sample. Measured airflow rates and air pressure in the sample are substituted into Darcy's law equation to estimate soil's air permeability. However, this method alters bulk density and soil structure and does not reflect the spatial variability of soil conditions in the field. Hence, this is not a recommended method to estimate the soil air permeability.

Barometric fluctuation test. This method monitors the changes in barometric pressure using piezometers installed within the vadose zone. Soil gas permeability is estimated by calculations that equate air flux between adjoining subsurface layers. However, normal diurnal barometric pressure changes interfere with the test if vadose zone is deep and the soil is very permeable. This test measures only vertical soil air permeability and is not recommended.

Air injection test. This is a field test in which a known volume of air is injected into the soil over a given time period. Pressure differential is measured before and after air injection and soil air permeability is calculated from Darcy's law equation. However, inaccurate (high) estimates result due to dispersion of soil particles when air is injected, and so this test is not recommended.

Pressure buildup test. This is a test developed in the petroleum industry in which the downhole pressure increase is monitored after a vapor extraction well is closed or shut in. This test was designed to test confined vapor deposits. Vapor pressures found in contaminated soil may not create significant pressure buildup when well is closed. This method is not recommended.

Pressure drawdown test. This method is the most accurate soil air permeability test and was developed in the petroleum industry. The method involves measuring drawdown or gauge pressure in a monitoring well at a given distance from a vapor extraction well, while air is extracted at a constant flow rate. Field measurements of gauge pressure are typically taken from at least three monitoring wells located at varying radial distances from the extraction well to compensate for lateral spatial variations. The test is conducted over a short time period—long enough to extract at least one to two pore volumes of air, yet short enough not to be hampered by variations in atmospheric pressure and soil moisture condensation.

After the tests have been completed, soil air permeability is estimated graphically using the field measurements. This involves plotting the vacuum measurements at the monitoring well vs. the natural logarithm of the time from the initiation of the test and then determining the slope of the best-fit regression line through the data. The equation that approximates expected pressure changes over time is[1,3]

$$P' = \frac{Q}{4\pi mK} \cdot \left\{ -0.5772 - \ln\left(\frac{r^2 \varepsilon \mu}{4KP_{atm}} \right) + \ln(t) \right\} \qquad (3.6)$$

where P' = vacuum measured at monitoring well
  Q = extracted flow rate
  m = stratum thickness
  K = soil air permeability
  μ = viscosity of vapor

r = distance of monitoring well where vacuum was measured
ε = vapor-filled void fraction (0.0–1.0)
P_{atm} = atmospheric pressure
t = time from start of test.

All of these parameters are obtained from the field and site data, except for the viscosity of air.

The slope (A) and y intercept (B) of the regression line of the pressure (negative) vs. natural log of time are given by the following equations:

$$A = \frac{Q\mu}{4\pi mK} \, , \tag{3.7}$$

$$B = \frac{Q\mu}{4\pi mK}\left[-0.5772 - \ln\left(\frac{r^2\varepsilon\mu}{4KP_{atm}}\right)\right]. \tag{3.8}$$

These equations can be used in two ways. If the flow rate from the extraction well (Q) and the thickness of the screened interval from which vapors are being extracted (m) are known, the soil air permeability (k) can be calculated by solving equation (3.7) as follows:

$$K = \frac{Q\mu}{4A\pi m} \, . \tag{3.9}$$

However, if Q and m are unknown, equation (3.7) is substituted into equation (3.8) and solved for K:

$$K = \frac{r^2\varepsilon\mu}{4P_{atm}\{\exp(B/A) + 0.5772\}} \, . \tag{3.10}$$

Air permeability of soil is also a function of the soil's intrinsic permeability (K_i) and liquid content. At contaminated sites, liquid(s) present in soil pores is a combination of soil moisture and nonaqueous phase liquids (NAPLs). Soil air permeability (K_a) can be estimated by multiplying the intrinsic permeability by the relative permeability (K_r).

$$K_a = K_i \cdot K_r \tag{3.11}$$

where K_r is a dimensionless ratio (0.0–1.0), describing the variation in air permeability as a function of air saturation.

The equation below is useful in estimating soil air permeability as a function of air saturation and liquid content.[4] The relative permeability is given by

$$K_r = \left(1 - S_e\right)^2\left(1 - S_e^{(2+\lambda/\lambda)}\right) \tag{3.12}$$

where S = θ / ε
S_e = $(S - S_r)/(1 - S_r)$
S = degree of saturation
θ = volumetric moisture content
ε = porosity of void space
S_r = residual saturation

S_e = effective saturation
λ = pore size distribution parameter.

When developing an estimate of relative permeability for a given soil matrix and liquid content, values of ε, S_r, S_e, and λ can be obtained from the literature.[5]

As noted earlier, there are factors other than soil air permeability that govern the airflow field during soil vapor extraction.

Porosity. Soils with higher porosities will allow a higher flow for the same induced vacuum. Porosity influences the intrinsic air permeability and is related to the particle size distribution of soil. Coarse textured soil will generally have higher porosities and thus higher permeabilities.

Subsurface conduits. The presence of subsurface features that are less restrictive to flow may result in preferential flow in directions different than expected based on site geologic conditions. Subsurface utility conduits (electric, telephone, sewer, and water) may be bedded in materials much more permeable than surrounding soils. Preferential flow paths will be formed if these conduits intersect the zone of vacuum influence.

The impact of screen interval and leakage from the surface will be discussed in later sections.

3.2.2 Contaminant Partitioning

Contaminant partitioning characteristics govern the performance and thus the eventual shutdown of the system at the end of a successful site cleanup. Organic contaminants can be present in the soil matrix in four basic ways (Figure 3.3):

- As an adsorbed film on soil particles and colloids
- As a residual immiscible liquid in the soil pore spaces (NAPL)
- As vapor in soil air present in the pore space
- As dissolved compounds in soil pore water, and groundwater
- In addition, as a floating product above the water table in the case of LNAPLs and as a pooled or sinking liquid in the case of DNAPLs.

When soil air remains undisturbed after contaminant spillage and infiltration, it becomes saturated by contaminant vapors volatilized from the liquid phase. This highly saturated vapor will be removed initially when soil vapor extraction is initiated. As this soil vapor is purged from soil pores, the concentration of contaminants in the extracted soil air begins to decrease as the process shifts away from equilibrium (Figure 3.4). At this stage, the interfacial mass transfer between the contaminant and flowing air controls the removal of contaminants from soil. Soil air, due to soil vapor extraction, moves much faster than the phase transfer taking place from the immiscible liquids, soil pore water, and the adsorbed film into the soil gas.

Optimization of the performance of a soil vapor extraction system, at a given site, requires a fundamental understanding of the mechanisms controlling contaminant removal from that particular soil matrix. Under actual field conditions, several mechanisms will occur simultaneously, and the contribution of each will shift along with the progress of remediation. Early in the treatment, direct volatilization from immiscible liquids and adsorbed film will dominate. Extraction of dissolved contaminants from the pore water and desorption from the mineral and organic fraction of the soil will take place only when the liquid mass of the contamination is gone. For a water-miscible volatile organic compound (VOC) such as acetone, extraction from the pore water will occur during the early phase of treatment.

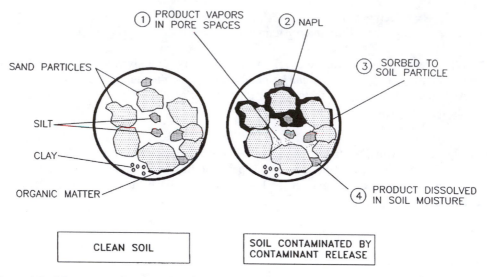

Figure 3.3 Phases of contaminants present in soil matrix.

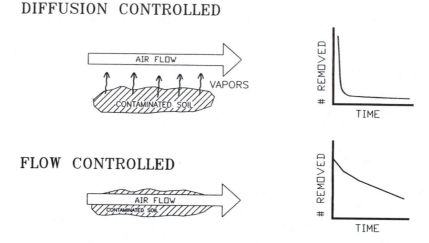

Figure 3.4 Decline in vapor concentrations under diffusion and flow-controlled regimes.

3.2.2.1   Contaminant Properties

The physical and chemical properties of a spilled compound control, to a great extent, the partitioning and thus the distribution into the various phases described earlier (vapor, dissolved in pore water, adsorbed to soil particles, and NAPLs).

The distribution of a compound among the four phases can be described by several parameters (Figure 3.5). The degree to which a compound partitions into the vapor phase is described by that compound's vapor pressure and Henry's law constant. The soil adsorption coefficient, k_d, describes the tendency of a compound to become adsorbed to soil. The solubility describes the degree to which a compound will dissolve into water.

3.2.2.1.1   Vapor Pressure

Vapor pressure is the parameter that can be used to estimate a compound's tendency to volatilize and partition into the gaseous phase. All solids and liquids possess a vapor pressure,

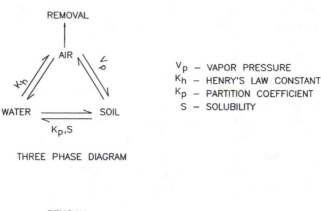

THREE PHASE DIAGRAM

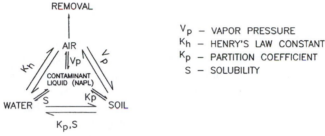

FOUR PHASE DIAGRAM

Figure 3.5 Distribution of contaminants into various phases.

and this parameter can be conceptualized as the solubility of the compound in air at a given temperature. More precisely, the vapor pressure is defined as the pressure exerted by the vapor at equilibrium with the liquid phase (NAPL) of the compound at a given temperature. Vapor pressure is typically expressed in terms such as millimeter of mercury (mmHg). Appendix C provides a listing of vapor pressures of the common environmental contaminants. Chemicals with vapor pressures greater than 0.5 to 1.0 mmHg (e.g., benzene, TCE) are expected to volatilize to a significant degree when released in the subsurface and will respond favorably to SVE technology.

For mixtures of compounds such as gasoline, the composition of the mixture also has a bearing on the vapor pressure according to the following relationship:

$$P_i = X_i A_i P_i^o \tag{3.13}$$

where P_i = equilibrium partial pressure of component i in the organic mixture
X_i = mole fraction of component i in the organic mixture
A_i = activity coefficient of component i in the organic mixture as a correction factor for compensating nonideal behavior[6,7]
P_i^o = vapor pressure of component i as a pure compound.

The above relationship, for example, states that the vapor pressure of benzene is related to the percentage (mole fraction) of benzene in gasoline.

Temperature also has a strong influence on the vapor pressure of a compound. It is stated that for most intermediate molecular weight organic compounds, vapor pressure increases three to four times for each 10°C increase in temperature.[6] In general, vapor pressure may be approximated by

$$P(T) = A \exp(-B/T) \tag{3.14}$$

where T = temperature in degrees Kelvin
       A, B = constants characteristic of the substance.

Based on the effects of temperature, expressed by the above equation, SVE effectiveness in the subsurface can be influenced by increase in temperature. However, seasonal fluctuations in soil temperature are dampened with depth of the vadose zone. Steam injection, soil heating, and radio-frequency heating are some techniques that may enhance the volatilization of the compounds. However, cost-effective application of these techniques requires much more research and development. In diffusion-limited conditions, increase in temperature may result in beneficial effects in increasing volatilization.

If the contamination present is not in the form of separate phase liquids (NAPLs), vapor pressure becomes a less accurate predictor of SVE effectiveness. Contaminant adsorption to soil, and solubility in soil moisture will determine the relative volatility.

3.2.2.1.2   Water Solubility

The solubility of a compound controls the degree to which the compound dissolves into pore water present in the vadose zone. Solubility also has an important impact on the partitioning fate and transport of the compound. Soluble compounds are likely to dissolve and move into the saturated zone due to infiltration. A fluctuating water table is another means of impacting contaminant concentration in the vadose zone and specifically in the capillary fringe.

Solubility of a compound can be defined as the maximum mass that can dissolve in pure water at a specific temperature (aqueous concentrations are usually stated in terms of weight per weight (ppm, ppb) or weight per volume (mg/l, μg/l)). For organic mixtures, such as gasoline, solubility is additionally a function of the mole fraction of each individual compound in the mixture. Appendix C presents the pure water solubilities for common environmental contaminants.

$$C_i = X_i A_i C_i^o \qquad (3.15)$$

where C_i = equilibrium concentration of compound i in the organic mixture
       X_i = mole fraction of compound i in the organic mixture
       A_i = activity coefficient of compound i in the organic mixture as a correction factor
              for compensating nonideal behavior[7]
       C_i^o = equilibrium solubility of compound i as a pure compound.

Under typical SVE scenario, vadose zone soils are relatively moist (10 to 14% by weight), and contaminants are generally dissolved in pore water. Solubility is also an important parameter for bioventing, since biodegradation of a compound is enhanced due to the increased availability for microbial uptake as a dissolved compound.

Cosolvation is another factor that should be taken into account when evaluating partitioning of contaminants in the subsurface. *Cosolvation* refers to a mobile phase consisting of multiple solvents that are miscible in one another. For example, the cosolvents can be pore water, benzene, toluene, and TCE present in NAPL form. The solubilitization of one contaminant into another will impact the partitioning and, thus, the removal efficiency under SVE operation.

3.2.2.1.3   Henry's Law Constant

The volatilization of a compound dissolved in water is governed by Henry's law, which describes the relative tendency for a compound in solution to exist in the vapor phase. It is somewhat analogous to the vapor pressure, which describes the partitioning behavior between a pure substance—instead of a compound in solution—and its vapor phase. The proportionality constant that relates concentration in solution to the partial pressure is known as Henry's law constant (K_H) and may be reported as "dimensionless" or as atm·m³/mol.

$$K_H = \frac{C_v}{C_l} \tag{3.16}$$

where K_H = Henry's law constant
         C_v = concentration in vapor phase at the water–vapor interface
         C_l = concentration in liquid at the water–vapor interface.

Henry's law constant is highly temperature-dependent and increases with increasing temperature. It is stated that Henry's law constant increases 1.6 times for an increase of 10°C in temperature.[7]

Appendix C lists the Henry's law constants for common environmental contaminants, and the relationship between vapor pressure, solubility and Henry's law constant is graphically presented in Figure 3.6. Under moist soil conditions, efficiency of SVE will be dependent on the Henry's law constant of the compound, similar to air stripping.[8] For example, although acetone is very volatile, efficiency of removal by SVE is very low due to its high solubility. Acetone tends to biodegrade very well and will be removed readily under bioventing.

3.2.2.1.4   *Soil Adsorption Coefficient*

Adsorption, absorption, and sorption are three terms that refer to similar phenomena.[9] *Adsorption* is defined as the accumulation occurring at an interface, *absorption* as the partitioning between two phases (e.g., accumulation from groundwater into soil organic carbon), and *sorption* as including both adsorption and absorption.[10,11] Often the terms adsorption and sorption are used interchangeably, as will be the case in this book.

In general, adsorption phenomena can be classified as either sorbent- or solvent-induced.[10] Sorbent-induced adsorption occurs when there is an attraction between the soil matrix (sorbent) and the contaminants (solute) and the contaminant accumulates at the surface due to the affinity of the surface for the contaminant (e.g., cation exchange at the mineral sites). Solvent-induced adsorption occurs when the contaminant is hydrophobic (nonpolar, water-disliking, and low solubility) and finds it energetically favorable to accumulate at an interface or partition into a nonpolar phase (e.g., partitioning onto the soil organic carbon) rather than remain in the water phase.

Adsorption of contaminants to soil particles is impacted by the two phenomena described above. The soil organic content and its mineral adsorptive surfaces are both capable of contaminant adsorption. Adsorption of contaminants at both these locations increases the immobility of the contaminant and decreases the relative volatility during SVE. Soil adsorption becomes particularly important (rate-limiting) under drier conditions. Soil moisture may decrease as airflows continuously, during SVE, due to the vapor pressure of water being 10 mmHg at typical soil temperatures. Figure 3.7 illustrates the difference in adsorption potential under wet and drier soil conditions.

Adsorption of a contaminant to soil and organic matter is described by the contaminant's soil adsorption coefficient, K_d. The total organic carbon (TOC) content, f_{oc}, strongly influences the adsorption capacity of nonionic organic compounds onto the soil matrix. Due to the nonpolar nature of most organic contaminants, there is little correlation between clay content and contaminant adsorption. While the adsorption coefficient, K_d, of a chemical has been observed to vary significantly from soil to soil (due to varying f_{oc}), it was observed that normalization of the K_d values by the respective f_{oc} values resolved in a parameter, k_{oc}, that

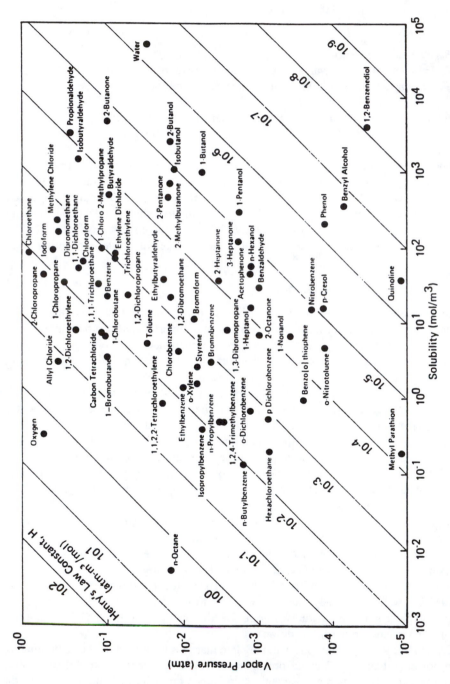

Figure 3.6 Solubility, vapor pressure, and Henry's law constant for selected chemicals. (After Lyman, W. J., Reehl, W. F., and Rosenblatt, D. H., *Handbook of Chemical Property Estimation Methods: Environmental Behavior of Organic Compounds*, McGraw-Hill, New York, 1982.)

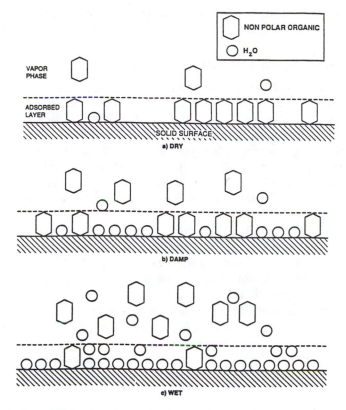

Figure 3.7 Illustration of VOC adsorption under three moisture regimes.

was much less variable (independent of the soil and a function only of the compound). Another way of expressing this relationship is defining K_{oc} as a proportionality constant.

$$K_d = f_{oc} \cdot K_{oc} \qquad (3.17)$$

where K_d = partitioning coefficient
f_{oc} = percentage of fraction of organic carbon in soils
K_{oc} = organic carbon partitioning coefficient.

Typical values of f_{oc} for subsurface soils have been reported in the range of 1 to 8%,[9] while alluvial sandy materials have been reported in the range of 0.02 to 1.0%. For soils with lower organic carbon content ($f_{oc} < 0.1\%$), adsorption to mineral grains may be dominant.

Researchers have evaluated a number of parameters for estimating K_{oc} values based on fundamental properties of the contaminant. One of the easiest way of obtaining K_{oc} is from the octanol–water partitioning coefficient (K_{ow}) of the contaminant. The K_{ow} of a contaminant describes the partitioning between water (a polar phase) and octanol (a nonpolar phase). The octanol–water partitioning is analogous to the partitioning in the soil matrix between the pore water and soil organic matter. There are many equations reported in the literature to describe the relationship between K_{ow} and K_{oc}.[9] An example is provided by the equation below:[12]

$$\log K_{oc} = 0.999 \log K_{ow} - 0.202 . \qquad (3.18)$$

Most organic contaminants are more easily adsorbed to the soil than they are desorbed. Therefore, it takes much more time and energy for the desorption than adsorption of a contaminant in the soil matrix. This phenomenon, known as *hysteresis*, tends to reduce the mass removal efficiency during SVE, particularly during the tail end of system operation.

3.2.2.1.5   Biodegradability of Contaminant

One of the most creative uses of SVE technology is in manipulating subsurface oxygen levels to maximize *in situ* biodegradation of aerobically biodegradable compounds in comparison to volatilization. This modification to conventional SVE is known as bioventing or bioenhanced soil venting. Bioventing can reduce effluent vapor treatment costs and can also result in the remediation of semivolatile organic compounds that are not readily volatilized.

Although microorganism populations in the vadose zone are sharply reduced in number and species diversity with depth, significant microbial degradation of some compounds occurs throughout the vadose zone. In bioventing, the natural biodegradation processes are enhanced by optimizing the system.

Most of the petroleum organic compounds such as benzene, toluene, ethylbenzene, and xylenes are very easily biodegradable under aerobic conditions. Some of the halogenated compounds such as methylene chloride and chlorobenzene will also biodegrade under aerobic conditions.[13,14] Most of the alcohols, ketones, and phenolic compounds will degrade under aerobic conditions (see Appendix D).

In contrast to the compounds discussed above, microbial degradation of volatile aliphatic chlorinated compounds, such as TCE and PCE, occurs primarily by reductive dehalogenation. Reductive dehalogenation of these compounds will take place only under reduced, anaerobic conditions. Hence, biodegradation of aerobically nonbiodegradable compounds during SVE is nonexistent or insignificant.

3.2.2.1.6   Weathering

The concept of weathering applies to SVE due to the effect SVE has on the composition of the organic contaminant mixture present in the subsurface. At a gasoline spill site, initially at system start-up, the compounds removed will be primarily the more volatile, lighter-end fractions. As SVE proceeds, the extracted vapor will likely have less of the lighter-end fractions and will be composed mostly of heavier compounds. This process is commonly described as weathering. Changes in organic contaminant mixture characteristics taking place due to natural volatilization, solubilization, and biodegradation of the readily mobile lighter compounds is also characterized as weathering. The degree of weathering will impact the overall effectiveness of SVE operation.

3.2.2.1.7   Other Contaminant Properties

There are several other contaminant properties that may influence the volatility and thus effectiveness of SVE operation. These properties are not as significant as the parameters described in the last few sections, and hence may be considered secondary in importance.

These properties are molecular weight, molecular structure, and polarity. Heavier and large molecules travel slowly to and from soil micropores and tend to adsorb very strongly to soil surfaces. During the tail end of SVE system operation, these properties may end up as the rate-limiting mass transfer properties. This may be a significant factor at petroleum spill sites and may influence the degree of mass removal and the residual mass left behind after a reasonable time of SVE operation.

3.2.2.1.8   Contaminant Partitioning Summary

When a volatile NAPL is present in the soil matrix, the bulk of the mass removed during SVE will be due to direct volatilization influenced by the vapor pressures of the contaminants.

This is analogous to a fan blowing past a pool of gasoline or TCE. It has been reported in the literature that more than 95% of NAPL can be removed after removing several hundred pore volumes of air.[15] In the presence of NAPL, mass removal rates are often linearly related to the airflow rates. When NAPL is not present in the subsurface, airflow requirements become very different, and are often governed by nonequilibrium rate-limiting conditions.

Under moist soil conditions and when there is no NAPL present, contaminant partitioning in the vadose zone can be described by the following equation.[16]

$$C_T = P_b C_A + w C_L + a C_G \qquad (3.19)$$

where C_T = total quantity of contaminant per unit soil volume
         C_A = adsorbed chemical concentration
         C_L = dissolved chemical concentration
         C_G = vapor concentration
         ρ_b = soil bulk density
         w = volumetric water content
         a = volumetric air content.

The equilibrium relationship between vapor concentration and the associated pore water concentration is given by Henry's law:

$$C_G = K_H \cdot C_L \qquad (3.20)$$

where C_G = vapor concentration
         K_H = Henry's law constant
         C_L = pore water concentration.

The relationship between equilibrium dissolved concentration and adsorbed concentration is given by

$$C_A = K_d \cdot C_L \qquad (3.21)$$

where C_A = adsorbed concentration
         K_d = adsorption coefficient
         C_L = dissolved concentration.

Again, $K_d = f_{oc} \cdot K_{oc}$.

3.2.2.2   Soil Properties

The soil environment, like the contaminant characteristics, has a significant impact on the effectiveness of SVE. Soil properties such as porosity, air permeability, water content, organic matter content, heterogeneity, and surface seals impact the movement of contaminant vapors.

3.2.2.2.1   Soil Porosity

Vapor migration in the subsurface occurs principally through air-filled pore spaces. Factors influencing vapor migration include the air-filled porosity (which is affected by the water content) of a soil and the orientation of the soil pores. Decreasing soil porosity will generally reduce the efficiency of SVE due to the decrease in diffusive mass transport from the soil matrix into the soil gas. Decrease in porosity also impacts SVE effectiveness due to the decrease in available cross-sectional area for airflow. Lower porosity soils also have higher available surface area for contaminant binding.

3.2.2.2.2   Water Content

Soil water content has competing effects on the air permeability. The primary effect of pore water is to reduce the air-filled porosity of soil. Research information is available to indicate that SVE would be more successful at lower water contents, since a greater percentage of the pore space will be air-filled and thus the induced airflow will be greater for a given vacuum.[15,17]

The water content also has a significant effect on the success of SVE through its effect on adsorption characteristics of organic compounds. Decrease in soil moisture allows for more contaminant adsorption onto the soil matrix to play a more prominent role in mass transfer during SVE. If the soil adsorptive capacities are strong, the benefits of soil dewatering may be partially offset due to the increased binding of contaminants.

Soil moisture may decrease as air is circulated through soil during SVE, since water has a vapor pressure of 10 mmHg at typical soil temperatures. As illustrated in Figure 3.7, under low soil moisture conditions, contaminants will adsorb directly on soil surfaces where fewer water molecules are competing for adsorption sites. A description of contaminant adsorption as a function of moisture content is also presented in Figure 3.8.[18] Adsorption was the highest during oven-dry conditions and decreased linearly with moisture content up to one mono layer. Water sufficient to form 1 to 5 mono layers exhibited complex behavior due to interactions between the contaminant and surface bound water. At moisture contents greater than five mono layers, there was a gradual increase in adsorption which was attributed to increase in contaminant resolution.

3.2.2.2.3   Soil Heterogeneity

Soil heterogeneity is caused by the structure, stratification, type, and size of soil particles that influence both contaminant migration and subsurface airflow pathways. An example of heterogeneous soil conditions is a coarse silty sand formation with interbedded clay lenses. Another example is coarse sand formation overlain by a silty sand formation, and the preferred extraction well configuration under this condition is shown in Figure 3.9.

Preferred flow path due to highly permeable macropores formed by fractures, cracks, root casts, or earthworms should also be taken into consideration. The presence of subsurface features, such as utility conduits, may also result in preferential flow pathways. Preferred flow pathways have major implications with regard to SVE effectiveness due to potential short-circuiting of induced airflow pathways. If short-circuiting occurs, contaminants adhering to soils through which limited airflow occurs may not be removed as rapidly as those present in zones through which a large volume of airflows. Removal of contaminants from soil particles in dead zones would therefore be more closely related to diffusion effects (Figure 3.10). Removal of contaminants from these dead zones may not be enhanced by increased vacuum or flow rates, but rather removal will be limited by diffusion from these zones to the more continuous flow paths.

3.2.2.2.4   Surface Seals and Air Inlet Wells

The airflow pathways will be influenced by the amount of airflow entering near the extraction well (where the pressure gradient is the greatest). An impermeable surface seal prevents air from entering near the injection well and forces air to be drawn from a greater distance and thus contact a greater volume of soil (Figure 3.11). Surface seals may also prevent infiltration of rainfall, and thus reduce the amount of water removed by the extraction well.

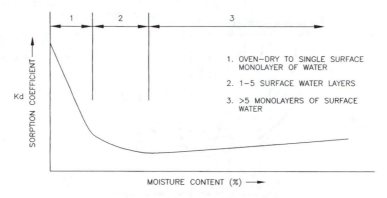

Figure 3.8 Adsorption of contaminants as a function of moisture content.

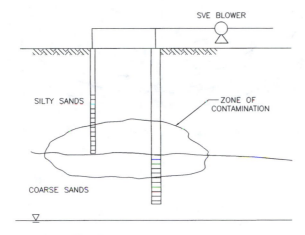

Figure 3.9 A cluster extraction well under varying geologic conditions.

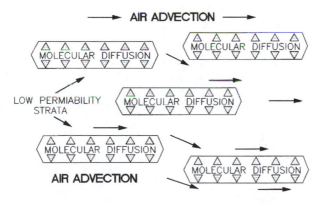

Figure 3.10 Advective/diffusive airflow schematic.

Depending on the characteristics of the site, different materials can be used as an impermeable seal. Existing paved surfaces, such as asphalt or concrete layers, or a rolled-out flexible membrane lining can be used as surface seals.

Injection or air inlet wells can also be located at numerous places at a SVE site depending on the needs. The function of air inlet wells is to control the subsurface airflow and overcome preferential pathways and dead zones (Figure 3.12). Air injection wells are passive and can be installed at the edge of a site, so as not to induce flow of contamination from an adjacent site. Typically, injection and inlet wells are similar in construction and can be interchanged during SVE operation.

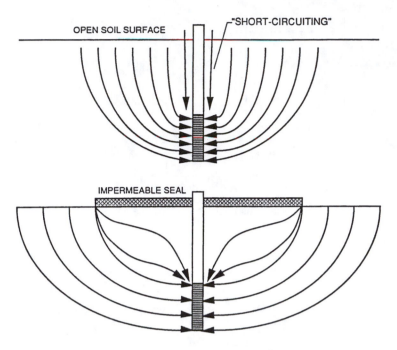

Figure 3.11 Effect of surface seal on vapor flow paths.

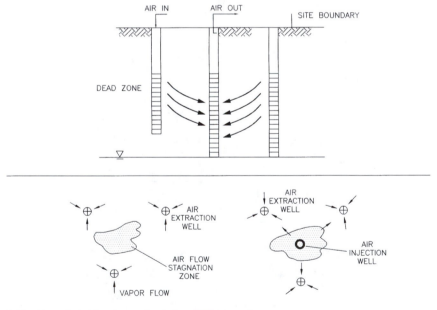

Figure 3.12 Use of air injection wells during SVE operation.

3.2.2.2.5   *Depth to Water Table*

If the SVE well penetrates the water table, the water table within the well will rise by an amount equal to the applied vacuum. A 60 in. water column vacuum at the well head will therefore result in a water table rise of 5 ft (Figure 3.13). Hence, if the screen of the SVE well is less than 5 ft, the rise in the water table will block the passage of air into the SVE

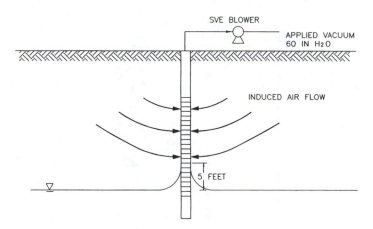

Figure 3.13 Water table rise during SVE operation.

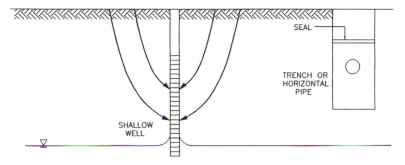

Figure 3.14 Limited radius of influence from vertical wells under shallow water table conditions and the use of horizontal extraction pipes.

well. Even if the screen length is more than 5 ft, the effective radius of influence will be reduced. This situation will be encountered more in conditions where the water table is shallow and the well is inappropriately designed. Installing the bottom of the SVE well screen above the water table will minimize the impact caused by water table uplift. Horizontal wells or pipes installed in trenches can be used in shallow water table conditions, thus expanding the available screen length for airflow and enabling operation of the system at reduced vacuum levels. Under this condition the water uplift will be minimized (Figure 3.14).

However, the fluctuation of the water table will create conditions of screen submergence during certain times of the year. The need to induce airflows at or near the capillary fringe will also require the bottom of the screen to be placed very close to the water table.

3.3 APPLICABILITY

There are many criteria that need to be evaluated before deciding the applicability of SVE at a given site. The applicability of this technology is greatly influenced by the information obtained from contaminant and site characterization activities (Figure 3.15). Contaminant characterization will include composition, type, age, concentration, phase, and distribution of contaminants in the subsurface. Site characterization will include geologic conditions, soil moisture content, manmade site conditions, topography, depth to water table, and horizontal and vertical extent of contamination.

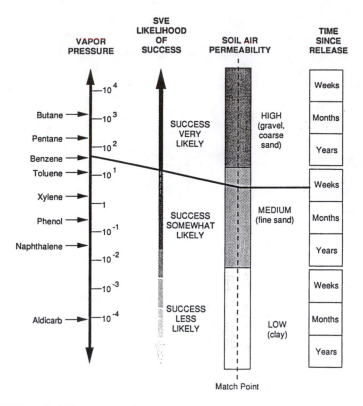

Figure 3.15 SVE applicability nomograph.

3.3.1 Contaminant Applicability

SVE is applicable when the contaminants present in the subsurface are volatile. As a simplified guideline, a compound or mixture of compounds is a likely candidate for SVE application if it has both of these characteristics:

- vapor pressure of 1.0 mmHg or more at 20°C
- Henry's law constant greater than 0.001 atm·m³/mol, or greater than 0.01 in the dimension less form of Henry's law constant.

Examples of contaminants amenable to SVE are listed in Table 3.1.[19]

Table 3.1 Contaminants Amenable to SVE

Contaminant	Henry's law constant (atm·m³/mol)	Vapor pressure (mmHg)
Benzene	0.00548 (25°C)	76 (20°C)
Toluene	0.00674 (25°C)	22 (20°C)
Trichloroethylene (TCE)	0.0099 (20°C)	57.8 (20°C)
Tetrachlorethylene (PCE)	0.00029	20 (26.3°C)

The impact of solubility is reflected by the Henry's law constant since the additive effects of solubility and vapor pressure are nonlinear. As noted in previous sections, the phase distribution, concentrations, and contaminant type will determine the mass removal efficiencies and the time required for site cleanup.

Evaluation of bioventing should be accompanied by the comparison of biodegradation rates to volatilization of the contaminants. If the contaminants are biodegradable, emphasis should be focused on the required mass removal rates and the time of cleanup. Presence of excessive levels of CO_2 and depletion of O_2 in the soil gas is an indication of ongoing natural biodegradation of contaminants in the vadose zone.

In most cases, the contaminant distribution is superimposed upon the geologic cross section of the area to be treated for the evaluation of SVE applicability. Appropriate laboratory methods should be used to analyze the volatile organic compounds (VOCs) present in the soil matrix.

In some instances, it becomes necessary to empirically measure the optimal system performance rather than predict from theoretical considerations. Field-scale pilot studies and bench-scale (laboratory) experiments are cost-effective means of achieving this objective. Field-scale pilot studies will be described in a later section. Laboratory experiments can be used to determine the mass removal rates and to optimize the required airflow, moisture content, and other control parameters.

Laboratory experiments can be conducted using soil columns. The effect of variations in flow rate and other parameters on microscopic scale mass transfer limitations can be studied this way. However, it must be recognized that laboratory experiments are conducted on disturbed soil samples that are inherently unrepresentative of actual field conditions. In addition, it must be remembered that the soil sample to be tested is only representative of a very localized area of a given site.

3.3.2 Site Characterization

SVE is typically more applicable in cases where the contaminated unsaturated zone is relatively permeable and homogeneous. Geologic parameters such as porosity, soil structure, and air permeability influence the extractability of induced airflow. Site surface topography can also influence the success of an SVE system. Ideally the site should be covered by an impermeable surface to minimize short-circuiting of airflow and infiltration. The presence of utility trenches (generally constructed with high permeability fill material) will also cause short-circuiting of airflow pathways.

Site characterization activities include

- geologic and hydrogeologic characterization
- contaminant distribution assessment (vertical and horizontal)
- soil gas surveys
- water table elevation and fluctuations
- barometric pressure changes
- presence of LNAPL on top of the water table or isolated globules ("ganglia") in the soil matrix
- groundwater concentrations.

Comprehensive geologic assessment will be complete only when information is collected on distinct geologic strata, subsurface conduits, depth to the water table, and seasonal fluctuation of the water table. Additionally, the estimated values of the physical properties (such as permeability, moisture content, organic carbon fraction) of each of the geologic strata should be collected. It is always important to develop detailed geologic cross sections in as many directions as possible. Most geologic cross sections are developed through analysis of drilling logs and site information available on tanks, pipes, etc. An example of a subsurface cross section used to evaluate the applicability of SVE is shown in Figure 3.16.

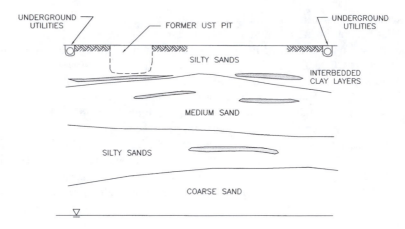

Figure 3.16 Typical cross section at a candidate site for SVE.

3.4 SYSTEM DESIGN

SVE technology has been implemented in the past using various design approaches which vary from an intuitive empirical approach to using sophisticated design models. The primary goals of SVE system design is to achieve the following objectives:

- Specification of system components such as vacuum blower, extraction wells, effluent air treatment unit, air/water separator, pipes, manifolds, etc.,
- Selecting operating conditions such as extraction vacuum levels, airflow rates, contaminant vapor concentrations, etc.,
- Estimation of cleanup levels, remediation time frame, residual concentrations, etc.,
- Meeting the constraints imposed by cost limits and permit limits, etc.

Prior to initiation of system design, a thorough evaluation of applicability of SVE should be completed. However, due to uncertainties and inherent limitations associated with natural heterogeneities, site characterization data, and accurate prediction of contaminant partitioning, system design objectives have to be realistic and flexible. Hence SVE system design has to incorporate modification of operating conditions even after system start-up. The additional cost required to build in adequate flexibility is usually very small compared to the potential benefits in the long term.

SVE system design activities can be grouped into two basic categories: selecting system operating parameters and selecting system components.

1. System operating parameters
    - extraction airflow rate
    - extraction vacuum
    - radius of influence
    - contaminant concentrations
2. System components and equipment
    - number of extraction wells
    - well locations
    - well construction (depth, screened interval)
    - extraction blower
    - air water separator
    - equipment manifolding and piping
    - instrumentation and control (measurement devices such as flow meters, pressure gauges, and control valves, etc.)
    - vapor treatment unit

Table 3.2 SVE System Design Criteria

	Design criteria	Significance
1.	The necessary vacuum to induce adequate airflow and radius of influence	The required vacuum dictates the type and size of vacuum pumps or blowers to utilize.
2.	Airflow rates to achieve adequate mass removal and closure within required time	Airflow requirements will dictate vacuum equipment and air emission control equipment sizing. Airflow requirments to be dictated by modeling, pilot testing, and closure levels.
3.	Number of vent wells to achieve closure within required time	The number of wells will be dictated by need to maintain adequate airflow in the subsurface. The pilot test in conjunction with the computer simulation will determine the number of required wells.
4.	Well screen positioning	Positioning of the well screens will be critical to successful design. Well screens will be placed within the water table to accommodate water table fluctuations and willl be screened at various intervals to ensure adequate flow throughout the soil. Screening intervals will be particularly important in ensuring that flow is induced in the most contaminated zone.
5.	Soil moisture content monitoring and manipulation	Soil moisture controls air permeability and equilibrium concentrations. Its proper manipulation may enhance system performance. High moisture in the extracted air stream requires separation before vapor treatment.
6.	Passive well placement	Passive wells provide influent air and should be located to minimize "dead," no-flow zones that may develop in multiple extraction point configurations.
7.	Off-gas treatment technology	Pilot-test data, cost analysis, and operation and maintenance logistics will enable the appropriate selection of off-gas treatment technologies.
8.	Operation and maintenance and remote monitoring	If the property is to be developed and the wells will not be accessible, remote system monitoring needs to be considered.
9.	No-failure piping and well design	Underground piping layout may consist of duplicate sections in sensitive areas to ensure nonfailure. All underground piping should be vacuum- or pressure-tested prior to burial.
10.	Closure SVE system management	Management strategies should be developed for the last months prior to closure. System pulsing, passive venting with natural wind-driven vacuum can be considered.

Table 3.2 presents the significance of some design criteria on SVE system performance.

3.4.1 Pilot Testing

It is prudent to perform a field pilot test at all SVE sites to determine whether the site is a suitable candidate for soil venting. Since the data collected during a pilot test involves collecting site-specific engineering design parameters, it is logical to call this test a field design test. Field design test activities are focused upon the collection of data to determine soil air permeability, zone of vapor extraction (radius of influence), extracted soil gas concentrations and composition, and the required flow rates and vacuum levels. Regardless of the size of the final system and site conditions, the field design test system should consist of the following components, at a minimum (Figure 3.17):

- vapor extraction test well
- vacuum blower to induce airflow
- at least three vacuum monitoring points, preferably at varying radial distances from the extraction well
- vapor treatment system (if required); vapor phase carbon may be sufficient for the field design test
- calibrated flow meter(s) or pitot tube(s)
- calibrated vacuum gauges
- sampling ports in the process lines

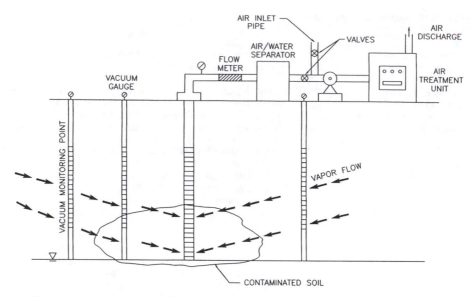

Figure 3.17 SVE pilot test schematic.

- sampling devices (sampling pumps, syringes)
- soil gas analytical instruments (PID, FID, field gas chromatograph, etc.)

A field design test usually consists of extracting contaminated air from the extraction well, at various flow regimes, and measuring the response to the applied vacuum in at least three monitoring points. The vacuum in the monitoring points is measured by installing vacuum seals and gauges on the surrounding monitoring wells and taking readings from them at different intervals during the various flow regimes. The extraction airflow rate from the venting wells is periodically measured, using flow meters or pitot tubes, at different time intervals of the various flow regimes. Air samples are taken from the extracted air at critical times of the test to analyze for the concentration and composition of contaminants. At least three air samples should be collected during each flow regime: one right after initiation of air extraction, one right before ending air extraction (after removal of at least one to two pore volume), and one during the middle of the test.

The extraction well location for the field design test should be selected within the area to be treated by the full-scale system. If the geologic conditions across the contaminated zone vary significantly, the need for more than one field testing location should be evaluated. The screened interval of the extraction well should be placed within the contaminated zone. Since induced airflow below the screened interval is generally not significant, the screen should be placed accordingly to attain the desired air distribution within the contaminated zone. Caution should be exercised to avoid any short circuit pathways (such as utility trenches) to obtain the best quality data.

The configuration of the typical SVE vertical well is shown in Figure 3.18. Horizontal wells or trenches should be considered for shallow water table conditions. In practice, existing groundwater monitoring wells have been used for SVE field design tests as vapor extraction wells. This is usually done to achieve cost savings, and caution must be provided to ensure that the screened interval of the groundwater monitoring well will provide the required air distribution.

A field design test will require at least three vacuum monitoring points in a homogeneous setting. The number of required monitoring points and the depths of installation will increase if the site geology is heterogeneous (Figure 3.19). Often one tries to locate the test in an area where existing groundwater monitoring points can be utilized as the vacuum monitoring

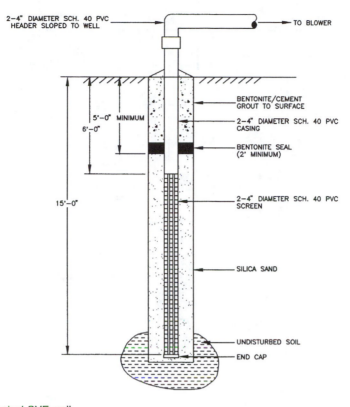

2–4" DIAMETER SCH. 40 PVC
HEADER SLOPED TO WELL

TO BLOWER

5'–0" MINIMUM
6'–0"

15'–0"

BENTONITE/CEMENT
GROUT TO SURFACE

2–4" DIAMETER SCH. 40 PVC
CASING

BENTONITE SEAL
(2' MINIMUM)

2–4" DIAMETER SCH. 40 PVC
SCREEN

SILICA SAND

UNDISTURBED SOIL

END CAP

Figure 3.18 Typical SVE well.

points. Again, caution should be exercised to avoid compromise on the quality of data collected. Vacuum monitoring points can also be installed very cheaply with small-diameter driven points with perforations at the bottom.

The capacity of the vacuum blower to be used for the field design test is usually picked based upon experience, geologic conditions, the expected flow rate, and the power supply needs. Availability of various types of vacuum pumps within specific flow and vacuum ranges is shown in Figures 3.20 and 3.21. Depending on the site geologic conditions, a high flow, low vacuum regime or a low flow, high vacuum regime may be required. Two vacuum blowers operated in series or parallel may be required for the field design test to attain the flow regime under certain site-specific conditions.

The assembly and connection of the vapor extraction test well, with the vacuum blower, flow meters, and pressure gauges are shown in Figure 3.17. Since most vacuum blowers are driven by fixed-speed motors, extraction flow rates are often controlled by installing flow control valves and an air inlet/dilution valve at the suction end of the vacuum blower. Partial opening of the air inlet valve provides the means to control the airflow rate from the extraction well and also the potential for overheating of the vacuum blower. It is very important to locate the flow meters and vacuum gauges before the air inlet valve for accurately measuring the flow characteristics of the extraction well (Figure 3.17).

During the field design test, two to three flow regimes can be achieved by manipulating the air inlet/dilution valve. Initially, the air inlet valve should be closed when the vacuum blower is activated. The vacuum distribution in the soil should be allowed to reach steady conditions. One can check whether the steady condition is reached by reading the changes in induced vacuum at the vacuum monitoring wells. Usually stabilization takes place within the first 15 to 30 min of pumping. Then the following parameters should be monitored:

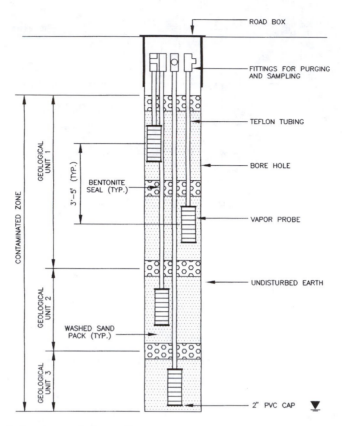

Figure 3.19 Nested probe monitoring well.

1. Vacuum at the extraction wellhead
2. Flow rate from the extraction well
3. Induced vacuum at the monitoring wells
4. Concentrations and composition of the extracted air samples

The next step is to repeat the above steps by opening the air inlet valve to get 20 to 25% decrease in vacuum at the extraction wellhead. Similarly, the inlet valve can be opened further to get a 40 to 50% decrease in vacuum at the wellhead.

As noted earlier, the screening of the applicability of SVE should be performed before and during the field design test. As a summary of the discussion in previous sections, Table 3.3 presents the variables that will impact the applicability, selection, and design of an SVE system.

3.4.2 Design Approaches

3.4.2.1 Empirical Approach

This is the most simplistic of all the design approaches and is based upon previous experience, a good understanding of the site conditions, site constraints, general guidelines, and intuition. The design is based on the basic geologic information and the selection of system components to meet the predicted operating conditions. Unfortunately, performance of the resulting system may not meet the remediation objectives.

In some cases the use of existing system or a skid-mounted system bought from a vendor could be made to match the design objectives. However, this approach also will be inefficient in meeting the remediation objectives. The initial cost savings may be offset by poor system performance and potential redesign of the whole system.

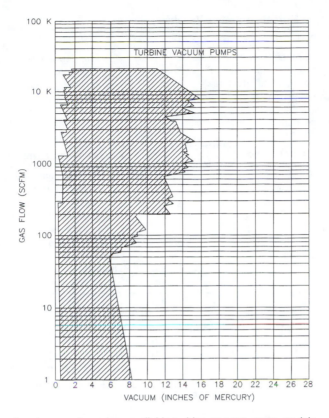

Figure 3.20 Operational ranges based on available turbine vacuum pump models.

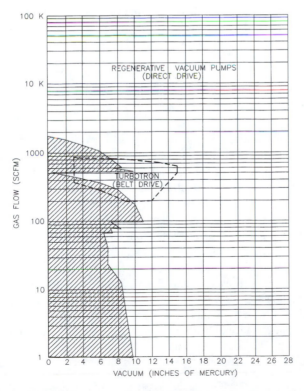

Figure 3.21 Operational ranges based on available regenerative vacuum pump models.

Table 3.3 Soil Vapor Extraction System Variables

Site conditions	Soil properties
Distribution of volatile organic compounds	Air permeability
Depth to groundwater table	Porosity
Location of heterogeneities and structures	Organic carbon content
Atmospheric pressure	Soil moisture content
Barometric changes	Particle size distribution
Temperature	Soil structure

Contaminant properties	Process control variables
Vapor pressure	Air extraction rate
Henry's law constant	Extraction vacuums
Solubility	Extraction well locations
Adsorption coefficient	Radius of influence
Density	Ground surface seals
	Contaminant levels and composition in the extracted air
	Extracted air temperature
	Extracted air moisture content
	Power usage

3.4.2.2   Radius of Influence Approach

This approach is the most widely used approach in the industry today.[20] The extraction flow rate vacuum, number of wells and their locations are selected based on the information obtained from a field pilot test. The pilot test is conducted to obtain steady state vacuum vs. distance relationships for a given site.

The steady state vacuum levels measured at monitoring wells located at varying radial directions are plotted on a semi-log graph paper, as shown in Figure 3.22. Vacuum levels are measured in various radial directions to compensate for the possible geologic variations in the subsurface. A best-fit line is then drawn through the points, as shown in Figure 3.22. The radius of influence of the extraction well is then determined as the distance at which a sufficient level of vacuum will be present to induce airflow. This "cut-off" vacuum level has been variously defined as 0.1 in. H_2O, 1.0 in. H_2O, or 10% of the applied vacuum at the extraction well. This arbitrary cut-off vacuum level is an empirical value based on prior experience and geologic conditions. The tailing effect of vacuum with distance is shown in Figure 3.23.

Once the radius of influence has been determined, overlapping circles with this radius are drawn on a site map over the zone of contamination (Figure 3.24). The number of wells and the locations of wells to cover the contaminated zone can be obtained this way. It must be noted that the vacuum vs. distance relationship may vary across the site and that the radius of influence may not be constant for all the extraction wells.

In this approach, the required screen lengths are estimated according to the depth of contamination as shown in a geologic cross section. Once the number of wells and the screen interval in each well has been selected, the total extraction airflow rate can be estimated using the curves shown in Figure 3.25. Selection of the appropriate vacuum blower can be done by matching the required airflow rate and the vacuum levels to the appropriate operating curve of various off-the-shelf blowers.

In practice, some sites may exhibit low flow, high vacuum conditions (less permeable geology) and other sites may exhibit high flow, low vacuum conditions (more permeable geology). Depending on the conditions encountered, it may be cost-effective to operate two smaller-capacity blowers in series or parallel instead of a single large-capacity blower (Figure 3.26).

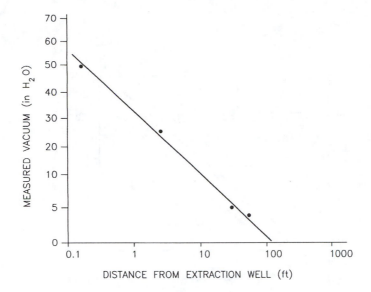

Figure 3.22   Vacuum vs. distance plot on a semi-log paper.

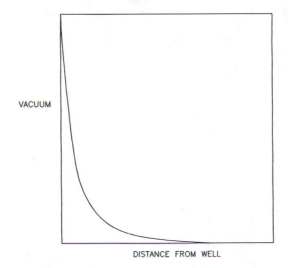

Figure 3.23   Tailing effect of vacuum away from an extraction well.

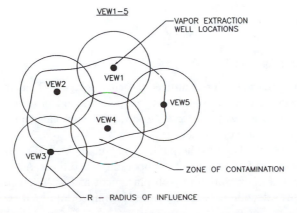

Figure 3.24   Determination of the required number of wells from the radius of influence.

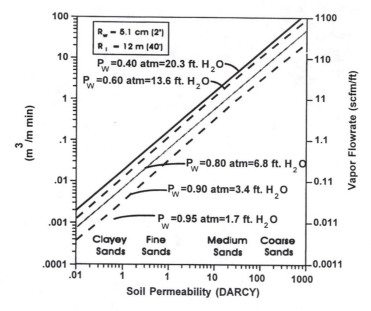

Figure 3.25 Airflow generation plot.

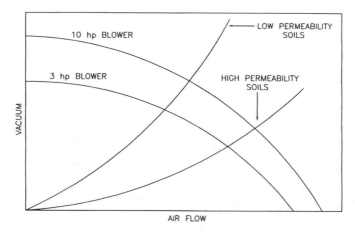

Figure 3.26 Use of blowers in series or parallel.

3.4.2.3   Modeling Approach

Computer simulation modeling is utilized to better design soil vapor extraction systems. This allows the model user to vary operating parameters and observe a simulation of the result. The type of models available for SVE system design can be classified into airflow models and multiphase transport models.

Flow-type models, available in the public domain, such as HYPER VENTILATE and AIR 3D, simulate only airflow under varying operating conditions. Airflow models therefore allow for selection and optimization of well placement and well screening within the contaminated zone. This selection is based on a predetermined, minimal acceptable airflow.

Multiphase transport modeling allows for simulating the SVE process over time based upon the selected well layout. The multiphase transport models are therefore often preceded by simple airflow models that locate extraction wells. The multiphase transport models simulate the subsurface perturbation caused by the SVE system to the contaminant partitioning and predict contaminant concentrations within the various media at different time frames.

3.4.2.4 Example of a Modeling Approach

The governing equations of groundwater flow provide a good approximation to gas flow in the vadose zone, particularly when used to represent pressure gradients encountered in most SVE systems.[21–24] The following are groundwater equations used for airflow in the vadose zone:

Theis Equation   *Jacob Equation*

$$s = \frac{114.6Q}{T} W(u) \qquad s = \frac{246Q}{T} \log\left[\frac{2.246Tt}{r^2 S}\right], \qquad (3.22)$$

$$u = \frac{r^2 S}{4Tt}, \qquad (3.23)$$

where s = drawdown in feet of air
T = gas transmisivity in ft²/day (= $K_g b$)
Q = flow rate in cfm
t = extraction time in days
r = distance from center of well in feet
S = storage coefficient
$W(u)$ = Theis well function of u
b = vadose zone thickness in feet

Theis equation is valid for all r, and Jacob equation is not valid for larger r (away from the extraction well). The limitations of both these equations are due to the assumption that there is no leakage of air from the surface and that airflows only in the horizontal direction.

The extrapolated logarithmic radius of influence R is given by the following equation, (since in the Jacob equation $s = 0$ at the limit of radius of influence):

$$R = \sqrt{\frac{2.246Tt}{S}} \qquad (3.24)$$

The straight-line approximation of the pressure drawdown data gives an underestimation of the actual radius of influence. Hence, the approach used in the radius of influence approach described in the previous section provides a conservative value for the radius of influence (Figure 3.27).

Substituting R into the flow equations, they become steady state equations as follows:

Theis (steady state approximation)   *Thiem* (steady state version of Jacob equation)

$$s = \frac{114.6Q}{T} \cdot W(u) \qquad s = 528Q \cdot \log\left(\frac{R}{r}\right) \qquad (3.25)$$

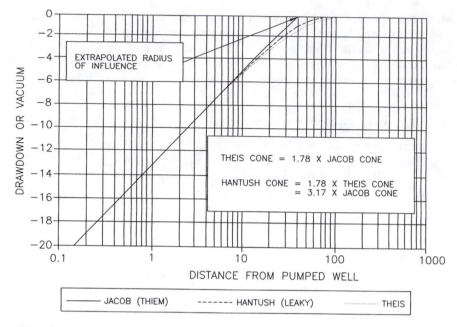

Figure 3.27 Comparison of hydraulics theories.

$$u = \frac{r^2}{1.781R^2}$$

Multiplying coefficients in equation (3.23) by $12 \cdot \rho_g$ to compute s in inches of water and for $\rho_g = 0.0012$:

<table>
<tr><td align="center">Theis</td><td align="center">Thiem</td></tr>
<tr><td align="center">$$s = \frac{1.65Q}{T} W(u)$$</td><td align="center">$$s = \frac{7.6Q}{T} \log\left(\frac{R}{r}\right)$$</td></tr>
<tr><td align="center">$$u = \frac{r^2}{1.781R^2}$$</td><td></td></tr>
</table>

In real life, significant amount of air leaks into the subsurface and the solution for the leaky vadoze zone (Figure 3.28) is given by the Hantush leaky equation. The Hantush leaky equation provides that

$$s = \frac{229 \cdot Q}{T} K_o\left(\frac{r}{B}\right), \text{ for } s \text{ in feet of air} \tag{3.26}$$

$$s = \frac{3.3 \cdot Q}{T} K_o\left(\frac{r}{B}\right), \text{ for } s \text{ in inches of water} \tag{3.27}$$

where K_o = a Bessel function
       B = leakage factor (.89 R).

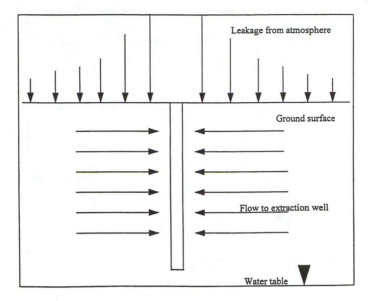

Figure 3.28 Airflow in the vadose zone with leakage.

The Hantush leaky equation provides a more realistic approximation of the pressure–drawdown relationship for large values or *r*. It is recommended to use the Thiem equation near the extraction well and the Hantush leaky equation away from the extraction well. The Hantush equation exhibits greater vacuum at greater distances. The reason for this is because the percentage of air that comes through leakage near the well is not significant. The Hantush leaky equation should be used at distances of $u > 0.05$ and at a radial distance of *r* which is 0.3 times the radius of influence.

The data requirements for simulating gas flow for SVE design include the following: steady state distance–drawdown data from a field pilot test (i.e., extraction rates, measured vacuums, well configuration); site map (showing buildings, paved and unpaved areas, pilot test well and monitoring well locations); and subsurface conformation (including boring logs, geologic cross sections, water table map or average depth to groundwater, and the extent of contamination). A typical approach to simulating gas flow might consist of the following steps:

1. Analysis of field pilot-test data—including plotting distance–drawdown data and applying the Theis or Hantush methods to estimate transmissivity and leakage
2. Calibration—reproducing the pilot test and adjusting model parameters to match observed vacuums
3. Simulation—using the calibrated model to examine the influence of various extraction well configurations; to examine the influence of vents, flow obstructions, paved areas, etc.; and to perform sensitivity analysis on key parameters such as air permeability and vertical leakage.

An example calculation of the above theory is described below. The vapor extraction test layout and data obtained from the pilot test is shown in Figures 3.29 and 3.30. The distance–drawdown data plotted is shown in Figure 3.31 on a semi-log paper.

Using the Thiem or Jacob equation and based on the semi-log analysis, determine the slope of the best-fit straight line, and use the equation

$$T = \frac{7.6}{\Delta s}Q \qquad\qquad (3.28)$$

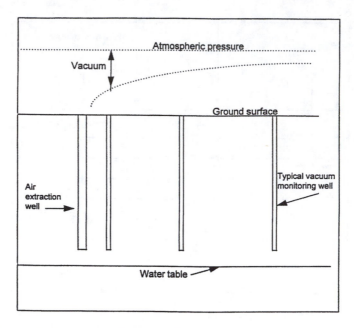

Figure 3.29 Vapor extraction test monitoring layout.

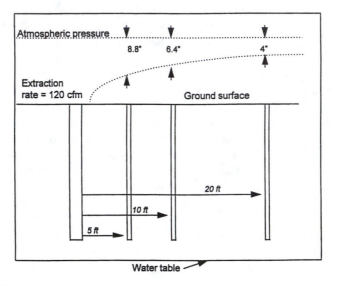

Figure 3.30 Typical pilot test monitoring results.

where T = gas transmissivity in ft²/day
        Q = flow rate in cfm
        Δs = change in vacuum over one log cycle in inches of water.

$$\Delta s = 8.1 \text{ from the plot (Figure 3.31)},$$

$$T = \frac{7.6 \cdot 120}{8.1} \text{ for the } Q \text{ of the test (120 cfm)}$$

$$= 113 \text{ ft}^2/\text{day}.$$

When the Q changes, the Δs will also change.

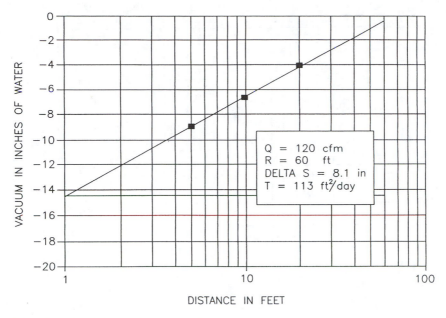

Figure 3.31   Semi-log distance–drawdown graph.

There is another more accurate and preferred method to do the same analysis which is known as the leaky log–log analysis using the Hantush leaky method. Using this procedure, a log–log plot of vacuum vs. distance from the extraction well is prepared. On a separate graph having the same scale as the data plot, a standard leaky type curve is prepared by plotting the Bessel function K_o against its argument r/B. To analyze the data, overlay the data plot on the type curve and, while keeping the coordinate axes of the two plots parallel, shift the data plot to align with the type curve, effecting a match position. Select and record the values of an arbitrary point referred to as the match point anywhere on the overlapping part of the plots (Figures 3.32 and 3.33). Record the match point coordinates: s, r, K_o, r/B.

Figure 3.32 shows the leaky type curve by plotting the Bessel function K_o against its argument r/B. K_o is the modified Bessel function of the second kind and of zero order. Tables of Values of this function are readily available.[25]

Figure 3.33 shows the data from the pilot test plotted on a log–log graph having the same scale as that used for the type curve. Figure 3.34 shows the data and type curve superimposed, effecting a position of best fit.

Gas transmissivity is computed from the formula shown here using the match point values for s and K_o.

$$T = \frac{3.3Q}{s}K_o$$

where T = transmissivity in ft²/day
   Q = flow rate in cfm
   K_o = match point value from type curve graph
   s = match point value from distance–drawdown graph.

The leakage factor B is computed from the equation shown here using match point values r and r/B.

$$B = \frac{r}{r/B}$$

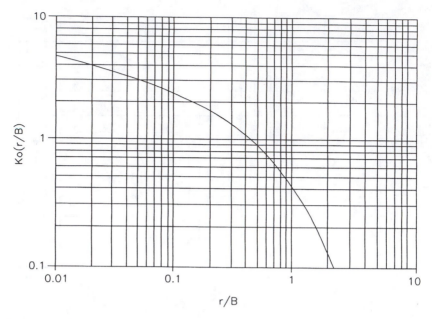

Figure 3.32 Leaky type curve.

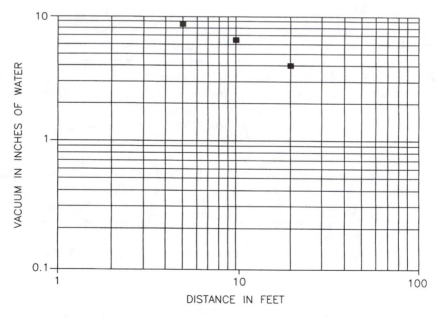

Figure 3.33 Log–log distance–drawdown graph.

where B = leakage factor in feet
        r = match point value from distance–drawdown graph
        r/B = match point value from type curve.

The log extrapolated radius of influence is computed from leakage factor.

$$R = \frac{B}{0.89}$$

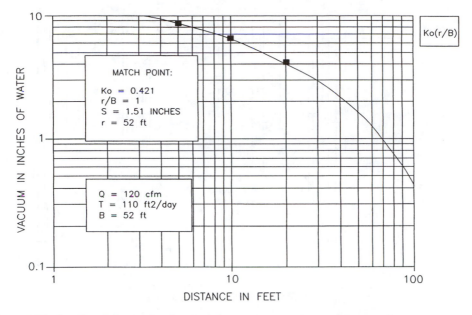

Figure 3.34 Log–log distance–drawdown graph.

The leakage is computed from the leakage factor and gas transmissivity. "Leakance", l, is defined as the ratio K'/b' where K' and b' are the gas conductivity and thickness, respectively, of the resistive layer separating the vadose zone from the atmosphere. All of the vertical resistance of the vadose zone sediments is assumed to be concentrated in a thin "leaky" layer at ground surface.

$$l = T/B^2$$

From the above exercise the obtained match point values and computed vadose zone parameters are presented below:

$$K_o = 0.421$$

$$r/B = 1$$

$$s = 1.51 \text{ in.}$$

$$r = 52 \text{ ft}$$

$$T = \frac{3.3 \cdot 120 \cdot 0.421}{1.51}$$

$$= 110 \text{ ft}^2 / \text{day}$$

$$B = \frac{52}{1} = 52 \text{ ft}$$

$$R = \frac{52}{0.89} = 58.4 \text{ ft}$$

$$l = \frac{T}{B^2} = \frac{110}{52^2}$$

$$= 0.041 \text{ day}^{-1}.$$

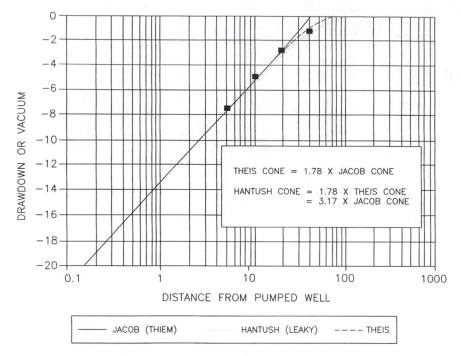

Figure 3.35 Comparison of hydraulics theories.

Because the Hantush equation is valid for all distances around the extraction well, the leaky analysis is preferred. The semi-log procedure can sometimes overestimate T and R, since the Jacob and Thiem equations are only valid near the well. If the pilot test incorporates vacuum monitoring wells beyond the valid range of the Jacob and Thiem equations, erroneous values will result.

Figure 3.35 shows vacuums observed in wells located 5, 10, 20, and 40 ft from the vapor extraction well. Data for the two wells nearest the extraction well fall on the straight-line logarithmic plot, whereas data from the two distant wells do not. If semi-log analysis is performed using the two closest wells, valid values for transmissivity and log-extrapolated radius of influence will be obtained. If the distant wells are used, however, the straight line of best fit is flatter, and both T and R are overestimated (Figures 3.36, 3.37, 3.38). Using leaky analysis procedures, all observed vacuums fall on the leaky type curve, so the correct T and R values are obtained regardless of which vacuum monitoring wells are used.

Application of the equations can be clarified with some examples given in the following five problems.

Problem 1: Given

- fine sand formation with estimated K_w of 3 ft/day
- water table at 40 ft
- extraction well has 12 in. bore hole (0.5 ft radius),

what flow rate Q can be expected from a blower having a vacuum of 80 in. of water?

Estimate gas conductivity:

$$K_g = 0.0656 \cdot K_w$$

$$= 0.0656 \cdot 3$$

$$= 0.20 \text{ ft/day.}$$

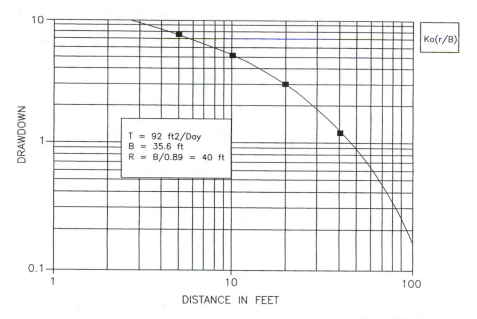

Figure 3.36 Vapor extraction test data analysis (leaky analysis gives correct *T* and *R* for all observation wells).

Calculate gas transmissivity:

$$T = K_g \cdot b$$

$$= 0.20 \cdot 40$$

$$= 8.0 \ \text{ft}^2/\text{day}.$$

In the absence of pilot test data, estimate radius of influence to be 40 ft.
Using Thiem equation to solve for Q:

$$Q = \frac{sT}{7.6 \log(R/r)}$$

$$= \frac{80 \cdot 8.0}{7.6 \log(40/0.5)}$$

$$= 44.3 \ \text{cfm}.$$

Problem 2: If subsequent pilot test data show $R = 27$ ft (by extending the straight line to $s = 0$), calculate the vacuum in an observation well 36 ft away from an extraction well producing 40 cfm.

Step 1 is to calculate the leakage factor from the radius of influence determined during the pilot test.

$$B = 0.89 \ R$$

$$= 0.89 \times 27$$

$$= 24 \ \text{ft}.$$

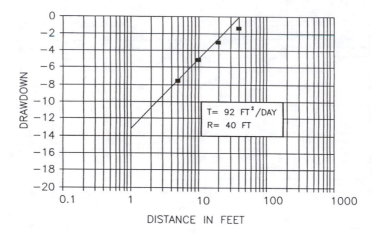

Figure 3.37 Vapor extraction test data analysis (near observation wells give correct *T* and *R* values).

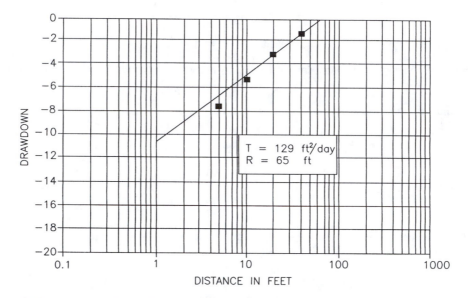

Figure 3.38 Vapor extraction test data analysis (observation wells give wrong *T* and *R* values).

Step 2 is to determine the ratio *r/B*

$$\frac{r}{B} = \frac{36}{24}$$

$$= 1.5.$$

Step 3 uses tables of Bessel function values to obtain the K_o value corresponding to the previously computed ratio *r/B* from tables:[25]

$$K_o[r/B] = K_o(1.5)$$

$$= 0.214.$$

Step 4 employs the Hantush leaky equation to compute vacuum at a distance from the extraction well.

$$s = \frac{3.3Q}{T} K_o(r/B)$$

$$= \frac{3.3}{8.0} \cdot 40 \cdot 0.214$$

$$= 3.5 \text{ in. of water column.}$$

Problem 3: At 40 cfm of airflow rate, what is the maximum distance of effective remediation of this well?

The first step is to establish a reasonable criterion for determining whether or not remediation will be adequate. A common procedure is to require a minimum of 0.1 or 0.2 in. of vacuum as the cut-off value at the point of concern. Another criterion is to require a minimum air velocity. For this example, 0.1 in. of vacuum will be used.

For Step 1, establish criterion, e.g., s required = 0.1 in.

For Step 2, use the Hantush equation and the known parameters to compute the value of the Bessel function, K_o. Find r for which $s = 0.1$ in. Solving the Hantush equation for K_o:

$$K_o[r/B] = \frac{sT}{3.3 \cdot Q}$$

$$= \frac{0.1 \cdot 80}{3.3 \cdot 40}$$

$$= 0.006.$$

Step 3 uses tables of Bessel function values to obtain the corresponding r/B ratio from which r can be computed. From Bessel function tables, the K_o value closest to 0.006 is 0.00623, which corresponds to $r/B = 2.5$. Thus,

$$r = 2.5 \cdot B$$

$$= 2.5 \cdot 24$$

$$= 60 \text{ ft.}$$

The basic equations discussed earlier are valid only under the following assumptions:

- air is incompressible
- the well is 100% efficient
- the well is fully penetrating.

If any of these assumptions are violated, the standard equations must be corrected to compensate this for deviation from the assumptions.

Corrections For Gas Expansion.   If the magnitude of vacuum exceeds 0.15 to 0.2 atm, the actual drawdown (vacuum) will deviate from the drawdown predicted by the equations, and a correction factor is required. Normally, vacuums this large are only observed at the extraction well itself and thus corrections are done only for the extraction well.

The equations shown here provide the correction factors needed to convert theoretical to actual vacuum and vice versa. At sea level, atmospheric pressure is 14.7 psi.

$$s_t = s_a - \frac{s_a^2}{2P_{atm}} \qquad (3.29)$$

$$s_a = P_{atm} - \sqrt{P_{atm}^2 - 2P_{atm} \cdot s_t} \qquad (3.30)$$

where s_t = theoretical drawdown (from equations)
s_a = actual drawdown observed in the field
P_{atm} = atmospheric pressure expressed in the same units as s_t and s_a.

At sea level,

$$P_{atm} = 14.7 \text{ psi, if } s \text{ is expressed in psi}$$

$$= 33.8 \text{ ft, if } s \text{ is expressed in ft}$$

$$= 406 \text{ in., if } s \text{ is expressed in inches.}$$

Problem 4: From the previous problem, $T = 8.0$ ft²/day and $B = 24$ ft. What would the flow rate be with an applied vacuum of 12 psi? This problem requires correcting for gas expansion, since the vacuum exceeds 0.2 atm.

For Step 1, the standard correction equation is used to compute an equivalent theoretical vacuum.

$$s_a = 12 \text{ psi}$$

$$s_t = s_a - \frac{s_a^2}{2P_{atm}}$$

$$= 12 - \frac{12^2}{2 \cdot 14.7}$$

$$= 7.1 \text{ psi}.$$

Expressed in inches,

$$s_t = 7.1 \cdot 2.3 \cdot 12$$

$$= 196 \text{ in.}$$

For step 2, the theoretical vacuum may be used in the Thiem equation to compute the anticipated airflow.

$$Q = \frac{sT}{7.6 \log(R/r)}$$

$$= \frac{196 \cdot 8.0}{7.6 \cdot \log(27/0.5)}$$

$$= \frac{196 \cdot 8.0}{7.6 \cdot 1.732}$$

$$= 119 \text{ cfm.}$$

The equations shown here have incorporated both the Thiem equation and the gas expansion correction into a single formula. The two equations are equivalent, one solving for vacuum, s, and the other solving for discharge, Q. When you find SVE airflow equations in the literature incorporating pressure squared as part of the formula, they are equivalent to the equations shown here. They are often written in terms if intrinsic permeability (K), gas density (ρ), and viscosity (μ), which tends to make them more confusing. The simplest approach is to use the standard airflow equations adapted from groundwater equations and apply the gas expansion correction as a separate step.

$$s = P_{atm} - \sqrt{P_{atm}^2 - \frac{15.2 \cdot P_{atm} \cdot Q \cdot \log(R/r)}{T}} \; , \tag{3.31}$$

$$Q = \frac{T}{7.6 \cdot \log(R/r)}\left(s - \frac{s^2}{2P_{atm}}\right). \tag{3.32}$$

It should be noted that P_{atm} is expressed in inches of water.

SVE well inefficiency, E, is demonstrated by a greater vacuum inside the borehole than outside (Figure 3.39). This difference in vacuum represents the head loss associated with drilling damage in the vicinity of the borehole. It is estimated by extending the drawdown curve left up to distance equal to borehole radius and comparing that value to the actual measured vacuum.

$$E = \frac{s_{outside}}{s_{inside}}. \tag{3.33}$$

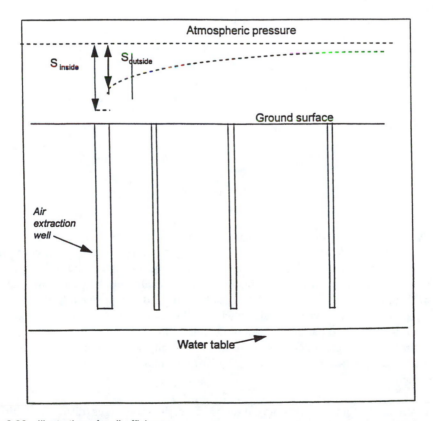

Figure 3.39 Illustration of well efficiency.

The theoretical equations use $s_{outside}$ not s_{inside}. Inefficiency is caused by pressure losses across the well screen, reduced permeability of sediments around the borehole, and other local factors. For a 100% efficient well, the drawdown inside the borehole is the same as the drawdown outside, so no correction is required. For a 60% efficient well, the vacuum outside the borehole will be 60% of that inside. For a 30% efficient well, the vacuum outside the borehole will be 30% of that inside.

Problem 5: Recompute Q in Problem 1 based on a T of 8.0 ft²/day, s of 80 in., the actual R of 27 ft (from the pilot test), and well efficiencies of 100%, 60%, and 30%.

For 100% efficiency, no correction is required.

$$Q = \frac{sT}{7.6\log(R/r)}$$

$$= \frac{80 \cdot 8.0}{7.6 \cdot \log(27/0.5)}$$

$$= 48.6 \text{ cfm.}$$

(We got 44.3 cfm when we assumed R = 40 ft.) For 60% efficiency:

$$s_{outside} = 0.6\, s_{inside}$$

$$= 0.6 \cdot 80$$

$$= 48 \text{ in.}$$

$$Q = \frac{48 \cdot 80}{7.6\log(27/0.5)}$$

$$= 29.2 \text{ cfm.}$$

For 30% efficiency:

$$s_{outside} = 24 \text{ in.}$$

$$\text{and } Q = 14.6 \text{ cfm.}$$

Correction for Partial Penetration. In partially penetrating systems, converging flow complicates the flow and drawdown distribution, thus resulting in different drawdowns than predicted by the standard equations. Before using the standard equations, the drawdowns must be corrected for partial penetration in a manner exactly analogous to procedures used for groundwater pumping tests. Pilot test data must be corrected to equivalent, fully penetrating values prior to analysis. Likewise, calculated drawdowns for the final remediation system must be corrected back to partially penetrating values.

As in groundwater pumping tests, the Hantush partial penetration equation may be used to provide the required correction factors. The Kozeny equation may also be used, but only at the extraction well.

- If screens do not fully penetrate the vadose zone, flow converges (distorts), resulting in complicated drawdown distribution.

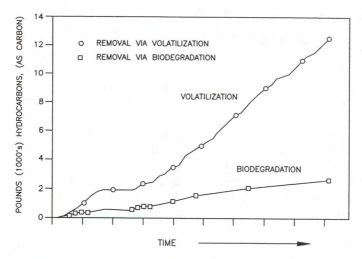

Figure 3.40 Biodegradation of contaminants during SVE of biodegradable VOCs.

- Pilot test data must be corrected for partial penetration prior to analysis.
- Calculated drawdown values must be corrected for partial penetration to predict actual well performance.
- Apply Kozeny correction factors to extraction well.
- Apply Hantush correction factors to observation wells as well as extraction well.

3.5 BIOVENTING

In most cases, SVE has been primarily applied for the removal of volatile organic compounds from the vadose zone utilizing volatilization as the primary mass transfer mechanism. However, circulation of air in the vadose zone soils can be utilized to enhance the degradation of volatile organic compounds which are aerobically biodegradable (Figure 3.40). Bioventing can reduce vapor treatment costs and can also result in the remediation of semivolatile organic compounds that cannot be removed by direct volatilization alone.[15,25]

In bioventing, the aerobic biodegradation processes are optimized to become the dominant process by modifying the venting system. Bioventing systems use the same blowers used in SVE systems to provide specific distribution and flux of air through the contaminated vadose zone to stimulate the indigenous microorganisms to degrade the contaminants to more benign compounds such as carbon dioxide, water, and biomass. For example, degradation of benzene takes place according to the equation

$$C_6H_6 + 7 \cdot 5\,O_2 \rightarrow 6CO_2\,6H_2O + \text{biomass.} \tag{3.34}$$

Based on the above equation about 3.5 g of O_2 are required per gram of benzene degraded. This method is relatively powerful, because atmospheric air contains 21% oxygen. During implementation of bioventing, soil gases present in the subsurface are generally monitored to ensure the presence of aerobic conditions. The gases monitored are O_2, CO_2, and CH_4. Increased levels of CO_2 (atmospheric level is 0.03%) and depletion of O_2 indicates a higher level of microbial activity. However, at several sites,[26] oxygen utilization has proven to be a more useful measure of biodegradation rates than carbon dioxide production. Abiotic factors such as soil pH and alkalinity may influence the soil gas O_2 concentrations. Higher pH and higher alkalinity soils will exhibit little CO_2 in the soil gas due to the formation of carbonates. The soil water content will also impact the measurable CO_2 in the soil gas due to the solubility of CO_2.

3.5.1   Laboratory Testing

A series of analyses are generally performed on site soil samples to evaluate the potential for microbial degradation of the contaminants. The evaluations are conducted to determine whether the site soil samples have (1) a generally healthy microbial population without being stressed by the contaminant concentrations, and (2) specific microbial species capable of degrading the contaminants present. In addition, an analysis should be performed to determine the availability of soluble N- and P-containing nutrients such as ammonium, nitrate, and phosphates. Environmental parameters such as pH and moisture content should also be evaluated.

Total heterotrophic microorganisms and specific degrader microorganisms population is enumerated using the plate count procedure. At least 10^5 colony forming units (CFU) per gram of soil should be present in the soil samples for bioventing to be feasible. Microbial population in the range of 10^7 to 10^8 CFU/gm of soil is considered very healthy for bioventing. Measurement of respiration rates can also be performed in the laboratory, using site soil samples, to evaluate the microbial activity.

In the past, biodegradability of specific contaminants was also evaluated in the laboratory. Currently, there is an abundance of data on biodegradability of common environmental contaminants (Appendix D).

3.5.2   Design of Bioventing Systems

The design of a bioventing system has three basic requirements. In order of importance, they are as follows:[15,20]

1. Maintain an oxygen flux through the contaminated soils matching the rates of active, aerobic biodegradation in those soils.
2. Maintain soil moisture within an optimal range for microbial activity.
3. Supply growth-limiting nutrients as needed to bolster microbial populations.

In order for these requirements to be satisfied, several operating parameters of a bioventing system need to be properly designed. These are discussed below.

3.5.2.1   Airflow Rate

It should be noted that there are different objectives for an SVE and a biovent system. An SVE system attempts to maximize airflow through the soil to maximize the rate of volatilization of contaminants into the air stream. A biovent system attempts to sufficiently manage the airflow rate to achieve maximum oxygen usage by the vadose zone microbial populations. Consequently, the biovent airflow rates are commonly an order of magnitude lower than SVE airflow rates. As a general rule of thumb, the air in the contaminated soil pore volume needs to be exchanged once every 1 to 2 days, depending on the air's oxygen content.

Viable aerobic bioactivity is typically found in contaminated sites where the oxygen levels in the soil are above 2%. When the oxygen content of the soil air falls below 2%, aerobic bioactivity becomes dormant or absent. At this point the soil air is dominated instead by carbon dioxide and methane. In examining field data on vadose zone air quality, keep in mind that the soil air is made up of a large number of small-scale zones, or micro-zones, arrayed like a three-dimensional jigsaw puzzle determined by the geology and contaminant distribution. Within each micro-zone, the air can either be (1) replenished relatively frequently and is thus aerobically active, (2) replenished relatively infrequently and is thus aerobically dormant with high carbon dioxide levels, or (3) largely stagnant and active only anaerobically with accumulating products such as methane and carbon dioxide. Most monitoring or extraction wells intersect several of these micro-zones. If it intersects only the first type or only the third type of micro-zones described above, the results will seem clear-cut. If the well

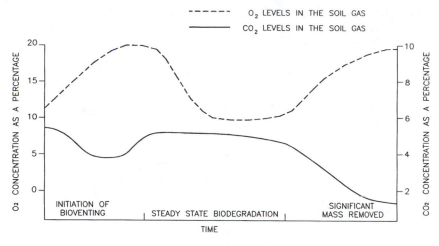

Figure 3.41 Typical O_2 and CO_2 percentages in the soil gas during bioventing.

intersects a mixture of micro-zone types, the results can be ambiguous and confusing. Typical O_2 and CO_2 percentages in the soil gas during bioventing are shown in Figure 3.41.

Well spacing is not the most important parameter for bioventing; rather, the airflow path is most important. The method is not reliant on the volatilization of contaminants for removal; rather, the introduction of oxygen for aerobic biodegradation to mineralize the contaminants. Consequently, the number of wells is often markedly fewer and in different locations than would be selected for an SVE system. These well locations are chosen to induce the flow of air with a high oxygen content across the area of contaminated soil.

3.5.2.2 Soil Moisture

For a bioventing system it is very important to maintain a moderate soil moisture, usually in the range of 40 to 60% of field capacity to maintain viable, aerobic bioactivity within the manipulated zone. If soil moisture increases appreciably above this range, the resultant pore sizes for air movement shrinks, decreasing the air permeability and flow rates. For these reasons, it is important to include in the design some means of adding moisture to the contaminated zone when necessary. It should be noted that impact of soil moisture on biodegradability is also influenced by the solubility of the contaminant.

Availability of O_2 is the rate-limiting step for bioventing in most cases. Optimum moisture content also plays a significant role, but not as dominant as the availability of oxygen. Macronutrients such as N and P and the micronutrients are generally available in the site soils. In rare cases, N and P containing soluble salts may have to be added to the site soils. Nitrogen and phosphorus initially consumed by the microbial mass are constantly recycled due to the lysis of dead microbial cells. Hence, the need and availability of N and P are critical only during the early stages of microbial growth.

3.5.2.3 Temperature

Temperature dependency of biodegradation rates have been reported to follow the van't Hoff–Arrhenius equation,[20,26,27] as well as the Phelps equation.[20,27] The van't Hoff–Arrhenius equation is as follows:

$$Y = Ae^{-Ea/RT}$$

(3.35)

where Y = temperature-corrected biodegradation rate
        A = baseline degradation rate
        E_a = activation energy
        R = gas constant
        T = absolute temperature.

The Phelps equation, which describes the relationship between respiration rate and temperature, is as follows:

$$R_t = R_7 \cdot \theta^{(t-7)} \tag{3.36}$$

where R_t = rate at temperature $t°C$
        R_7 = rate at 7°C
        θ = thermal coefficient.

The design criteria of a bioventing system is summarized in Table 3.4.

Table 3.4 Design Criteria for a Bioventing System

	Parameter	Impact
1.	Lateral and vertical extent of contamination	Contaminant mass estimate
2.	Geology of the contaminated zone	Manipulation of airflow in various zones and thus required airflow rates and vacuum levels
3.	Air residence time and pore volume exchange	Biodegradation rates, required airflow rates, and thus extraction blower capacities
4.	Location of air extraction wells, and air recharge conditions and the need for air injection wells	Design of the bioventing system
5.	Soil moisture content and background nutrient levels	Design of moisture and nutrient addition systems
6.	Continuous monitoring of O_2, CO_2, and other soil gas concentrations	Focused monitoring and optimization of bioventing performance

3.5.3 *In Situ* Respiration Test

An *in situ* respiration test is perhaps the best method currently available to assess the rates of biodegradation that can be sustained by a bioventing system. This section describes the steps that are generally necessary for successfully carrying out an *in situ* respiration test.

3.5.3.1   Equipment

The equipment needed for carrying out an *in situ* respiration test can be fairly simple. If an SVE pilot test is also to be carried out at the site, the test can make use of that pilot test equipment which is comparable to those listed below. Also, should a more rigorous test be required, some or all of the optional equipment can also be used. The basic equipment needed is listed below:

- Small air compressor or blower (a minimum of 1 to 2 cfm capacity). Examples of easily available equipment include car tire inflation pumps or air mattress inflation pumps. These can be plugged into a car cigarette lighter or other electric outlet.
- Oxygen (O_2) field monitoring device for air, with ± 0.5% accuracy or less. Examples: Firite meter, explosimeter, Gastech meter.
- Carbon dioxide (CO_2) field monitoring device for air with ± 0.5% accuracy or less.
- In-line sampling and field monitoring devices.

3.5.3.2   Test Procedures

1. Monitor the candidate wells for O_2, CO_2, CH_4, and total volatile organics concentrations in the adjacent vadose zone. Use the wells with at least a portion of their screen extending above the water table. Throughout these test procedures, remove several well casing volumes prior to collecting and analyzing samples.

2. Select a well for the test in a moderately contaminated area with predominantly aerobic bioactivity. Soil contaminant concentrations should be greater than 50 mg/kg. It is important to remember that, regardless of the conditions currently prevailing at the site, whether anaerobic, aerobic, or both, bioventing will ultimately convert the site to active, aerobic biodegradation. We must select those contaminated subareas of the site that can quickly and easily be converted to active, aerobic bioactivity during the test. Failure to select a proper test site will likely result in negative results.

3. Prior to beginning the test, measure the O_2, CO_2, CH_4, and total volatile organics concentrations in the test well.

4. Inject at least 1000 ft^3 of air into the test well. Air can be injected in up to five wells. It is generally advisable to mix 1 to 2% He with the injected air as a tracer gas. Detection of He implies that the air sampled is the same that was injected and the changes in its makeup can be attributed to microbial activity. A volume of 1000 ft^3 is deemed necessary to avoid boundary effects from interfering with the test results. Regardless of the capacity of the air pump used, this volume of air should be injected within 24 h. It usually takes about 24 h for the soil microbial populations to convert from their dormancy to active respiration. If a higher capacity pump is used, monitoring may optionally be carried out during the remainder of the 24-h period. However, the respiration rates monitored are typically distinctly lower than the respiration rates subsequently measured. At the least it provides a testing period for assessing the accuracy of the monitoring devices.

5. Begin periodic monitoring of O_2 and CO_2 concentrations in the test well 24 h following the start of air injection. Samples should be analyzed at least several times per day and, at most, hourly. The actual frequency should be selected in the field so that the data collected show a definite trend in O_2 and CO_2 concentrations in spite of the scatter caused by meter error.

6. Continue the monitoring of the test well until O_2 and CO_2 concentrations have changed at least 5%. Preferably, the test would be conducted until the O_2 and CO_2 concentrations have changed over 10%. This duration is necessary to define a definite trend through much of the range in O_2 concentration that active, aerobic biodegradation occurs. This will likely require a test duration of at least 1 day and possibly 2 days.

7. The data should be analyzed for both zero-order and first-order relationships. For zero-order analysis, compare the O_2 and CO_2 concentrations with respect to time (e.g., $[O_2]$ vs. time). For first-order analysis, compare the ratio of the test O_2 and CO_2 concentrations to the initial concentrations with respect of time. Determine the linear regression equation for each analysis as well as the coefficient of determination (R^2). Select the order of analysis that has the higher R^2 values. Previous studies have empirically found both approaches to be appropriate. Usually (but not always), the O_2 depletion trends have been found to be more reliable than CO_2 production trends. However, this is partly a function of the monitoring devices used and the geochemistry of the given site.

3.5.4   Modified Applications of Bioventing

3.5.4.1   Closed Loop Bioventing

It should be noted that 23.9 lb of oxygen can be introduced into the vadose zone in a day by introducing air at a flow rate of 1 scfm. At this rate of oxygen introduction into the vadose zone, all the oxygen present in the air will not be consumed by the microorganisms for contaminant biodegradation. As noted earlier, aerobic biodegradation of hydrocarbons require only about 3.5 g of oxygen per gram of hydrocarbons.

A closed-loop bioventing system, as shown in Figure 3.42, can be operated to maximize the percentage of contaminant mass biodegraded during system operation. Conceptually this

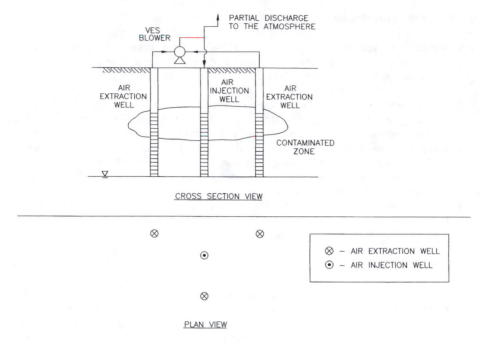

Figure 3.42 Closed-loop bioventing.

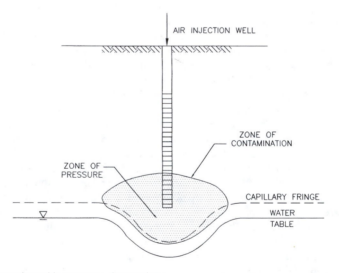

Figure 3.43 Bioventing with pressure dewatering.

configuration utilizes the vadose zone soils as a bioreactor to degrade the contaminants. Continuous operation of the closed-loop system will eventually decrease the level of oxygen in the air extracted and reinjected. Hence, a partial discharge of the extracted air periodically or continuously will help in replenishing the amount of oxygen consumed by the microorganisms. The rate of this partial discharge does not have to exceed 10% of the total flow of the extracted air.

3.5.4.2   Bioventing with Pressure Dewatering

In this modified form of bioventing, air is injected into the vadose zone under pressure.[28] The introduction of air pressure, locally, just above the water table (Figure 3.43), depresses the water table, and this phenomenon is called pressure dewatering. The depression of the water table and subsequent gravity drainage, in turn, exposes the smear zone for increased levels of oxygen and thus faster rates of biodegradation in the capillary fringe. Faster removal of contamination in the capillary fringe and smear zone also helps in improving groundwater quality without direct groundwater remediation.

In addition the oxygen present in the injected air will enhance the biodegradation in the vadose zone. Caution should be provided to ensure the control of the migration of contaminant vapors beyond the zone of contamination, specifically into underground utilities and vaults.

3.5.4.3   Intrinsic Bioventing

This approach is discussed only at a conceptual level in this section. As noted earlier, a very low airflow rate will be sufficient to introduce the 3.5 g O_2 required per gram of hydrocarbons biodegraded. It is conceivable to design a system in which the positive pressure difference between the atmospheric pressure and the subsurface pressure caused by barometric changes (during the daytime) could be utilized to induce the airflow into the vadose zone.[29] An automatic control system to induce the airflow into the vadose zone and close the reverse flow during nighttime will provide sufficient O_2 for enhanced biodegradation.

3.6   MONITORING REQUIREMENTS

As noted earlier, the performance of an SVE system has to be optimized and fine-tuned continuously as the system operation proceeds. Monitoring data are used to assess system performance and calibration of models, and to guide necessary operational changes and equipment modifications.

The simplest way of assessing SVE process performance is to monitor the flow, vacuum responses and the concentration, and composition of the contaminants in the extracted air. This is the minimum monitoring required to identify mass removal rates and any changes in the subsurface conditions impacting the flow characteristics.

Table 3.5 presents an expanded list of monitoring requirements and the impacts on system performance.

3.7   VAPOR TREATMENT TECHNOLOGIES

The most common technologies for treating air streams laden with volatile organic compounds (VOCs) are described briefly below. Types of common VOCs that require treatment when present in an air stream are as follows:

- aliphatic hydrocarbons
- aromatic hydrocarbons
- chlorinated hydrocarbons
- alcohols, ethers, and phenols
- ketones and aldehydes

Applicability of various vapor treatment technologies to the above contaminant types is shown in Table 3.6.

3.7.1   Thermal Oxidation

A tried and true technology for VOC control is oxidation, either catalytic or thermal. Oxidation units can destroy nearly 75% of the VOC and toxic emissions targeted by the Clean

Table 3.5   SVE System Performance Monitoring Requirements

Data required	Impact(s)
• Flow rate with time	• Pore volumes removed per day
	• Subsurface changes with respect to air permeability
	• Subsurface air distribution
• Applied vacuum with time	• Subsurface changes in air permeability, moisture content
	• Induced air distribution and zone of influence
• Extracted air concentration with time	• Rate of mass removal
	• Typically declines with time
	• Cumulative mass removed
	• Decline to low mass removal rates can occur well before soil cleanup standards are reached
	• Volatilization phase shifting to diffusive phase
	• Vapor treatment technologies
• Extracted air composition with time	• Weathering of contaminants
	• Rate of mass removal
	• Microscopic scale phenomena of contaminant partitioning
	• Volatilization phase shifting to diffusive phase
	• Ability to reach soil cleanup standards
	• Aerobic vs. anaerobic conditions
	• O_2/CO_2 concentration ratios will be an indicator of biodegradation activity in the subsurface
	• Vapor treatment technologies
• Monitoring well vacuum measurements	• Areal extent of vacuum coverage
	• Induced airflow distribution pattern

Table 3.6   Applicability of Selected Vapor Treatment Technologies to Different Types of VOCs

	VOC type				
Treatment technology	Aliphatic HC	Aromatic HC	Chlorinated HC	Alcohols, ethers, phenols	Ketones, aldehydes
Thermal oxidation	×	×	×	×	×
Catalytic oxidation	×	×	*	×	×
Adsorption	×	×	×	×	
Condensation	×	×	×	×	×
Biofiltration	×	×	*	×	×
Membrane filtration	×	×	×	×	×

* Sometimes applicable.

Air Act Amendments of 1990. Thermal oxidation, also known as thermal incineration, operates on a simple premise: sufficiently heating a VOC in the presence of oxygen will convert the VOC to harmless end products.[30]

During thermal oxidation, the VOC-laden air is captured by a ventilation system, preheated, thoroughly mixed, and combusted at high temperatures to form carbon dioxide and water. A thermal oxidation unit typically consists of a fan to move VOC-laden air; a filter–mixer to mix the VOC-laden air; a fan to supply combustion air (if required); a combustion unit consisting of a refractory-lined chamber and one or more burners; heat-recovery equipment; and a stack for atmospheric release of the treated exhaust. A continuous monitor is recommended to measure the combustion temperature to ensure complete oxidation.

Frequently, thermal oxidation systems require the use of supplemental fuel, such as natural gas or oil, to sustain the combustion temperature, which is nominally about 1200°F to 1600°F. Low-concentration VOC streams may not possess the oxidation energy required to maintain the combustion temperature; therefore, oxidation of these streams would require supplemental fuel or electrical energy. On the other hand, if exhaust streams contain very high concentrations of VOCs, dilution air may be required to prevent explosion.

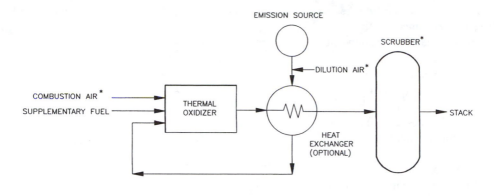

* REQUIRED UNDER SPECIFIC CONDITIONS.

Figure 3.44 Schematic diagram of a recuperative thermal oxidizer.

Heat-recovery equipment may be installed with a thermal oxidation system to preheat the VOC-laden air stream prior to combustion. Preheating the incoming stream reduces the amount of supplemental fuel that would be required to maintain the combustion temperature.

In addition to combustion temperature, two other parameters that influence the VOC destruction efficiency of a thermal oxidation system are *residence time* and *degree of mixing*. Residence time is the amount of time required for complete oxidation of VOC. Generally residence time varies from 0.5 to 1 s.[31] Longer residence times may be required if certain VOCs such as chlorinated organic compounds are present. The residence time is affected by the degree of mixing of VOC-laden stream prior to combustion. More thoroughly mixed VOCs require less residence time for complete oxidation.

There are three types of thermal oxidation systems: after-burners, recuperative, and regenerative. They are differentiated by the equipment used for heat recovery.

Common *after-burners* utilize a direct-fired burner(s) in an insulated combustion chamber sustaining a temperature of 1500°F for at least 0.5 s. They have no heat-recovery equipment. The burning chamber will be sized to allow for the given residence time at a certain airflow velocity. This system has the lowest capital cost but is the most expensive to operate. High VOC concentrations will help minimize auxiliary fuel consumption.

Recuperative systems (Figure 3.44) use a heat exchanger, typically crossflow, counterflow, or concurrent flow, for heat recovery. The heat exchanger is the heart of this technology. The VOC-laden air stream is preheated to a maximum of approximately 80% of the 1500°F combustion temperature. This system is sized based on the flow rate, and the cost will go up in proportion to the efficiency and number of passes being made in the exchanger. The recuperative system is best suited for moderately high VOC concentrations and can tolerate a broad mix of constituents. But if the design VOC concentration is reduced or if there are swings in incoming VOC concentrations, the auxiliary fuel consumption will go up dramatically.

Regenerative systems use ceramic material for heat recovery. The ceramic material is stored in separate beds that feed to a central combustion chamber. They generally use less supplemental fuel than the other systems because of the superior heat transfer capability of the ceramic material. As a result, these systems may be attractive for controlling low VOC streams.

During thermal oxidation of chlorinated hydrocarbons, hydrochloric acid (HCl) fumes are formed in addition to CO_2 and H_2O. As a result, removal of the HCl formed is necessary before the treated air is discharged to the atmosphere. A scrubber installed before the stack will facilitate the removal of the acid present in the effluent air stream prior to discharge. Caustic soda or potash can be used as the absorbent solution in the scrubber. Less than perfect combustion of the VOCs may lead to solubilization of the residual VOCs in the scrubbing solution. Care should be taken to dispose the neutralized scrubbing solution properly under these circumstances.

In summary, the design criteria for thermal oxidation is governed by

- influent airflow rate
- VOCs concentration
- VOCs stream composition
- influent stream fuel value
- combustion temperature
- residence time
- degree of mixing
- oxygen content of flue gas

3.7.2 Catalytic Oxidation

Catalytic oxidation is very similar to thermal oxidation and combines a conventional-type heat exchanger with a catalyst. A catalyst inside the combustion unit lowers the activation energy for combustion; thus, combustion occurs at a lower temperature than for thermal oxidation. The catalyst, either precious or base metal, will allow oxidation to occur at a fairly low temperature of about 500°F to 700°F. As a result, fuel costs for catalytic oxidation are usually much lower than for an equally applicable thermal oxidation system. In the absence of centralized gas supply or under safety restrictions for flame combustion, electric power may be a convenient means to heat the influent air stream.

Typical catalyst materials include platinum, palladium, and metal oxides such as chrome-alumina, cobalt oxide, and copper oxide–manganese oxide.[31] The average catalyst lifetime is 2 to 5 years, after which deactivation by inhibitors, blinding by particle entrainment, and thermal aging render the catalyst ineffective. Catalytic materials may be inserted into the combustion unit in either a monolithic or beaded configuration.

Because of its sensitivity to VOC-laden air streams and process operating characteristics, the catalyst dictates the optimum operating conditions for a catalytic oxidation system (Figure 3.45). There is an inverse relationship between conversion efficiency and process flow rates. The higher the conversion efficiency desired, the lower the flow rate that can be processed. Higher flow rates require the installation of multiple catalysts.

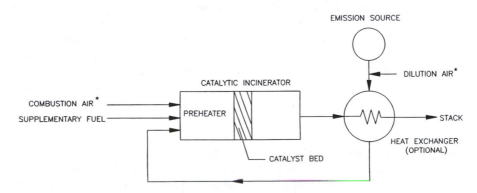

* REQUIRED UNDER SPECIFIC CONDITIONS.

Figure 3.45 Schematic diagram of a catalytic oxidizer.

Catalytic oxidation systems are not effective in streams containing lead, arsenic, sulfur, silicone, phosphorus, bismuth, antimony, mercury, iron oxide, tin, zinc, and other catalyst deactivators. These compounds have a tendency to mask or poison the catalyst's cell structure. When this occurs, the catalyst's ability to react chemically with the hydrocarbon decreases, requiring more fuel to rise to a temperature where proper oxidation will take place. Masking of the catalyst may also cause airflow restrictions whereby more horsepower could be needed to push the air through the system. In order to prolong the lifetime of the catalyst, it has to be cleaned periodically to remove deactivators and particulates.

Catalytic oxidation systems are typically applied to low-VOC-concentration streams, since high VOC concentrations and associated high heat contents can generate enough heat of combustion to deactivate the catalyst. Dilution air may be required when the influent VOC concentrations are high. The temperature and pressure across the catalyst bed should be monitored to ensure catalyst viability. The temperature rise across the catalyst indicates the extent of VOC oxidation; a decrease in the temperature rise across the catalyst indicates that VOC oxidation is incomplete. Since excessive heat can deactivate most catalysts, the inlet temperature to the catalyst bed must be kept sufficiently low to preserve catalyst activity. Similarly, the pressure drop across the catalyst is also an indication of the catalyst bed viability. The substantial amount of process control required to operate a catalytic oxidizer is illustrated in Figure 3.46.

Design criteria for catalytic oxidation systems can be summarized as follows:

- influent airflow rate
- influent stream composition
- influent VOCs concentration
- operating temperature
- catalyst properties
- space velocity and retention time
- influent stream fuel value
- oxygen content of flue gas
- presence of impurities and poisoning compounds
- type of heat exchanger
- availability of energy source (electricity vs. natural gas)

3.7.3 Adsorption

Adsorption refers to the process where gaseous VOC molecules contact a solid adsorbent and bond via weak intermolecular forces. Activated carbon is the most common adsorbent in use today for VOC removal. Other adsorbents include silica gel, alumina, and specialized resins. Activated carbon is derived from wood, coal, or other carbonaceous raw materials such as coconut shells. Granular activated carbon is currently the most common type of carbon used for VOC abatement, because the granules contain a significant amount of available surface area. Powdered activated carbon is generally cheaper and of lower quality than granular activated carbon, and when used in packed columns may cause unacceptably high pressure drops. Additionally, powdered activated carbon is nonregenerable, and must be disposed once it is spent. Granular activated carbon prepared from coconut shells can be prepared in large particle sizes necessary to help minimize pressure drop and is extremely hard, which leads to low attrition even under rough handling and high gas velocity conditions.

The adsorption of VOCs on the micropore surfaces of carbon is mainly a physical process involving van der Waals type forces and involves the liberation of from 2 to 5 kcal/mol of heat. Since these interactions are weak, the VOC is in dynamic equilibrium with the carbon surface and is being constantly adsorbed and desorbed. This adsorption/desorption

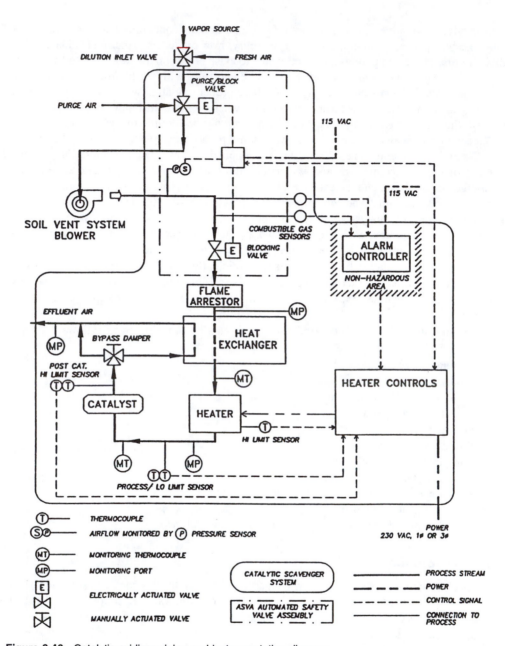

Figure 3.46 Catalytic oxidizer piping and instrumentation diagram.

process causes the VOC to be retained in the carbon pore structure. The more the VOC is "like" the carbon surface, the stronger it will interact with the surface. For example, molecules consisting mainly of aromatic or aliphatic moieties will adsorb more strongly than oxygenated molecules.

Vapor-phase granular activated carbon is generally used in a fixed bed, and the contaminated air is passed through the adsorbent bed. The adsorption of VOCs from air by a carbon bed is a continuous process, but for convenience of presentation can be envisioned as taking place in layers. Thus one can envision that when a contaminant stream passes through the bed, some contaminant is removed by layer "a" at the entrance of the bed, leaving a lower contaminant level in the stream to contact the next layer "b". Layer "b" then in turn removes more of the contaminant. The reduction of contaminant level as the stream contacts further layers of unused carbon continues until finally layer "z" is reached. The layers of "a" to "z" are known as the adsorption zone or the mass transfer zone (MTZ). As the flow continues, the MTZ progressively advances and propagates until breakthrough occurs. The following factors play important roles in adsorption dynamics and the length and shape of MTZ.

- Type, size, and macro- and micro-pore surface area of the carbon
- Depth of the adsorbent bed and empty bed contact time
- Gas velocity
- Temperature of the influent air stream
- Concentration of contaminants to be removed
- Moisture content and relative humidity
- Pressure of the system
- Vapor pressure of the contaminants to be adsorbed
- Possible decomposition or polymerization on the carbon surface.

Adsorption decreases with increasing temperature. Because the equilibrium capacity of adsorbents is lower at higher temperatures, the dynamic or breakthrough capacity will also be lower, and the MTZ proportionately changes with temperature. It should be also noted that the adsorption process is exothermic. As the adsorption front moves through the bed, a temperature front also proceeds in the same direction, and some of the heat is imparted to the gas stream. When the gas leaves the adsorption front, the heat exchange will reverse and the gas will impart heat to the bed. This increase in temperature in downgradient zones in the bed decreases the capacity in that zone. As a result, the maximum inlet concentration to a carbon bed should be limited to less than 10,000 ppmv.

The relative humidity and the moisture content of the influent stream also have a significant impact on the adsorption capacity of the bed. Small quantities of moisture actually enhance the adsorption process, as the heat of adsorption is carried with the moisture. However, relative humidities in excess of 50% tend to lessen the effectiveness of the bed. Moisture knockout systems or in-line heaters installed upstream of the bed will help to alleviate this problem. Moisture content plays an important role when treating effluent air streams of an air stripper. Polymeric adsorbent resins which are significantly more hydrophobic than carbon have shown to be able to adsorb at least 10 times more VOC mass than activated carbon.

When using activated carbon for air treatment, the spent carbon must be replaced or regenerated. If the replacement option is chosen, as soon as the carbon is spent it is removed and sent to an off-site reprocessing facility to be regenerated. Removal of the spent carbon from the bed and introduction of new carbon is accomplished as a wet slurry. This option is preferred when the total mass of VOCs to be removed during the life of the project is not high.

When the contaminant mass to be removed is high, either during short-term or long-term operations, fixed regenerative beds may be the preferred mode. On-site regenerative beds consist of two or more beds of activated carbon. Continuous operation of the system is maintained by the concurrent adsorption in at least one bed and desorption by the other beds. A schematic diagram of a fixed regenerative carbon bed is shown in Figure 3.47.

Desorption of the carbon bed refers to the process or regenerating the carbon to restore its adsorbing capabilities and preserve its useful life (usually 2 to 5 years). The desorption process normally lasts 1 to 2 h and consists of the following three steps: regeneration of the carbon, bed drying, and returning the bed to its operating temperature. Carbon regeneration is accomplished by volatilizing the adsorbed compounds either by raising the temperature of the carbon bed via steam (steam desorption) or lowering the temperature in the bed to vacuum conditions to increase the vapor pressure of the adsorbed VOCs (vacuum desorption). The adsorption time of each bed is dependent on the influent mass

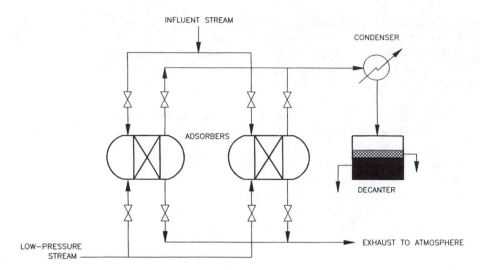

Figure 3.47 Schematic diagram of a fixed regenerative bed carbon adsorber.

flux rate and is usually finalized during the installation and start-up period. The automatic cycling of beds between adsorption and desorption modes is controlled by an onboard programmable logic controller.

When chlorinated organic compounds are present, HCl may be produced during steam regeneration of the carbon beds. Accumulation of HCl in the beds may corrode the container and require periodic replacement. Under such circumstances, use of a container that has a corrosion-resistive internal surface such as Teflon or Haztalloy should be used. Another method of overcoming the same concern is to use heated nitrogen as a carrier gas under vacuum desorption conditions. Nitrogen also helps in maintaining an inert environment in the bed. The nitrogen containing all the purged VOCs is condensed, and the VOCs are separated as a liquid before reusing the nitrogen gas.

When the influent air stream flow rates are small and the total VOC mass is less, smaller-sized carbon canisters that are usually of 55 gal in volume and contains approximately 200 lb of carbon can be used. Once the carbon becomes saturated, these canisters are disposed and replaced with a new canister.

Carbon adsorption is not effective for controlling highly volatile VOCs (such as vinyl chloride), which do not adsorb well. Similarly, highly nonvolatile VOCs (compounds with very high molecular weight) which do not desorb well are also not good candidates for removal by carbon adsorption.

Design criteria for carbon adsorption systems can be summarized as follows:

- influent airflow rate
- influent stream composition
- influent VOCs concentration and total mass
- adsorption capacity
- influent air temperature
- influent air moisture content and relative humidity
- adsorption equilibria related to waste mix

3.7.4 Condensation

Condensation is the process of removing VOCs from a noncondensable gas stream. Condensation can occur by lowering the gas stream temperature at constant pressure or increasing the gas stream pressure at constant temperature (or a combination of both).[30]

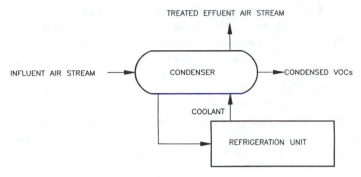

Figure 3.48 Schematic diagram of a refrigerated condenser.

There are two popular types of condensers: surface and direct contact. *Surface condensers* are generally shell and tube heat exchangers where coolant flows inside the tubes to condense the VOCs in the gas stream flowing outside the tubes. *Contact condensers* operate by spraying a cool liquid directly into a gas stream to cool it and condense the VOCs. In both types of condensers, the VOCs may be reused or disposed. Excessive influent moisture content will impact the process by ice formation. Removal of moisture prior to condensation may be required under those circumstances.

Coolants used to condense VOCs include chilled water, brine solutions, chlorofluorocarbons (CFCs), and cryogenic fluids.[30] Chilled water is an effective coolant down to approximately 45°F. Brine solutions are effective coolants down to approximately −30°F. CFCs are effective coolants down to approximately −90°F. However, the production and use of most CFCs are expected to be eliminated by the year 2000. Cryogenic fluids, mainly liquid nitrogen or carbon dioxide, can be effective down to temperatures as low as −320°F. Figure 3.48 is a schematic diagram of a typical condensation system.

Design criteria for condensation systems are as follows:

- influent airflow rate
- stream composition
- required condensation temperature
- mixture dew point
- moisture content in influent stream

3.7.5 Biofiltration

Biofiltration, a well-established technology in Europe, is just starting to get noticed in the U.S. Biofiltration harnesses the natural process of contaminant degradation with immobilized microorganisms. In a biofilter, the microorganisms grow on materials such as soil, compost, peat, or heather, supplemented sometimes with synthetic materials including activated carbon (to adsorb certain VOCs) and polystyrene, which provides bulking and structural stability. Recent designs tend to favor mineral soils and synthetic mixtures designed for durability and good retention of moisture and structural characteristics.[31] Biofiltration beds may be open to the atmosphere or enclosed. Single-bed or multiple-stack configurations are available. Figure 3.49 presents a schematic diagram of a closed, single-bed biofiltration system.[32]

In a biofiltration system, the VOC-laden airstream is dedusted, cooled, or humidified as necessary and then transported via a blower and a network of perforated piping into the beds which contain immobilized microorganisms. As the VOCs enter the biofilter, they diffuse into the biofilm, where the microorganisms oxidize them to carbon dioxide water and chloride (in the case of chlorinated of VOCs). The oxygen present in air also diffuses into the biofilm to facilitate the aerobic microbial metabolism. The residence time of the air stream depends on such factors such as the composition of the waste stream and its flow rate. The process takes place entirely in the biofilm: no contaminants are permanently transferred to the filter material. The microorganisms at the heart of biofiltration can be native to the soils used in the bed or specially cultured microorganisms adapted to the specific VOCs. Most biofiltration systems can be acclimatized to the target air stream in about 2 to 3 weeks, either with native microorganisms or specialized microbial species. Biofilters are an especially good candidate to treat air streams which contain biodegradable compounds such as alcohols, ketones, ethers, esters, organic acids, and lighter end petroleum compounds. Some of chlorinated organic compounds (such as methylene chloride and chlorobenzene) can also be treated with the practical residence times available.

As mentioned earlier, pretreatment of the VOC-laden air stream is important for preserving the longevity of the filter bed. Dust particles in the air stream can clog the piping and the filter bed, thus increasing the pressure drop and reducing available sites for biofilm growth. Cooling of the air stream to the optimum operating temperature of approximately 30°C to 40°C is required to prevent deactivation of the microorganisms. Humidification of the incoming air stream is important to maintain the required moisture content.

A consistent moisture level (the correct level depends upon the type of filter material) is essential for biofilter effectiveness. The microorganisms require a moist environment to maintain a viable biofilm and, clearly, blowing air on damp filter material is going to have a drying effect. In addition, the filter must be consistently kept moist to prevent cracking due to repeated drying and wetting. Cracks formed in the bed would permit escape of untreated VOCs to the atmosphere. Moisture can be controlled by humidification as a pretreatment step or by moisturizing the filter bed with spray nozzles (Figure 3.49).

Design criteria for biofiltration systems are summarized below:

- influent airflow rate
- stream contaminant composition (biodegradability of contaminants)
- influent contaminant concentrations
- required pretreatment
- available residence time
- influent temperature
- space requirements
- filter size
- filter bed material

3.7.6 Membrane Filtration

Membrane filtration technology refers to the use of a semipermeable membrane to separate VOCs from an air stream. Membrane technology has been found to be effective in removing some VOCs that traditionally have been difficult to recover, such as chlorinated hydrocarbons and CFCs. In this process, VOC-laden air contacts one side of a membrane that is permeable to organic vapors but is relatively impermeable to air. A partial vacuum, applied on the other side, draws the organic vapors through the membrane. The permeate vapor is then compressed and condensed to recover the organic fraction. The purified air stream is removed on the feed side.[33]

The membrane unit, shown in Figure 3.50, is a spirally wound module comprised of a perforated pipe bound within the wound membrane and spacers.[32] In the system shown in Figure 3.50, VOCs in a compressed air stream enter the first of two membrane stages. The first stage concentrates most of the VOCs into the permeate stream. The permeate is recompressed and condensed, normally producing only water present in the air stream as moisture. The bleed stream leaving the condenser enters the second membrane stage, reducing the VOC content

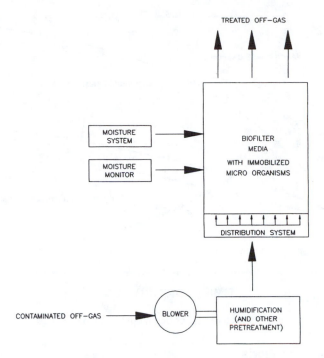

Figure 3.49 Components of a biofiltration system. (After Vembu, K. and Walker, C. S., Biofiltration holds VOCs, odors at bay, *Environ. Prot.*, February 1995.)

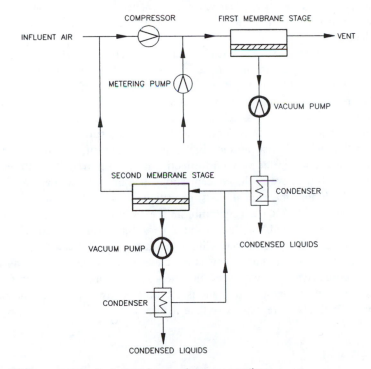

Figure 3.50 VOC removal in air streams by membrane separation.

further. The permeate from this stage is concentrated enough with VOCs to allow for its condensation. The bleed stream from this condenser is recirculated back to the first stage membrane unit. In its entirety, the system separates the VOC-laden air stream into a VOC-depleted stream and condensed liquid VOCs.

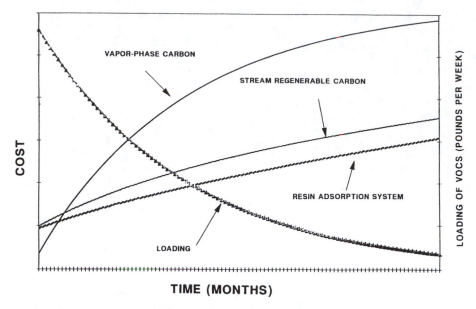

Figure 3.51 Comparison of costs for three vapor treatment technologies.

3.7.7 Cost Considerations

Capital and annualized costs of vapor treatment technologies are fundamental to the selection of a specific technology. Clearly, a treatment technology that proves to be overwhelmingly expensive for the amount of VOCs to be treated will be eliminated early in an equipment selection process.

As fundamental as costs are to equipment selection, they are also very unit-specific, because of their dependence on unique stream characteristics such as flow rate, VOCs composition and concentration, and operating temperature. It is presumptuous to conclude that a cost estimate for a particular treatment technology based on a specific set of operating conditions can be applied universally to every other possible set of operating conditions.

Another important aspect to consider during design of subsurface remediation system is the expected decline of VOC concentration in the air stream during the life of the project. A particular technology that may be very cost-effective during the initial, high-concentrations phase may not be cost-effective as the concentrations decline. A break-even analysis should be performed by including the capital costs, operating costs, and the impact on operating costs due to the decline of VOCs concentration in the air stream to be treated. An example of such an analysis performed for three technologies for a specific stream is shown in Figure 3.51. The other technologies were eliminated during the initial screening process.

Energy required to operate the systems is the major component contributing toward the annual operating costs. Availability of a specific energy source may have a significant impact on the operating cost. For example, propane is 1.5 times more expensive than natural gas, and electricity is 3.5 times more expensive than natural gas.

REFERENCES

1. U.S. Environmental Protection Agency, Forced Air Ventilation for Remediation of Unsaturated Soils Contaminated by VOC, Robert S. Kerr Environmental Research Laboratory, Ada, OK, EPA/600/S2-91/016, July 1991.

2. U.S. Environmental Protection Agency, Remedial Action, Treatment, and Disposal of Hazardous Waste, Proceedings of the Seventeenth Annual RREL Hazardous Waste Research Symposium, EPA/600/9-91/002, 1991.

3. Bear, J., *Hydraulics of Groundwater*, McGraw-Hill, New York, 1979.

4. DiGiulio, D.C., et al., Conducting field tests for evaluation of soil vacuum extraction application, *Proc. Fourth Natl. Outdoor Action Conference on Aquifer Restoration, Groundwater Monitoring and Geophysical Methods*, Las Vegas, Nevada, May 1990.

5. Rawls, W. J., Brakensiek, D. L., and Saxton, K. E., Estimation of soil water properties, *Trans. Am. Soc. Agric. Eng.*, 1316, 1982.

6. Jury, W. A., Winer, A. M., Spencer, W. F., and Focht, D. D., Transport and transformations of organic chemicals in the soil-air-water ecosystem, *Rev. Environ. Contamination Toxicol.*, 99, 120, 1987.

7. Lyman, W. J., Reehl, W. F., and Rosenblatt, D. H., *Hardbook of Chemical Property Estimation Methods: Environmental Behavior of Organic Compounds*, McGraw-Hill, New York, 1982.

8. Munz, C. and Roberts, P. V., Air-water phase equilibria of volatile organic solutes, *J. AWWA*, 75 (5), 62, 1987.

9. Knox, R. C., Sabatini, D. A., and Canter, L. W., *Subsurface Transport and Fate Processes*, Lewis Publishers, Boca Raton, FL, 1993.

10. Weber, W. J., Jr., *Physiochemical Processes for Water Quality Control*, John Wiley & Sons, New York, 1972.

11. Weber, W. J., Jr., McGinley, P. M., and Katz, L. E., Sorption phenomena in subsurface systems: concepts, models and effects on contaminant fate and transport, *Water Res.*, 25, 499, 1991.

12. Hassett, J. J., Means, J. C., Banwart, W. L., and Wood, S. G., Sorption Properties of Sediments and Energy Related Pollutants, U.S. Environmental Protection Agency, EPA/600/3-80-041, 1980.

13. Sulfita, J. M., Microbial Ecology and Pollutant Biodegradation in Subsurface Ecosystems, Chapter 7, Transport and Fate of Contaminants in the Subsurface, U.S. Environmental Protection Agency, EPA/625/4-89/019, 1989.

14. Kobayashi, H. and Rittman, B. E., Microbial removal of hazardous organic compounds, *Environ. Sci. Technol.*, 16, 170A, 1982.

15. U.S. Environmental Protection Agency, Soil Vapor Extraction Technology, Reference Handbook, EPA/540/2-91/0-003, 1991.

16. Sims, R. C., Soil remediation technologies at uncontrolled hazardous waste sites: A critical review, *J. Air Waste Manage. Assoc.*, 40, 704, 1990.

17. Stonestrom, D. A. and Rubin, J., Air permeability and trapped air content in two soils, *Water Resour. Res.*, 25(a), 1959, 1989.

18. Ong, S. K. and Lion, L. W., Trichloroethylene vapor sorption onto soil minerals, *Soil Sci. Soc., Am. J.*, 55, 1559, 1991.

19. Montgomery, J. H. and Welkom, L. M., *Groundwater Chemicals and Desk Reference*, Lewis Publishers, Boca Raton, FL, 1990.

20. William C. Anderson, Ed., *Vacuum Vapor Extraction: Innovative Site Remediation Technology Series*, Vol. 8, American Academy of Environmental Engineers, Annapolis, MD, 1993.

21. Massmann, J. W., Applying groundwater flow models in vapor extraction system design, *J. Environ. Eng. Div. ASCE*, 115(I), 129, 1989.

22. Shan, C., Falta, R. W., and Javandel, I., Analytical solutions for steady state gas flow to a soil vapor extraction well, *Water Resour. Res.*, 28 (4), 1105, 1992.

23. Strack, O. D. L., *Groundwater Mechanics*, Prentice-Hall, Englewood Cliffs, NJ, 1989.

24. Schafer, D., personal communication, 1995.

25. Kruseman, G. P. and deRidder, N. A., *Analysis and Evaluation of Pumping Test Data*, 2nd ed., Publication 4T, International Institute for Land Reclamation and Improvement, Wageningen, The Netherlands, 1972.

26. Leeson, A., et al., Statistical Analyses of the U. S. Air Force Bioventing Initiative Results, *Third Internatl In Situ and On Site Bioreclamation Symposium*, San Diego, April 1995.

27. Simpkin, T. J., et al., The Influence of Temperature on Bioventing, *Third Internatl. In Situ and On Site Bioreclamation Symposium*, San Diego, April 1995.

28. Reisinger, H. T., et al., Pressure Dewatering: An Extension of Bioventing Technology, *Third Internatl. In Situ and On Site Bioreclamation Symposium*, San Diego, April 1995.

29. Vidumsky, J., personal communication, 1995.

30. American Institute of Chemical Engineers, *Reducing and Controlling Volatile Organic Compounds*, Center for Waste Reduction Technologies, New York, 1993.

31. McHale, C., personal communication, 1995.

32. Vembu, K. and Walker, C. S., Biofiltration holds VOCs, odors at bay, *Environ. Prot.*, February 1995.

33. U.S. Environmental Protection Agency, Volatile Organic Compound Removal from Air Streams by Membrane Separation, Emerging Technology Bulletin, EPA/540/F-94/503, 1994.

4

IN SITU AIR SPARGING

4.1 INTRODUCTION

In situ air sparging is a remediation technique that has been used since about 1985, with varying success, for the remediation of volatile organic compounds (VOCs) dissolved in the groundwater, sorbed to the saturated zone soils, and trapped in soil pores of the saturated zone. This technology is often used in conjunction with vacuum extraction systems (Figure 4.1) to remove the stripped contaminants, and has broad appeal due to its projected low costs relative to conventional approaches.

The difficulties encountered in modeling and monitoring the multiphase air sparging process (i.e., air injection into water saturated conditions) have contributed to the current uncertainties regarding the process(es) responsible for removing the contaminants from the saturated zone. Even today, engineering design of these systems is largely dependent on empirical knowledge and experience of the design engineer. Thus, air sparging should be treated as a rapidly evolving technology with a need for continuous refinement of optimal system design and mass transfer efficiencies. The mass transfer mechanisms during *in situ* air sparging relies on the interactions between complex physical, chemical, and microbial processes, many of which are not well understood.

A typical air sparging system has one or more subsurface points through which air is injected into the saturated zone. At the technology's inception, it was commonly perceived that the injected air travels up through the saturated zone in the form of air bubbles;[1–3] however, it is more realistic that the air travels in the form of continuous air channels.[4–6] The airflow paths will be influenced by pressure and flow rate of the injected air and depth of injection; however, structuring and stratification of the saturated zone soils appear to be the predominant factors.[4–6] Significant channeling may result from relatively subtle permeability changes, and the degree of channeling will increase as the size of the soil pore throats get smaller. Research shows that even minor differences in permeability due to stratification can impact sparging effectiveness.[5]

In addition to conventional air sparging, in which injection of air is as shown in Figure 4.1, many modifications of the technique to overcome the geologic/hydrogeologic limitations to the technology's success will also be discussed in this chapter.

4.2 GOVERNING PHENOMENA

In situ air sparging is potentially applicable when volatile and/or aerobically biodegradable organic contaminants are present in water-saturated zones, under relatively permeable conditions. The *in situ* air sparging process can be defined as injection of compressed air at controlled pressures and volumes into water-saturated soils. The contaminant mass removal

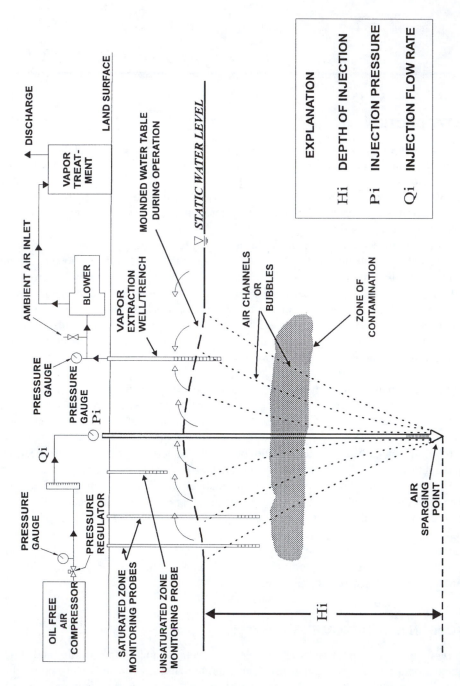

Figure 4.1 Air sparging process schematic.

processes that occur during the operation of air sparging systems include (1) *in situ* air stripping of dissolved VOCs; (2) volatilization of trapped and adsorbed phase contamination present below the water table and in the capillary fringe; and (3) aerobic biodegradation of both dissolved and adsorbed phase contaminants.

It was found that during *in situ* air sparging of petroleum hydrocarbon sites, in the short term (weeks/months), stripping and volatilization account for much more removal of hydrocarbons than does biodegradation.[7] Biodegradation only becomes more significant for mass removal with long-term system operations.

4.2.1 *In Situ* Air Stripping

Among the three contaminant removal mechanisms discussed above, *in situ* air stripping may be the dominant process for some dissolved contaminants. Henry's law constant provides a qualitative assessment of the potential removal efficiencies of dissolved VOCs during air sparging. Compounds such as benzene, toluene, xylenes, ethylbenzene, trichloroethylene, and tetrachloroethylene are considered to be very easily strippable (See Appendix C for a full list of Henry's law constants). However, a basic assumption made in analyzing the air stripping potential during air sparging is that Henry's law applies to the volatile contaminants and that all the contaminated water is in close communication with the injected air. In-depth evaluation of these assumptions exposes the shortcomings and complexities of interphase mass transfer during air sparging.

First of all, Henry's law is valid only when partitioning of the dissolved contaminant mass has reached equilibrium at the air–water interface. However, the residence time of air traveling in discrete channels may be too short to achieve the equilibrium due to the high air velocities and short travel paths. Another issue is the validity of the assumption that the contaminant concentration at the air–water interface is the same as in the bulk water mass. Due to the removal of contaminants in the immediate vicinity of the air channels, it is safe to assume that the contaminant concentration is going to be lower around the channels than outside the channels. To replenish the mass lost from the water around the air channel, mass transfer by diffusion and convection must occur from water away from the air channels. Therefore, it is likely that the density of air channels will play a significant role in mass transfer efficiencies by minimizing the distances required for a contaminant "molecule" to encounter an air channel. In addition, the density of air channels will also influence the interfacial surface area available for mass transfer.

The literature suggests that the air channels formed during air sparging mimic a "viscous fingering" effect, and that two types of air channels are formed: large-scale channels and pore-scale channels.[8] The formation of both types of channels enhances the channel density and the available interfacial surface area.

It has been proposed that *in situ* air sparging also helps to increase the rate of dissolution of the sorbed phase contamination, and eventual stripping below the water table. This is due to the enhanced dissolution caused by increased mixing and the higher concentration gradient between the sorbed and dissolved phases under sparging conditions.

4.2.2 Direct Volatilization

The primary mass removal mechanism for VOCs present in the saturated zone during pump and treat operations is resolubilization into the aqueous phase and the eventual removal with the extracted groundwater. During *in situ* air sparging, direct volatilization of the sorbed and trapped contaminants is enhanced in the zones where airflow takes place. Direct volatilization of any compound is governed by its vapor pressure, and most VOCs are easily removed through volatilization. Figure 4.1 is a schematic of an air channel moving through an aquifer containing sorbed or trapped (NAPL) contamination. In the regions where the soil is predominantly air-saturated or the air channel is next to the zone of trapped contaminants, the process is similar to soil vapor extraction or bioventing, albeit in a microscopic scale.

Where significant levels of residual contamination of VOCs or NAPLs are present in the saturated zone, direct volatilization into the vapor phase may become the dominant mechanism for mass removal in areas where air is flowing. This may explain the significant increase in VOC concentrations typically observed in the soil vapor extraction effluents at many sites.[7]

4.2.3 Biodegradation

In most natural situations, aerobic biodegradation of biodegradable compounds in the saturated zone is rate-limited by the availability of oxygen. Biodegradability of any compound under aerobic conditions is dependent on its chemical structure and environmental parameters such as pH and temperature. Some VOCs are considered to be easily biodegradable under aerobic conditions (e.g., benzene, toluene, acetone, etc.) and some are not (e.g., trichloroethylene and tetrachloroethylene).

Typical dissolved oxygen (DO) concentrations in uncontaminated groundwater are less than 4.0 mg/l. Under anaerobic conditions induced by the natural degradation of the contaminants, DO concentrations in groundwater are often less than 0.5 mg/l. Dissolved oxygen levels can be raised by air sparging up to 6 to 10 mg/l under equilibrium conditions.[1,7,9] This increase in the DO levels will contribute to enhanced rates of aerobic biodegradation in the saturated zone. This method of introducing oxygen for enhanced biodegradation rates is one of the inherent advantages of *in situ* air sparging. However, the oxygen transfer into the bulk water is a diffusion-limited process. The diffusion path lengths for transport of oxygen through groundwater are defined by the distances between air channels. Where channel spacing is large, diffusion alone is not sufficient to transport adequate oxygen into all areas of the aquifer for enhanced aerobic biodegradation. The pore-scale channels formed and the induced mixing during air sparging enhances the rate of oxygen transfer.[8]

4.3 APPLICABILITY

4.3.1 Examples Of Contaminant Applicability

Contaminant type is a major variable affecting air sparging design and contaminant mass removal rate. Based on the discussion in the previous section, Table 4.1 describes the applicability of air sparging for a few selected contaminants based on the properties of strippability, volatility, and aerobic biodegradability. In order for air sparging to be effective, the VOCs must transfer from the groundwater into the injected air, and oxygen present in the injected air must transfer into the groundwater to stimulate biodegradation.

In practice, the criterion for defining contaminant strippability is based on the Henry's law constant being greater than 1×10^{-5} atm·m^3/mol. In general, compounds with a vapor pressure greater than 0.5 to 1.0 mmHg can be volatilized easily; however, the degree of volatilization is also limited by the flow rate of air. The half-lives presented in Table 4.1 are estimates in groundwater under natural conditions without any enhancements to improve the rate of degradation.

Many of the constituents present in heavier petroleum products such as no. 6 fuel oil will not be amenable to either stripping or volatilization (Figure 4.2). Hence, the primary mode of remediation, if successful, will be due to aerobic biodegradation. Required air injection rates under such conditions will be lower and influenced only by the requirement to introduce sufficient oxygen into the saturated zone.

Figure 4.2 qualitatively describes different mass removal phenomena in a simplified version under optimum field conditions. The amount of mass removed by stripping and volatilization have been grouped together, due to the difficulty in separating them in a

Table 4.1 Examples of Contaminant Applicability for *In Situ* Air Sparging

Contaminant	Strippability	Volatility	Aerobic biodegradability*
Benzene	High ($H = 5.5 \times 10^{-3}$)	High ($V_P = 95.2$)	High ($t_{1/2} = 240$)
Toluene	High ($H = 6.6 \times 10^{-3}$)	High ($V_P = 28.4$)	High ($t_{1/2} = 168$)
Xylenes	High ($H = 5.1 \times 10^{-3}$)	High ($V_P = 6.6$)	High ($t_{1/2} = 336$)
Ethylbenzene	High ($H = 8.7 \times 10^{-3}$)	High ($V_P = 9.5$)	High ($t_{1/2} = 144$)
TCE	High ($H = 10.0 \times 10^{-3}$)	High ($V_P = 60$)	Very low ($t_{1/2} = 7704$)
PCE	High ($H = 8.3 \times 10^{-3}$)	High ($V_P = 14.3$)	Very low ($t_{1/2} = 8640$)
Gasoline constituents	High	High	High
Fuel oil constituents	Low	Very low	Moderate

Note: H = Henry's law constant (atm·m³/mol); V_P = vapor pressure (mmHg) at 20°C; $t_{1/2}$ = half-life during aerobic biodegradation, hours.

* It should be noted that the half-lives can be very dependent on the site-specific subsurface environmental conditions.

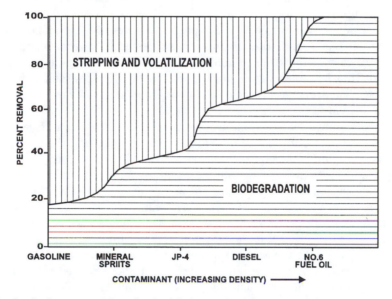

Figure 4.2 Qualitative presentation of potential air sparging mass removal for petroleum compounds.

meaningful manner. However, the emphasis should be placed on total mass removal, particularly of mobile volatile constituents, and closure of the site regardless of the mass transfer mechanisms.

4.3.2 Geologic Considerations

Successful implementation of *in situ* air sparging is greatly influenced by the ability to achieve significant air distribution within the target zone. Good vertical pneumatic conductivity is essential to avoid bypassing or channeling of injected air horizontally, away from the sparge point. It is not an easy task to evaluate the pneumatic conductivities in the horizontal and vertical direction for every site considered for *in situ* air sparging.

Geologic characteristics of a site are very important when considering the applicability of *in situ* air sparging. The most important geologic characteristic is stratigraphic homogeneity or heterogeneity. Presence of low permeability layers under stratified geologic conditions will impede the vertical passage of injected air. Laboratory-scale studies[5] illustrate the impact of geologic characteristics on air channel distribution. Under laboratory conditions, injected air may accumulate below the low permeability layers and travel

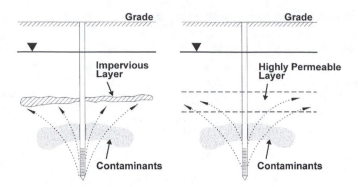

Figure 4.3 Potential situations for the enlargement of a contaminant plume during air sparging.

in a horizontal direction, thus potentially enlarging the contaminant plume (Figure 4.3). High permeability layers may also cause the air to preferentially travel laterally, again potentially causing an enlargement of the plume (Figure 4.3). Horizontal migration of injected air limits the volume of soils that can be treated by direct volatilization due to the inability to capture the stripped contaminants. Horizontal migration can also cause safety hazards if hydrocarbon vapors migrate into confined spaces such as basements and utilities. Hence, homogeneous geologic conditions are important for the success and safety of *in situ* air sparging.

Both vertical permeability and the ratio of vertical to horizontal permeability decrease with decreasing average particle size of the sediments in the saturated zone. The reduction of vertical permeability is directly proportional to the effective porosity and average grain size of the sediments.[10] Hence, based on the empirical information available, it is recommended that the application of conventional *in situ* air sparging be limited to saturated zone conditions where the hydraulic conductivities are greater than 10^{-3} cm/s.[4,11]

It is unlikely that the design engineer will encounter homogeneous geologic conditions across the entire cross section at most sites. In fact, the optimum geologic conditions for air sparging may be where the permeability increases with increasing elevation above the point of air injection. Decreasing permeabilities with elevation above the point of air injection will have the potential to enlarge the plume due to lateral movement of injected air.

4.4 DESCRIPTION OF THE PROCESS

4.4.1 Air Injection Into Water-Saturated Soils

The ability to predict the performance of air sparging systems is limited by the current understanding of airflow in the water-saturated zone and limited performance data. There were two schools of thought in the literature describing this phenomenon. The first, and the widely accepted one, describes that the injected air travels in the vertical direction in the form of discreet air channels. The second school of thought describes injected air travel in the form of air bubbles. Airflow mechanisms cannot be directly observed in the field. However, conclusions can be reached by circumstantial evidence collected at various sites, and laboratory-scale visualization studies.

Sandbox model studies performed[5,6] tend to favor the "air channels" concept over the "air bubbles" concept. In laboratory studies simulating sandy aquifers (grain sizes of 0.075 to 2 mm), stable air channels were established in the medium at low injection rates. Under laboratory conditions simulating coarse gravels (grain sizes of 2 mm or larger), the injected air rose in the form of bubbles. At high air injection rates in sandy, shallow, water-table aquifers, the possibility for fluidization (loss of soil cohesion) around the point of injection exists,[4,6] and thus the loss of control of the injected air may occur.

4.4.2   Mounding of Water Table

When air is injected into the saturated zone, groundwater must necessarily be displaced. The displacement of groundwater will have both a vertical and lateral component. The vertical component will cause a local rise in the water table, sometimes called water table mounding. Mounding has been used by some as an indicator of the "radius-of-influence" of the sparge well during the early stages of development of this technology.[1,2,9,12,13] Mounding is also considered to be a design concern because it represents a driving force for lateral movement of groundwater and dissolved contaminants and can therefore lead to spreading of the plume. The magnitude of mounding depends on site conditions and the location of the observation wells relative to the sparge well. Mounding can vary from a negligible amount to several feet in magnitude.

Simulations of the flow of air and water around an air sparging well were performed with a multiphase, multicomponent simulator (TETRAD) originally developed for the study of problems encountered during exploitation of petroleum and geothermal resources.[14,15] The simulations were performed by defining two primary phases of transient behavior that lead to a steady-state flow pattern (Figures 4.4 and 4.5). The first phase is characterized by an expansion in the region of airflow (Figure 4.4). During this phase, the rate of air injection into the saturated zone exceeds the rate of airflow out of the saturated zone into the vadose zone. It is during this transient expansion phase that groundwater mounding first develops and reaches its highest level, where it extends from near the injection well to beyond the region of airflow in the saturated zone. When injected air breaks through to the vadose zone, the region of airflow in the saturated zone begins to collapse or shrink (Figure 4.5). During this second transient phase of behavior, the preferred pathways of higher air permeability from the point of injection to the vadose zone are established. The air distribution zone shrinks until the rate of air leakage to the vadose zone equals the rate of air injection. During this collapse phase, mounding near the sparge well dissipates. When steady state conditions are reached, little or no mounding exists. This behavioral pattern has also been observed in the field[4,13,15] (Figure 4.6). This behavior reflects the building and decay of the groundwater mound at a sparging location.

The transience of groundwater mounding at most sites has important implications for the risk of lateral movement of the contaminant plume. Because the water table returns close to its presparging position during continuous air injection, the driving force for lateral movement of groundwater caused by air injection becomes very small.

An important aspect of groundwater mounding is that it is not a direct indicator of the physical presence of air in the saturated zone. Water table mounding at a given place and time may or may not be associated with the movement of air in the saturated zone at the same location. Some mounding will occur beyond the region of airflow in the saturated zone. Additionally, a transient pressure increase without water table mounding commonly occurs beyond the limits of airflow, especially where airflow is partially confined. Because of its transient nature and the fact that the water table is displaced ahead of injected air, water table mounding can be a misleading and overly optimistic indicator of the distribution of airflow within the saturated zone.

4.4.3   Distribution of Airflow Pathways

It is often envisioned that airflow pathways developed during air sparging form an inverted cone with the point of injection being the apex. This would be true only if soils were

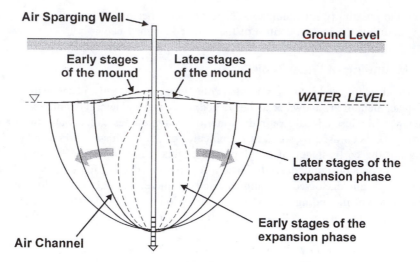

Figure 4.4 The first transient behavior after initiation of air injection into the saturated zone.

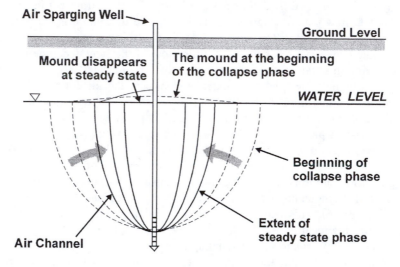

Figure 4.5 The second transient behavior before reaching steady state during air sparging.

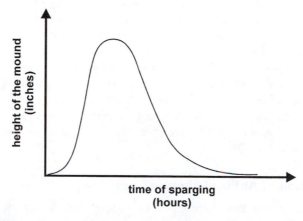

Figure 4.6 Appearance and disappearance of groundwater mound during *in situ* air sparging.

perfectly homogeneous or comprised of coarse-grained sediments, and the injected airflow rate was low. During laboratory experiments using homogeneous media with uniform grain sizes, symmetrical airflow patterns about the vertical axis were observed.[5] However, media simulating mesoscale heterogeneities yielded nonsymmetrical airflow patterns.[5] The asymmetry apparently resulted from minor variations in the permeability and capillary air entry resistance, which resulted from pore-scale heterogeneity. Hence, under natural conditions, it is realistic to expect that symmetric air distribution will never occur.

These same experiments also indicated that the channel density and thus the interfacial surface area increased with increased airflow rates, since higher volumes of air occupy an increased number of air channels. Assuming that the air channels are cylindrical in shape and that the number of channels and air velocity in the channel remains the same even for a change in airflow rate, the interfacial surface area will be increased by a ratio $(Q_{final}/Q_{initial})^{0.5}$, where Q is airflow rate.

It is reported in some literature that, at low sparge pressures, air travels 1 to 2 ft horizontally for every foot of vertical travel.[1] However, it should be noted that this correlation was not widely observed. It was also reported that as the sparge pressure is increased, the degree of horizontal travel increases.[2,6,14] Field observations have indicated that airflow channels extend 10 to 40 ft away from the air injection point, independent of flow rate and depth of sparge point.[7,10,14]

4.4.4 Groundwater Mixing

Mixing of groundwater during air sparing is an important mechanism to overcome the diffusional limitation for contaminant mass transfer and provide adequate oxygen transport into the aquifer. Groundwater mixing during air sparging may significantly reduce the diffusion limitation for mass transfer without generating any changes in the bulk groundwater flow. It has been shown that nonsteady-state mixing mechanisms induced in opposite directions at different times as a result of pulsed sparging operations will enhance the mass removal efficiencies.[8,13]

There are many possible mechanisms for groundwater mixing during air sparging.[8] Several possible mechanisms are as follows:

- physical displacement by injected air
- capillary interaction of air and water
- frictional drag by flowing air
- water flow in response to evaporative loss
- thermal convection
- migration of fines

Groundwater is physically displaced by air as it moves through the saturated zone soil during sparging. This process occurs during nonsteady-state airflow conditions, where the percentage of air saturation changes with time until the formation of spatially fixed air channels. The amount of mixing due to this physical displacement is dependent upon the amount of groundwater displaced and the duration of nonsteady-state flow conditions. The rate of water displacement is permeability limited and, therefore, the duration of these effects is generally greater in low-permeability soils. The process will take place over both microscopic distances (inches) and site-scale distances. Pulsed sparging will frequently create nonsteady-state conditions and enhance groundwater mixing.[8]

While physical displacement of water by air involves changes in fluid saturation, capillary fluid interactions during sparging can cause groundwater movement without a change in air saturation. This process can be expected to be more pronounced during nonsteady-state conditions, when higher air injection pressures can be maintained.[8,13,15] Pulsed sparging may enhance this mixing mechanism by increasing the time during which conditions are in a nonsteady state.

Frictional drag on groundwater can be induced by transfer of shear stresses from flowing air to pore water during non-Darcy airflow conditions.[8] For fluid flow in a porous medium, a critical value of Reynolds number (Re) for non-Darcy flow is 1,[8] which corresponds to an air velocity of 0.015 to 0.15 m/s for fine sands to coarse sands.

Evaporative loss of water to the injected air stream can result in water inflow to the sparged zone to maintain volume balance. This volume balance approach must consider changing air saturations and is very sensitive to the degree of air saturation and relative humidity of the injected air and their effects on the rate of evaporation. This is also a thermodynamic process, where heat lost to evaporation cools the groundwater, leading to downward density-driven flow. This flow would be opposite to that induced by frictional drag (for upward airflow).[8]

Thermal convection can occur through density-driven flow of cooled groundwater as indicated above, or through heating of groundwater by injecting heated gases. This process is sensitive to the air saturation developed by its effect on heat transfer. The heat capacity of air is much less than that of water, potentially limiting the warming of groundwater.[8]

The migration of fine sediments has been shown to significantly reduce the permeability of petroleum reservoirs by "sealing" pore throats.[8] Fines migration also has been observed during sparging in both laboratory sand tank studies and in field studies. Airflow paths may be destabilized by changes in air permeability caused by fines migration, and the resulting redirection of airflow may cause groundwater mixing as the water is displaced by or displaces air.

Based on the above discussion, physical displacement of water and capillary interactions seem to be relevant primarily during nonsteady-state conditions. Frictional drag, evaporative loss, thermal convection, and fines migration may also cause groundwater mixing after steady state conditions are reached, but the magnitude of mixing resulting from these processes may be less than that which occurs during the nonsteady state.[8]

Groundwater mixing is important during air sparging to effectively transport dissolved oxygen for *in situ* bioremediation. Groundwater mixing can be effective if it occurs at the pore scale as well as over site-scale distances, since either process can reduce the diffusion limitation of sparging. Theory and field measurements indicate that mixing does occur, is most pronounced during nonsteady-state conditions, and is enhanced by pulsed sparging. This mixing is commonly bidirectional, which may prevent development of a discernible site-scale flow pattern. Because sparging without groundwater mixing will be of limited effectiveness, the increased volatile organic compound removal and DO addition that occurs during sparging and is enhanced by pulsing provides strong indirect evidence that mixing does occur.[8]

4.5 SYSTEM DESIGN PARAMETERS

In the absence of readily available and reliable models for the *in situ* air sparging process, empirical approaches are used in the system design process. The parameters which are of significant importance in designing an *in situ* air sparging system are listed below.

- Air distribution (zone of influence)
- Depth of air injection
- Air injection pressure and flow rate
- Injection mode (pulsing or continuous)
- Injection wells construction
- Contaminant type and distribution

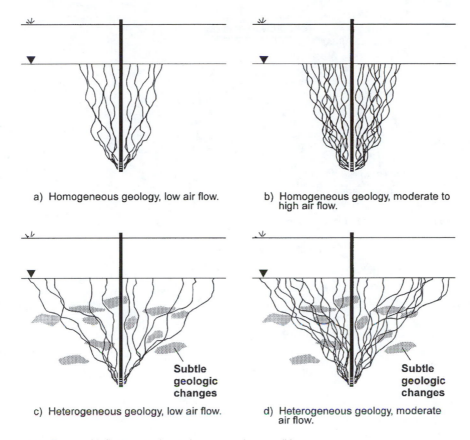

a) Homogeneous geology, low air flow.

b) Homogeneous geology, moderate to high air flow.

c) Heterogeneous geology, low air flow.

d) Heterogeneous geology, moderate air flow.

Figure 4.7 Zones of influence under various operating conditions.

4.5.1 Air Distribution (Zone of Influence)

During the design of air sparging systems, it may be very difficult to define a radius of influence in the same manner it is used in pump and treat and/or soil venting systems. Due to the asymmetric nature of the air channel distribution and the variability in the density of channels, it is safer to assume a "zone of influence" rather than a radius of influence[4,6,16] (Figure 4.7).

It becomes necessary to estimate the "zone of influence" of an air sparging point, similar to any other subsurface remediation technique, to design a full-scale air sparging system consisting of multiple points. This estimation becomes an important parameter for the design engineer to determine the number of required sparge points. The zone of influence should be limited to describing an approximate indication of the average distance traveled by air channels from the sparge point in the radial directions, under controlled conditions.

The zone of influence of an air sparging point is assumed to be an inverted cone (Section 4.4.3); however, it should be noted that this assumption implies homogeneous soils of moderate to high permeability, which is rarely observed in the field. As noted earlier during a numerical simulation study on air sparging,[14] three phases of behavior were predicted following initiation of air injection (Figures 4.4 and 4.5). These are (1) an expansion phase in which the vertical and lateral limits of airflow grow in a transient manner; (2) a second transient period of reduction in the lateral limits (collapse phase); and (3) a steady state phase, during which the system remains static as long as injection parameters do not change. The zone of influence of air sparging was found to reach a roughly conical shape during the steady state phase.

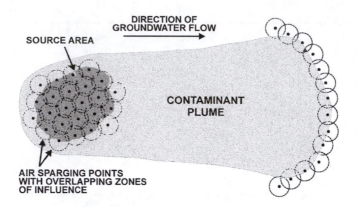

Figure 4.8 Air sparging point locations in a source area and in a curtain configuration.

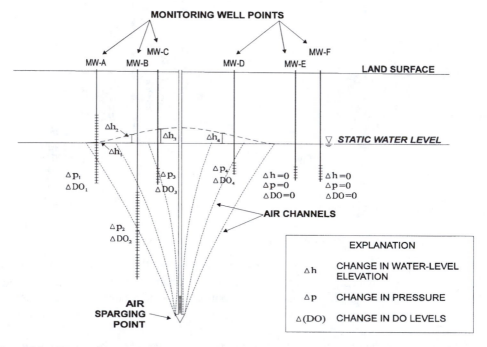

Figure 4.9 Air sparging test measurements.

Based on the inverted cone airflow distribution model, many air sparging system designs are performed based on the zone of influence measured during a field design test. When a hot spot or source area is under consideration for cleanup, it is prudent to design the air sparging system in a grid fashion (Figure 4.8). The grid should be designed with overlapping zones of influence that provide complete coverage of the area under consideration for remediation. If an air sparging curtain is designed to contain the migration of dissolved contaminants, the curtain should be designed with overlapping zones of influence in a direction perpendicular to the direction of groundwater flow (Figure 4.8).

A properly designed pilot test can provide valuable information. The limitations of time and money often restrict field evaluations to short duration single-well tests. Potential measuring techniques (Figure 4.9) of the zone of influence have evolved with this technology during the last few years.

- *Measurement of the lateral extent of groundwater mounding in the adjacent monitoring wells.*[1,2,9,13]

  This was the earliest technique used during the very early days of implementation of this technology. However, it did not take long to realize that the lateral extent of the mound is only a reflection of the amount of water displaced and does not correspond to the zone of air distribution.

- *Measurement of the increase in dissolved oxygen (DO) levels and redox potentials in comparison to presparging conditions.*[2,3,12,17–20]

  These parameters should be measured in the monitoring well itself by using field probes. Oxygen transfer could take place during sample collection and handling, and may bias the results of the analysis.

  This concept lost its value when it was realized that the injected air travels in the form of channels rather than bubbles. Increases in DO levels in the bulk water due to diffusion-limited transport of oxygen will be noticeable only during a long-term pilot study. In most cases, the increased DO levels observed during short-duration pilot tests were due to the air channels directly entering the monitoring wells and not due to overall changes in dissolved oxygen levels in the aquifer.

- *Measurement of soil gas pressures.*

  This technique involves the measurement of any increase in the soil gas pressure above the water table due to the escape of the injected air into the vadose zone. The escaped air will quickly equilibrate in the vadose zone, and may spread over a larger area than the zone of air distribution in the vadose zone. As a result, during the combined operation of soil venting and air sparging, measurement of this parameter may be totally misleading.

- *Increase in head space pressure within sealed saturated zone monitoring probes (piezometers) which are perforated below the water table only.*

  This technique is widely used, and currently considered to be the most reliable in terms of detecting the presence of air pathways at a specific distance in the saturated zone. When an air channel enters a monitoring probe via the submerged screen, the head space pressure could increase up to the hydrostatic pressure at the point of entry. However, the actual distribution of air channels may extend beyond the furthest monitoring probe.

- *The use and detection of insoluble tracer gases, such as helium and sulfur hexafluoride.*[4,16,20]

  Initial monitoring of the tracer gas in the vadose zone is typically performed while the SVE system is off. The potential to balance the mass of the injected tracer and the amount of recovered tracer raises the level of confidence in the estimation of the capture rate of injected air. The use of sulfur hexafluoride as a tracer gas has the advantage due to its solubility being similar to that of oxygen. Hence the detection of sulfur hexafluoride in bulk water will be an indicator for the diffusional transport of oxygen.

  This technique will also provide information on vapor flow paths and vapor recovery efficiencies during air sparging.

- *Measurement of the electrical resistivity changes in the target zone of influence as a result of the changes in water saturation due to the injection of air (electrical resistivity tomography (ERT) method).*

  Tomography is a method of compiling large amounts of one-dimensional information in such a way as to produce a three-dimensional image (CAT scans, MRIs, and holograms make use of tomography). ERT is a process in which a three-dimensional depiction of air saturation within the saturated zone is generated by measuring the electrical resistance of the soil between electrodes placed at various locations on wells during installation. Other tomographic methods include vertical induction profiling (VIP) and geophysical diffraction tomography (GDT).

VIP is similar to ERT, except that ERT uses a direct current (DC) potential while VIP uses a 500 Hz alternating current (AC). The use of AC makes it possible to detect electrical field strength by induction, so existing PVC wells can be used without the requirement for subsurface electrode installation.

GDT is another three-dimensional subsurface imaging technology that could be used to characterize air movement through the subsurface during air sparging. GDT is a high-resolution acoustic technique that provides quantitative subsurface imaging by measuring variations in acoustic velocity between various locations on the ground surface and depths in monitoring wells.

ERT may be the most reliable method among all the techniques discussed in this section. The high drilling costs, associated with installing large numbers of electrodes in the subsurface, preclude it from being used widely.

- *Measurement of moisture content changes within the target zone of influence using time domain reflectrometry (TDR) technique.*
  TDR is a well-established and accurate means to measure the moisture content of soils and has been widely used in the agricultural industry. When injected air travels within the zone of influence, moisture content will decrease due to the displacement of water. TDR data, collected from probes placed in the aquifer, can accurately reflect the changes in moisture content.

- *Neutron probe technique to measure the changes in water saturation.*[21]
  This technique utilizes a neutron probe to measure changes in water saturation (thus air saturation) below the water table during air sparging. The neutron probe detects the hydrogen in water and thus translates into a "water" saturation value. The water saturation values can be converted to air saturation values. The neutron probe can also detect the hydrogen in petroleum hydrocarbons and hence can bias the "fluid" saturation values.

- *The actual reduction in contaminant levels due to sparging.*
  This evaluation gives an indication of the extent of the zone of influence in terms of contaminant mass removal, but the test has to be run long enough to collect reliable data.

Since cost and budgetary limitations influence how a field design test is performed, availability of resources will determine the type of method used. The most reliable method is the one which measures the changes in electrical resistivity due to changes in air/water saturation. The most cost-effective method is the one which determines the head space pressure within the saturated zone probes.

4.5.2   Depth of Air Injection

Among all the design parameters, the depth of air injection may be the easiest to determine, since the choice is very much influenced by the contaminant distribution. It is prudent to choose the depth of injection at least a foot or two deeper than the "deepest known point" of contamination. However, in reality, the depth determination is influenced by soil structuring and the extent of stratigraphic layering, since injection below low permeability zones should be avoided. Current experience in the industry is mostly based on injection depths of less than 30 to 60 ft below the water table.[1,7,20]

The depth of injection will influence the injection pressure and the flow rate. The deeper the injection point is located, the greater the zone of influence will be expanded, and thus more air will be required to provide a reasonable percentage of air saturation within the zone of influence.

4.5.3 Air Injection Pressure and Flow Rate

Injected air will penetrate the aquifer only when the air pressure exceeds the sum of the water column's hydrostatic pressure and the threshold capillary pressure, or the *air entry pressure*. The air entry pressure is equal to the minimum capillary entry resistance for the air to flow into the porous medium. Capillary entry resistance is inversely proportional to the average diameter of the grains and porosity.[10,14]

The injection pressure necessary to initiate *in situ* air sparging must be able to overcome the following:

1. The hydrostatic pressure of the overlying water column at the point of injection.
2. The capillary entry resistance to displace the pore water; this depends on the type of sediments in the subsurface.

The capillary pressure can be quantitatively described,[16] under idealized conditions, by the following equation:

$$P_c = \frac{2s}{r} \tag{4.1}$$

where P_c = capillary pressure
  s = the surface tension between air and water
  r = the mean radius of curvature of the interface between fluids.

This equation reveals that as r decreases, the capillary pressure increases. Generally, r will decrease as grain size decreases. Therefore, the required pressure to overcome capillary resistance increases with decreasing sediment size.

The pressure of injection (P_i) in feet of water could be defined as

$$P_i = H_i + P_a + P_d \tag{4.2}$$

where H_i = saturated zone thickness above the sparge point (feet of water)
  P_c = air entry pressure of formation (feet of water)
  P_d = air entry pressure for the screen and packing.

The air entry pressure for a formation is heavily dependent on the type of geology. In reality, the air entry pressure will be higher for fine-grained (1 to 10 ft of water column pressure) than coarse-grained media (1 to 10 in. of water column pressure). When H_i is significantly greater than P_c and P_d combined, it is likely that air will enter the formation primarily near the top of the injection screen.

The notion that excessively high pressures and flow rates correspond to better air sparging performance is not true. Increasing the injection rate to achieve a greater flow and wider zone of influence must be implemented with caution.[4,14] This is especially true during the start-up phase due to the low relative permeability to air attributable to low initial air saturation. The danger of pneumatically fracturing and thus creating secondary permeability in the formation under excessive pressures should also be taken into consideration in determining injection pressures. As such, it is very important to gradually increase the pressure during system start-up. Refer to Chapter 9 to estimate the maximum pressure that can be safely applied without causing any fracturing in the formation.

The typical values of injected airflow rates reported in the literature range from 1 cfm to 15 cfm per injection well.[4,11] Injection airflow determinations are also influenced by the ability to recover the stripped contaminant vapors through a vapor extraction system, thus containing the injected air within a controlled air distribution zone.

4.5.4   Injection Mode (Pulsing and Continuous)

Direct and speculative information available in the literature indicates that the presence of air channels impedes, but does not stop, the flow of water across the sparging zone of influence. The natural groundwater flow through a sparged zone of an aquifer will be slowed and diverted by the air channels due to changes in water saturation and thus relative hydraulic permeability. This potentially negative factor could be overcome by pulsing the air injection and thus minimizing the decrease in relative permeability due to changes in water saturation.

An additional benefit of pulsing will likely be due to the increased mixing of groundwater resulting from air channels formation and collapse during each pulse cycle. This should also help to reduce the diffusional rate limitation for the transport of contaminants in the bulk water phase toward the air channels, due to the cyclical displacement of water during pulsed air injection. As noted earlier, the expansion phase during air sparging (Figures 4.4 and 4.5) appears to have a greater zone of influence than under the steady state conditions; therefore, pulsing may improve the efficiency of air sparging by creating cyclical expansion and collapse of the zone of influence.

4.5.5   Injection Wells Construction

Injection wells must be designed to accomplish the desired distribution of airflow in the formation. Conventional design of an air sparging well under shallow "sparge depth" conditions (less than 20 ft) and deeper sparge depth conditions (greater than 20 ft) are shown in Figures 4.10 and 4.11. Schedule 40 or 80 PVC (polyvinyl chloride) piping and screens in various diameters can be used for the well construction. In both configurations, the sparge point should be installed by drilling a well to ensure an adequate seal to prevent short-circuiting of the injected air up the well bore. At large sites where many wells are required, the cost of installing multiple sparge points may prohibit the consideration of air sparging as a potential technology.

Injection well diameters range from 1 to 4 in. The performance is not expected to be affected significantly by changes in well diameter. Economic considerations favor smaller-diameter wells (1 to 2 in.), since they are less expensive to install. However, as the diameter of the well is reduced, the pressure drop due to the flow-through piping increases and may become significant, especially at deeper depths.

Driven air sparge well points made out of small-diameter cast iron (3/4 in. to $1\frac{1}{2}$ in.) flush-jointed sections (Figure 4.12) may help in making this technology more cost effective under some conditions. However, the absence of a sand pack around the sparge points may allow clogging of the sparge points to develop over a long period, particularly under pulsing conditions. Specifically, continuous expansion and collapse of the soils around the sparge point during the pulse cycles will create a "sieving" action, thus allowing finer sediments to accumulate around the sparge points and eventually clog them.

The well screen location and length should be chosen to maximize the flow of injected air through the zone of contamination. At typical injection flow rates, most of the air will escape through the top 12 in. of the screen. A 10-slot PVC screen is normally used for air sparging applications.

4.5.6   Contaminant Type and Distribution

Volatile and strippable compounds will be most amenable to air sparging, though it is anticipated that nonvolatile but aerobically biodegradable compounds can also be addressed by this technique. There is no limit on the dissolved concentrations of contaminants treatable by air sparging. For air sparging to be effective, the air saturation percentage and the radius and density of air channels are important factors for mass transfer efficiencies of both contaminants and oxygen. The rates of stripping and biodegradation are both limited by diffusion through water. It is not possible to optimize them separately.[22]

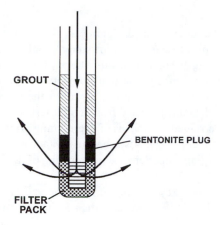

Figure 4.10 Schematic showing conventional design of an air sparging point for shallower applications.

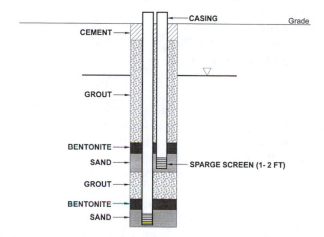

Figure 4.11 Diagram of a nested sparge well for deeper applications.

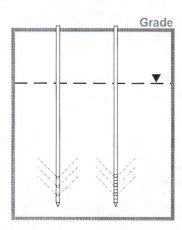

Figure 4.12 Small-diameter air sparging well configuration.

4.6 PILOT TESTING

Because the state-of-the-practice in designing an *in situ* air sparging system has not progressed beyond the "empirical stage," a pilot study should be considered only to prove the effectiveness of air sparging at a particular site. The pilot study could be most appropriately defined as a field design study, since the primary objective is to obtain site-specific design information. However, due to the still unknown nature of the mechanics of the process and mass transfer mechanisms, the data collected from a pilot test should be treated with caution. The collected data should be valued as a means of overcoming prior concerns, if any, regarding the implementation of this technology. Also, because vapor extraction is a complimentary technology to *in situ* air sparging, simultaneous pilot testing of the integrated system is highly recommended.

Short-term pilot tests play a key role in the selection and design of *in situ* air sparging systems. Most conventional pilot tests are less than 24 to 48 h in duration and consist of monitoring changes in

- pressure buildup in sealed piezometers screened below the water table
- dissolved oxygen levels
- water levels in wells
- soil gas pressures
- contaminant concentrations in soil gas
- presence and capture of tracer gases

These parameters are assumed to be indicators of air sparging feasibility and performance and are also used in the design of full-scale systems. As noted earlier, the understanding regarding the value and usefulness of each of the parameters listed above is improving. However, it is suggested that collection and comparison of as many parameters as possible will provide valuable insight on site-specific applicability of air sparging as a remediation technique.

It is very important to perform a preliminary evaluation of the geologic and hydrogeologic conditions for the applicability of *in situ* air sparging prior to the pilot study. In addition, a thorough examination of the degree and extent of contamination should be performed.

Evaluating the site-specific parameters listed in Table 4.2 prior to designing a pilot test will enhance the quality of data that would be collected. A typical equipment setup used for an air sparging pilot test is shown in Figure 4.1. The data that should be collected during the pilot study, to be used for the design of a full-scale system, include the following engineering parameters:

Table 4.2 Evaluation of Site-Specific Parameters as a Preliminary Screening Tool

Condition	Favorable conditions	Impact
Saturated zone soil permeability (horizontal)	10^{-3} cm/s	Applicability, flow rate vs. pressure mass removal efficiencies vs. transport rate
Geologic stratification and anisotropy	Sandy, gravelly soils, homogenous	Applicability, air distribution
Aquifer type	Unconfined	Recovery of injected air
Depth of contamination below the water table	Less than 40–50 ft	Sparging depth (injection pressure)
Type of contaminant	High volatility, high strippability, high aerobic biodegradability	Applicability — volatility/strippability/ biodegradability
Extent of contamination	No separate phase contamination	Applicability and mass removal efficiency (multiple sparge points)
Soil conditions above the water table	More than 5 ft of vadose zone, permeable soils	Ability to capture the stripped contamination by vapor extraction Vapor flow paths

- *Zone of Air Distribution:* For any subsurface remediation system, the zone of air distribution is the key design parameter, since it would determine the number of injection points required. The zone of influence under various pressure and flow combinations should be measured. Methods to measure the zone of influence were described in Section 4.5.1 and Figure 4.9.
- *Injection Air Pressure:* Injection air pressure is significantly influenced by the depth of injection and the subsurface geology. The required baseline pressure during the pilot test should be equal to or just above the value necessary to overcome the sparging depth. The impact of any additional required pressure should be evaluated carefully in incremental steps, because excessive pressures may fracture the soils around the point of injection.
- *Injection Flow Rate:* The injection flow rate should provide an adequate percentage of air saturation within the zone of air distribution. The greater the sparging depth, the higher will be the flow rate required to achieve a percentage of air saturation. Evaluation of the injection flow rate should also be governed by the ability to capture the stripped contaminant vapors and the net pressure gradient in the vadose zone. At a minimum, the airflow rate should be sufficient to promote significant volatilization rates and/or maintain dissolved oxygen levels greater than 2 mg/l. Typical injection flow rates are in the range of 1 to 15 cfm per injection point, depending on the type of geology and the sparging depth.
- *Mass Removal Efficiency:* Another key objective during the field study should be to demonstrate the mass removal efficiency of the *in situ* air sparging process. This can be accomplished by measuring the net increase in contaminant levels in the effluent of the vapor extraction system after the initiation of the air sparging system. To evaluate the net increase in contaminant levels in the effluent, the field test should be conducted as a sequential test in two phases. In the first phase, perform the vapor extraction test until 1.5 to 2 pore volumes are removed from the unsaturated zone and the concentrations in the extracted air reach pseudo-"steady" state conditions. Then initiate the air sparging during the second phase and monitor the contaminant levels in the vapor extraction system air stream. An increase in the contaminant level along with the duration of this spike would indicate the short-term mass removal efficiency due to air sparging (Figure 4.13). The second phase of the test should be continued until a decline in contaminant concentrations in the effluent air stream is observed. Once the decline is observed (Figure 4.13), the effect of pulsed sparging on mass removal rates should be evaluated. If the dissolved contaminant concentrations within the zone of air distribution are low, the vapor phase contaminant spike described in Figure 4.13 may go unnoticed, especially without frequent sampling. Conversely, the spike may be very noticeable in the presence of trapped contaminants in the saturated zone.

Determination of the increase in contaminant levels due to air sparging is important to the evaluation of safety considerations associated with implementing this technology. Continuous removal of the contaminants transferred into the vadose zone by soil vapor extraction is very important. Buildup of these contaminants to explosive levels must be avoided at any cost. The air injection and air extraction rates should be controlled in order to maintain a net negative pressure within the target area.

4.7 MONITORING CONSIDERATIONS

In situ and aboveground monitoring data should be used to assess the performance of operating conditions to determine whether system adjustments or expansions are necessary. Table 4.3 lists the various parameters that can be utilized to monitor the system performance.

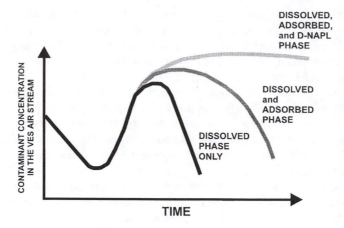

Figure 4.13 Contaminant removal efficiencies during a pilot test.

Table 4.3 *In Situ* Air Sparging System Monitoring Parameters

In situ parameters	Measurement
Groundwater quality improvement	Obtaining periodic groundwater samples from monitoring wells after shutting down air injection
Dissolved oxygen levels/temperature	Field probes in the monitoring wells after shutting down air injection
Redox potential/pH	Field probes in the monitoring wells after shutting down air injection
Biodegradation byproducts such as CO_2	Groundwater samples obtained with a flow-through cell
Soil gas concentrations	FID, PID, explosimeter or field gas chromatograph or laboratory air samples
Soil gas pressure/vacuum	Pressure/vacuum gauge or manometer
Groundwater level	Water level meter

System operating parameters	Measurement
Injection well pressure	Pressure gauge or manometer
Soil vapor extraction well vacuum	Vacuum gauge or manometer
Injection well flow rate	Airflow meters
Soil vapor extraction flow rate	Airflow meters
Extraction vapor concentrations	FID, PID, explosimeter, field GC, or laboratory air samples
O_2, CO_2, N_2, CH_4	Laboratory analysis
Pulsing frequency	Timer

In situ field parameters obtained during air sparging are often subject to a wide range of interpretations. Data obtained from groundwater monitoring wells while air is being injected will be always questioned with respect to the potential of an air channel bubbling up through the water in the well. Therefore, it is recommended that groundwater quality samples be collected after air injection is shut down and sufficient time is provided for equilibrium to be reached. Dissolved oxygen, redox, and CO_2 levels are recommended to be measured with continuous flow through cells due to the potential for these parameters to change during conventional groundwater sampling and handling procedures.

4.8 PROCESS EQUIPMENT

Successful implementation of an air sparging system is dependent on the proper selection of the process equipment. Primary components of an *in situ* air sparging system are

- air compressor or blower
- vacuum blower
- fittings and tubing to connect the compressor to the well(s)
- air filters
- pressure regulator
- flow meters
- pressure gauges
- air drying unit

4.8.1   Air Compressor or Air Blower

Selection of an air compressor or air blower will depend on the required pressure rating at which air has to be injected (obtained from the pilot test). Air blowers (positive displacement) can be used only when the required pressure rating is less than 12 to 15 psi. There are various types of air compressors available, such as

- positive displacement compressors
  - reciprocating piston
  - diaphragm
  - rocking piston
  - rotary vane
  - rotary screw and lobed rotor
- nonpositive displacement compressors
  - centrifugal
  - axial flow
  - regenerative

Positive displacement compressors generally provide the most economic solution for systems that require relatively high pressures. Disadvantages include low flow rates, oil removal in some cases, and the heat generated during operation.

Unlike the positive displacement compressor, a nonpositive displacement compressor does not provide a constant volumetric flow rate over a range of discharge pressures. The most important advantage of nonpositive displacement compressors is their ability to provide high flow rates. Table 4.4 and Figure 4.14 provide a summary of air compressor characteristics. It should be noted that the size and capacity requirements of an air compressor required for an air sparging site will be typically below 100 hp and 150 psi.

Table 4.4   Summary of Compressor Characteristics

Class	Category	Type	Power range (hp)	Pressure range (psi)	Advantages
Positive displacement compressors	Reciprocating	Piston air-cooled	1/2 – 500	10 – 250	Efficient, light-weight
		Piston water-cooled	10 – 500	10 – 250	Efficient, heavy-duty
		Diaphragm	10 – 200	10 – 250	No seal, contamination-free
	Rotary	Sliding vane	10 – 500	10 – 150	Compact, high-speed
		Screw (helix)	10 – 500	10 – 150	Pulseless delivery
		Lobe, low-pressure	15 – 200	5 – 40	Compact, oil-free
		Lobe, high-pressure	7½ – 200	20 – 250	Compact, high-speed
Nonpositive displacement compressors	Rotary	Centrifugal	50 – 500	40 – 250	Compact, oil-free, high-speed
		Axial flow	1,000 – 10,000	400 – 500	High-volume, high speed
		Regenerative peripheral blower	1/4 – 20	1 – 5	Compact, oil-free, high volume

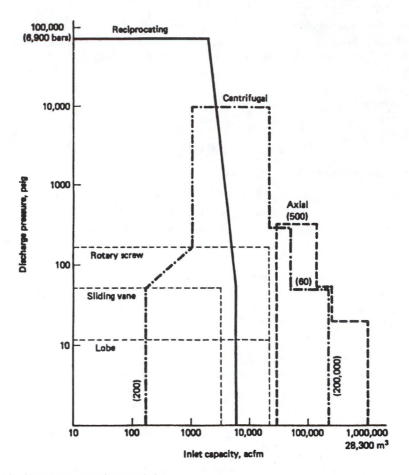

Figure 4.14 Air compressor characteristics.

Oil contamination in the injected air can affect the *in situ* air sparging system performance. A variety of filters have been developed to filter out the contained oil. An alternative is to use an oil-less compressor. Higher capital and maintenance costs are typical of oil-less equipment as compared to their oil-lubricated counterparts.

Some pneumatic systems cannot tolerate moisture formed by the cooling of air caused by compression. While a mechanical filter removes most of the solid and liquid particulates from the air, it is not very effective for removing water and oil vapors. The moisture may later condense and freeze in the pipes downstream when very low temperatures are encountered. An air dryer prevents condensation by reducing the humidity of the air stream. A practical type of dryer is the desiccant unit, which uses a moisture-absorbing chemical, usually in pelletized form. Water vapor also can be removed by condensation by passing the air through a chilling unit. Coalescing filters are also effective in removing mists of tiny water droplets. Less costly options include heat tracing of the piping/manifold or the use of a receiver tank with a manual or automatic drain to remove the condensation.

4.8.2 Other Equipment

Selection of the vacuum blower will depend on the required airflow rate and vacuum levels necessary for efficient subsurface vapor recovery. Depending on the geologic conditions encountered, high vacuum, low flow vs. low vacuum, high flow combination has to be evaluated. For further description of vacuum blowers, see Chapter 3.

Valves and control devices in compressed-air systems fall into three general categories: those that control pressure, those that control direction of airflow, and those that control flow rate. Flow control appurtenances such as flow meters, valves, and pressure gauges are described in detail in Appendix A.

4.9   MODIFICATIONS TO CONVENTIONAL AIR SPARGING APPLICATION

For the purposes of discussion in this book, conventional application of air sparging is defined as shown in Figure 4.1. Due to the geologic and hydrogeologic conditions encountered at many sites across the country, this form of application may have to be limited to only 25% of the remediation sites.[7] However, the concept of using air as a carrier for removing contaminant mass still remains a very attractive and cost-effective alternative as compared to currently available options.

Several modifications to conventional air sparging are described in the following sections.

4.9.1   Horizontal Trench Sparging

Trench sparging was developed to apply air sparging under less permeable geologic conditions when depth of contamination is less than 30 ft. When the hydraulic conductivities (in the horizontal direction) are less than 10^{-3} cm/s, it is prudent to be cautious when injecting air directly into the water-saturated formations. This technique is generally applicable where there is a shallow depth to groundwater and the formation is fine grained.

Trench sparging includes (1) placement of a single or parallel trench(es) perpendicular to groundwater flow; (2) injection of air through lateral or vertical pipes at the bottom of the trench; and (3) extraction of air from lateral pipes in the trench above the water table. Figure 4.15 shows a typical trench sparging system.

The primary focus in this modified approach is to create an artificially permeable environment in the trench(es) in which the distribution of injected air is controlled. As the contaminated groundwater travels through the trench, the strippable VOCs will be removed from the groundwater and captured by the vapor extraction pipe placed above the water table (Figure 4.15B).

Due to the extremely slow groundwater velocities under conditions where this technique may be preferred, the residence time of the moving groundwater in the trench will be high. For this reason, the injection of the air does not have to be continuous, and a pulsed mode of injection can be implemented. When biodegradable contaminants are present, the trench can be designed to act like an *in situ* fixed film bioreactor, and the rate of injection of air can be further decreased. The need for effluent air treatment may be avoided under these circumstances. Injection of nutrients, such as nitrogen and phosphorus, can enhance the rate of biodegradation in the trench (Figure 4.15B).

The treated groundwater leaving the trench will be saturated with dissolved oxygen and nutrients (if added) and can enhance the degradation of dissolved and residual contaminants downgradient of the trench. If the primary focus of remediation is containment only, this concept can be implemented as a low-cost containment technique with just one downgradient trench. Depending on the need to clean up the site faster, this variation on air sparging can be implemented as shown in Figure 4.15A with multiple trenches.

The biggest limitation of this technique will be the total depth to which the trench has to be dug. Total depths beyond 30 to 35 ft may preclude the implementation of this technique due to shoring costs, site accessibility, and the potential need to deal with a large volume of contaminated soils.

When the depths of a sparge trench are limited to less than 35 ft, air injection into the trench can be accomplished with a blower instead of a compressor. The extracted air can then be treated with a vapor treatment unit (probably vapor-phase granular activated carbon

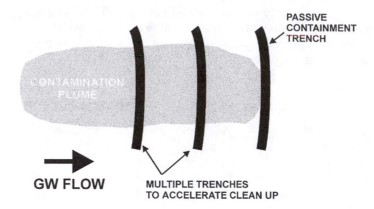

Figure 4.15A   Horizontal trench sparging. (Plan view).

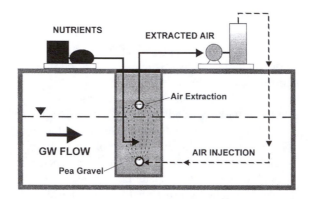

Figure 4.15B   Horizontal trench sparging. (Section view).

(GAC) due to the low levels of mass expected) and reinjected back into the trench as shown in Figure 4.15B. This configuration will eliminate the need to take regular air samples for regulatory purposes.

4.9.2   In-Well Air Sparging

This modification was developed as a means to use air as the carrier of contaminants and to overcome the difficulties of injecting air into "non-optimum" geologic formations. In-well air sparging, shown in Figure 4.16, can also overcome the depth limitations and overall difficulties of installing trench(es) described in the previous section.

The injection of air into the inner casing (Figure 4.16) induces an "air lifting effect," which is limited only to the inner casing. The water column inside the inner casing will be lifted upward (in other words, water present inside the inner casing will be pumped) and will overflow over the top of the inner casing, as shown in Figure 4.16. As a result, contaminated water will be drawn into the lower screen from the surrounding formation and will be continuously "air lifted" in the inner tube.

Due to the "mixing" of air and contaminated water, as the air–water mixture rises inside the inner tube, strippable VOCs can be air stripped and captured for treatment as shown in Figure 4.16. Treated, clean water which spills over the top of the inner casing will be reinjected back into the formation via the top outer screen. This approach can completely eliminate the need for extracting water for above-ground treatment under some conditions.

An added benefit to in-well air sparging is that the reinjected water, saturated with dissolved oxygen, can enhance the biodegradation of aerobically biodegradable contaminants

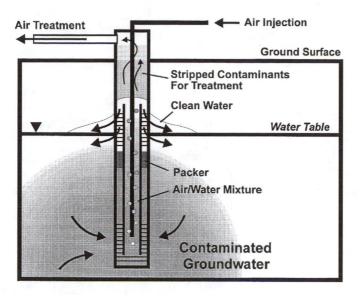

Figure 4.16 In-well air sparging.

present in the saturated zone. The need to inject nutrients inside the well can be evaluated on a site-specific basis.

4.9.3 Biosparging

As discussed in previous sections, injection of air into water-saturated formations has a significant benefit in terms of delivery of oxygen to the microorganisms for *in situ* bioremediation. If delivery of oxygen for the biota is the primary objective for air injection, the volume of airflow does not have to be at the same level required to achieve stripping and volatilization. Control of air channel formation and distribution and capturing of the stripped contaminants also become less significant under these circumstances. Application of this technique to remediate a dissolved plume of acetone, for example, which is a nonstrippable but extremely biodegradable compound, will be appropriate.

Injection of air at very low flow rates (0.5 cfm to less than 2 to 3 cfm per injection point) into water-saturated formation to enhance biodegradation is defined as biosparging. Limitations caused by geological formations also become less significant, since the path of air channels can be allowed to follow the path of least resistance. However, it has to be noted that the time required to increase the dissolved oxygen levels in the bulk water depends on the time required for the diffusion of O_2 from the air channels into the water surrounding the channels. It is estimated that only 0.5% of the oxygen present in the injected air will be transferred into the dissolved phase during air sparging.[6,7,13] Therefore, caution must be exercised in terms of evaluating the changes in dissolved oxygen (DO) levels after the initiation of biosparging. It is common practice to assume that the observed increase in DO levels in monitoring wells is due to the changes in the bulk water. Direct introduction of air into the monitoring wells due to an air channel being intercepted could also be a reason for increased DO levels in monitoring wells.

4.9.4 Vapor Recovery via Trenches

Trench vapor recovery is a minor modification to conventional air sparging that involves the recovery of stripped vapors from fine-grained formations (Figure 4.17). Trench vapor recovery can be used when there is a shallow depth to groundwater, and overlying fine-grained formation extends from the surface to below the water table. This geologic situation would

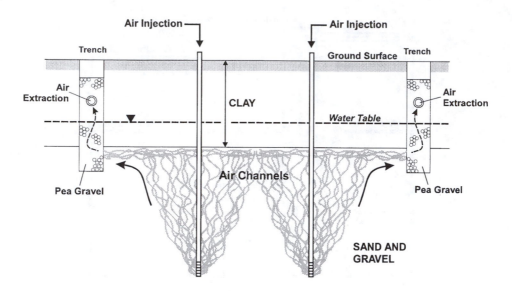

Figure 4.17 Modified air sparging with vapor recovery through trenches.

normally inhibit extraction of stripped vapors by vapor extraction wells. Saturated zone mass transport and removal rates and mechanisms are very similar to those of conventional air sparging, with the exception of the capillary fringe area. If contaminants are adsorbed to the fine-grained formation matrix in the capillary zone, trench sparging may be ineffective in transferring and removing the contaminant mass from these areas. Trench vapor recovery systems are most effective when only dissolved-phase contaminants need to be addressed.

4.9.5 Pneumatic Fracturing for Vapor Recovery

The application of *in situ* air sparging using pneumatic fracturing to enhance vapor recovery is applied to sites with fine-grained formations that extend below the water table and depths to water that prohibit trenching (Figure 4.18). Pneumatic fracturing increases hydraulic and vapor flow conductivity near the top of the water table and in the overlying unsaturated zone, while allowing stripped contaminants to be collected without spreading out laterally. Balancing injection flow rates is critical when using this method of vapor recovery. Mass transport and recovery have limitations similar to trench vapor recovery. Mass transfer and removal of adsorbed contaminants near the top of the water table are limited in areas between fractures.

4.10 CLEANUP RATES

To date, there are no reliable methods for estimating groundwater cleanup rates. A mass removal model for *in situ* air sparging has been reported using air stripping as the only mass-transfer mechanism.[3–17] However, this model was based on the premise that injected air travels in the form of bubbles. Because it has been established that the primary mode of travel of injected air at most sites is in the form of air channels, the reliability of this model may be questionable.

In the presence of air channels, the rate of mass transfer will be limited by either kinetics of the mass transfer at the interface, or by the rate of transport of the contaminant through the bulk water phase to the air–water interface. Based on these assumptions, reaching nondetectable levels may be possible only with biodegradable contaminants (Figure 4.19).

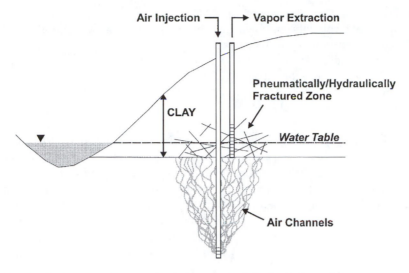

Figure 4.18 Modified air sparging combined with pneumatic/hydraulic fracturing.

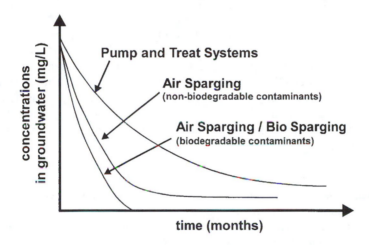

Figure 4.19 Cleanup rates for various contaminants during *in situ* air sparging.

Cleanup times of less than 12 months to 3 years have been achieved in many instances. Reports in the literature indicate that sites that have implemented air sparging have often met groundwater cleanup goals in less than 1 year.[2,9,11,18,19] However, it should be noted that, at most of these sites, the cleanup goal was approximately 1 mg/l for total benzene, toluene, ethyl benzene, and xylenes (BTEX) and that BTEX compounds are very biodegradable.

The air saturation and the size of the air-filled regions have the greatest effect on the mass transfer rates. Specifically, a high volume of air-filled spaces that are finely distributed will promote a faster mass transfer. Numerical analyses show that the air saturation must be greater than 0.1% and the size of air channels must be on the order of 0.001 m in order for air sparging to be successful.[22] At sites with optimum geologic conditions, the above criteria will be met easily.

The required cleanup times for a site will depend on the following:

- Target cleanup levels
- Extent and phases of contamination:
  - Contaminant mass present in the saturated zone and the capillary fringe
  - Extent of dissolved and sorbed phase contamination

- The presence and absence of a DNAPL
- Strippability, volatility, and biodegradability of contaminants present
- Solubility, partitioning of the contaminants
- Geologic conditions:
  - Percentage of air saturation
  - Density of air channels
  - Size of the air channels

Figure 4.19 shows the effectiveness of air sparging on decreasing the dissolved groundwater contaminant levels in comparison to a pump and treat system. These data were summarized from operational history obtained from approximately 40 *in situ* air sparging systems.[7] Implementation of air sparging at sites with biodegradable contaminants led to more rapid attainment of cleanup standards. At sites with nonbiodegradable contaminants, specifically at sites with chlorinated organic compounds, an asymptotic concentration level was reached. However, this asymptote was at a lower concentration and was reached in less time than what could have been accomplished with a pump and treat system.

4.11 LIMITATIONS

At first glance, *in situ* air sparging appears to be a simple process: injection of air into a contaminated aquifer below the water table with the intent of volatilizing VOCs and providing oxygen to enhance biodegradation. Previous discussions in this chapter have included the applicability of the *in situ* air sparging process. There are also discussions in the literature regarding the rates of mass transfer of contaminants and oxygen between the air channels and the relatively stagnant surrounding water. This section summarizes the conditions under which the conventional application of this technology is not recommended.

- Tight geologic conditions with hydraulic conductivities less than 10^{-3} cm/s: The vertical passage of the air may be hampered, the potential for the lateral movement of contaminants will be increased, and there is the potential for inefficient removal of contaminants. The conventional form of air sparging should be evaluated with extreme caution under the above conditions.
- Heterogeneous geologic conditions, with the presence of low permeability layers overlying zones with higher permeabilities: Again, the potential for the enlargement of the plume exists due to the inability of the injected air reaching the soil gas above the water table.
- Contaminants present are nonstrippable and nonbiodegradable (see Appendix B and C).
- Mobile free product has not been removed or completely controlled: Air injection may enhance the uncontrolled movement of this liquid away from the air injection area.
- Air sparging systems that cannot be integrated with a vapor extraction system to capture all the stripped contaminants: In some instances, the stripped contaminants can be biodegraded in the vadose zone, if optimum conditions are available. Thicker vadose zones and very low injection rates are more appropriate to implement this than shallower depths.
- The structural stability of nearby foundations and buildings may be in jeopardy due to the potential of soil fluidization or fracturing.
- Potential for uncontrolled migration of vapor contaminants into nearby basements, buildings, or other conduits.

4.12 KNOWLEDGE GAPS

The following recommendations are provided for further research of this technology so that *in situ* air sparging systems in the future can be designed on a scientific basis instead of using an empirical approach.

- Clarify the understanding of the mode and behavior of air travel. Determine the influence of saturated zone soil structuring on the mode of air travel.
- Develop and refine field methods for estimation of the percentage of air saturation and the size and density of air channels in a deterministic way.
- Optimize optimum pressure, flow, and distribution of airflow relationships in relation to soil type and structure.
- Develop further understanding of mass transfer mechanisms during air sparging.
- Develop reliable models of the hydraulic behavior and the mass transfer mechanisms, thus simplifying the process of designing the system and estimating cleanup times.
- Design enhancements to overcome the geologic and hydrogeologic limitations.
- Eliminate the need to capture all the stripped contaminants if they are biodegradable by enhancing the biodegradation rates in the vadose zone to meet the mass removal rates due to air sparging.

4.13 SUMMARY OF CASE STUDIES IN THE LITERATURE

As of the writing of this book, there is limited information available in the literature regarding successful case studies. This section quickly summarizes the site conditions and the type of contaminants at these sites where *in situ* air sparging was successfully implemented. Variations exist among the sites surveyed with respect to contaminants treated, soil type, geologic features, complimentary technologies used, and many other factors.[7,11,26,27]

Contaminants treated include

- Gasoline constituents—benzene, toluene, ethylbenzene and xylenes, total petroleum hydrocarbons (TPHs)
- Chlorinated solvents—trichloroethylene (TCE), tetrachloroethylene (PCE), trichloroethane (TCA), dichloroethene (DCE), dichloroethane (DCA), etc.
- Acetone, methyl ethyl ketone (MEK), cyclohexane (under biosparging conditions)

Initial contaminant concentrations have ranged from 300 mg/l to less than 1 mg/l. In the case of petroleum compounds, cleanup levels of less than 5 ppb have been reached in many cases.[7,26] In spite of the considerable questions surrounding the short-term and long-term effectiveness of air sparging, many case studies indicate that *in situ* air sparging can achieve a substantial and permanent decrease in groundwater concentrations.[7,26]

Most of the successful sites reported have permeable soil types such as sand and gravel.[7,26,27] Some successful cleanups have been reported under silty soils for both petroleum and chlorinated compounds.[7,26] However, it should be noted that there may be many unsuccessful cases due to the improper application of this technology, not reported in the literature.

Minimum sparging depth reported for successful sparging is 8 ft,[4,11] and the maximum depth is 60 ft below the water table.[7] There are many reports in the literature claiming successful closures of sites in less than 6 months of operation.[26] The typical range reported in the literature seems to be in the range of 9 to 30 months.[7,11,26,27]

Other conclusions from available case studies in the literature follow.[7,26]

- Sparging was especially effective at sites where the contaminants were present only in the dissolved phase (no NAPLs). *In situ* air sparging was effective in reducing dissolved groundwater contaminant levels by 1 to 4 orders of magnitude.
- Rebound of contaminants takes place during the 6 to 12 months after the system is shut down. The rebound effects appeared to be minimized by a high density of sparge wells, close spacing, and high flow addressing the entire source area.
- Rebound effect was significant at sites where NAPLs were present.

REFERENCES

1. Brown, R. A., Air sparging: a primer for application and design, *Subsurface Restoration Conf.*, U.S. Environmental Protection Agency, 1992.
2. Brown, R. A., Treatment of Petroleum Hydrocarbons in Groundwater by Air Sparging, Section 4, Research and Development, Wilson, B., Keeley, J., and Rumery, J. K., Eds., RSKERL, U.S. Environmental Protection Agency, 1992.
3. Sellers, K. and Schreiber, R., Air sparging model for predicting groundwater cleanup rate, *Proc. Petroleum Hydrocarbons and Organic Chemicals in Groundwater: Prevention, Detection, and Restoration*, Houston, TX, 1992.
4. Johnson, R. L., Johnson, P. C., McWhorter, D. B., Hinchee, R. E., and Goodman, I., An overview of in situ air sparging, *Groundwater Monitoring Remediation*, Fall 1993.
5. Wei, J., Dahmani, A., Ahlfeld, D. P., Lin, J. D., and Hill, E., III, Laboratory study of air sparging: air flow visualization, *Groundwater Monitoring Remediation*, Fall 1993.
6. Johnson, R. L., Center for Groundwater Research, Oregon Graduate Institute, Beaverton, Oregon, personal communication.
7. Geraghty & Miller, Inc., Air Sparging Projects Data Summary, 1995.
8. Clayton, W. S., Brown, R. A., and Bass, D. H., Air sparging and bioremediation: The case for in situ mixing, *Third Internatl Symp. In Situ and On Site Bioreclamation*, San Diego, April 1995.
9. Brown, R., Herman, C., and Henry, E., The use of aeration in environmental cleanups, Presented at *HAZTECH Internatl. Pittsburgh Waste Conf.*, Pittsburgh, PA, 1991.
10. Bohler, J. B., Hotzl, H., and Nahold, M., Air injection and soil air extraction as a combined method for cleaning contaminated sites—observations from test sites in sediments and solid rocks, in *Contaminated Soil*, Arendt, F., Hinsevelt, M., and van der Brink, W. J., Eds., Springer-Verlag, Berlin, 1990.
11. U.S. Environmental Protection Agency, Evaluation of the State of Air Sparging Technology, Report 68-03-3409, Risk Reduction Engineering Laboratory, Cincinnati, OH, 1993.
12. Kresge, M. W. and Dacey, M. F., An evaluation of in situ groundwater aeration, *Proc. Ninth Ann. Hazardous Waste Mater. Manage. Conf. Internatl.*, Atlantic City, NJ, 1991.
13. Boersma, P. M., Diontek, K. R., and Newman, P. A. B., Sparging effectiveness for groundwater restoration, *Third Internatl. Symp. In Situ On-Site Bioreclamation*, San Diego, April 1995.
14. Lundegard, P. D. and Andersen, G., Numerical simulation of air sparging performance, *Proc. Petroleum Hydrocarbons Organ. Chem. Groundwater: Prevention, Detection, Restoration*, Houston, TX, 1993.
15. Lundegard, D. D., Air sparging: Much ado about mounding, *Third Internatl. Symp. In Situ On-Site Bioreclamation*, San Diego, April 1995.
16. Ahlfeld, D. P., Dahmani, A., and Wei, J., A conceptual model of field behavior of air sparging and its implications for application, *Groundwater Monitoring Remediation*, Fall 1994.
17. Marley, M. C., Li, F., and Magee, S., The application of a 3-D model in the design of air sparging systems, *Proc. Petroleum Hydrocarbons Organ. Chem. Groundwater: Prevention, Detection, Restoration*, Houston, TX, 1992.
18. Marley, M. C., Walsh, M. T., and Nangeroni, P. E., Case study on the application of air sparging as a complimentary technology to vapor extraction at a gasoline spill site in Rhode Island, *Proc. HMCRI 11th Ann. Natl. Conf.*, Washington, DC, 1990.
19. Ardito, C. P. and Billings, J. F., Alternative remediation strategies: The subsurface volatilization and ventilation system, *Proc. Petroleum Hydrocarbons Organ. Chem. Groundwater: Prevention, Detection, Restoration*, Houston, TX, 1990.

20. Marley, M. C., Bruell, C. J., and Hopkins, H. H., Air sparging technology: A practice update, *Third Internatl. Symp. In Situ On-Site Bioreclamation*, San Diego, April 1995.

21. Acomb, L. J., et al., Newtron probe measurements of air saturation near an air sparging well, *Third Internatl. Symp. In Situ On-Site Bioreclamation*, San Diego, April 1995.

22. Mohr, D. H., Mass transfer concepts applied to in situ air sparging, *Third Internatl. Symp. In Situ On-Site Bioreclamation*, San Diego, April 1995.

23. Rollins, J. P., Ed., *Compressed Air and Gas Handbook*, 5th ed., Compressed Air and Gas Institute, Cleveland, OH, 1989.

24. Howard, P. H., et al., *Handbook of Environmental Degradation Rates*, Lewis Publishers, Boca Raton, FL, 1991.

25. Lyman, W. J., Reehl, W. F., and Rosenblatt, D. H., *Handbook of Chemical Property Estimation Methods*, McGraw-Hill, New York, 1992.

26. Bass, D. H. and Brown, R. A., Performance of air sparging systems—a review of case studies, Presented at the *Superfund 1995 Conf.*, Hazardous Materials and Control Research Institute, Washington, DC, November 1995.

27. American Petroleum Institute, *In Situ Air Sparging: Evaluation of Petroleum Industry Sites and Considerations for Applicability, Design and Operation*, API Pub. No. 4609, April 1995.

5

IN SITU BIOREMEDIATION

5.1 INTRODUCTION

Biological processes, which take place in the natural environment, can modify organic contaminant molecules at the spill location or during their transport in the subsurface. Such biological transformations, which involve enzymes as catalysts, frequently bring about extensive modification in the structure and toxicological properties of the contaminants. These biotic processes may result in the complete conversion of the organic molecule to innocuous inorganic end products, cause major changes that result in new organic products, or occasionally lead to only minor modifications. The available body of information suggests that the major agents causing the biological transformations in soil, sediment, surface water, and groundwater are the indigenous microorganisms that inhabit these environments.

Biodegradation can be defined as the microbially catalyzed reduction in complexity of chemicals. In the case of organic compounds, biodegradation frequently, although not necessarily, leads to the conversion of much of the carbon, nitrogen, phosphorus, sulfur, and other elements in the original compound to inorganic end products. Such a conversion of an organic substrate to inorganic end products is known as mineralization. Thus, in the mineralization of organic C, N, P, S, or other elements, CO_2 or inorganic forms of N, P, S, or other elements are released by the organisms and enter the surrounding environment. Few nonbiological reactions in nature bring about comparable changes.

Natural communities of microorganisms present in the subsurface have an amazing physiological versatility. Microorganisms can carry out biodegradation in many different types of habitats and environments, both under aerobic and anaerobic conditions. Communities of bacteria and fungi can degrade a multitude of synthetic compounds and probably every natural product.

In situ bioremediation is the application of biological treatment to the cleanup of hazardous chemicals present in the subsurface. The optimization and control of microbial transformations of organic contaminants require the integration of many scientific and engineering disciplines.

Hazardous compounds persist in the subsurface because environmental conditions are not appropriate for the microbial activity that results in biochemical degradation. The optimization of environmental conditions is achieved by understanding the biological principles under which these compounds are degraded, and the effect of environmental conditions on both the responsible microorganisms and their metabolic reactions. The "biodegradation triangle" (Figure 5.1) for understanding the microbial degradation of any natural or synthetic organic compound consists of knowledge of the microbial community, environmental conditions, and structure and physicochemical characteristics of the organic compound to be degraded.

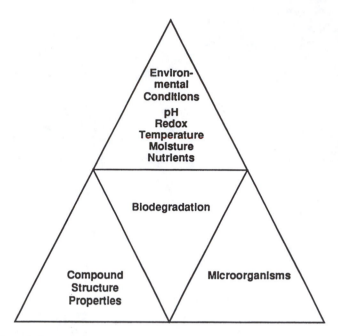

Figure 5.1 Biodegradation triangle.

5.2 MICROBIAL METABOLISM

During the process of *in situ* bioremediation, microorganisms use the organic contaminants for their growth. In addition, compounds providing the major nutrients such as nitrogen, phosphorus, and minor nutrients such as sulfur and trace elements are also required for their growth. In most cases, an organic compound that represents a carbon and energy source is transformed by the metabolic pathways that are characteristic of heterotrophic microorganisms. It should be stressed, however, that an organic compound need not necessarily be a substrate for growth in order for it to be metabolized by microorganisms. Two categories of transformations exist. In the first, biodegradation provides carbon and energy to support growth, and the process, therefore, is growth-linked. In the second, biodegradation is not linked to multiplication, but to obtaining the carbon for respiration in order for the cells to maintain their viability. This maintenance metabolism may take place only when the organic carbon concentrations are very low. Cometabolic transformations also fall into the second category.

It has been observed that the number of microbial cells or the biomass of the species acting on the compound of interest increases as degradation proceeds.[1] During a typical growth-linked mineralization brought about by bacteria, the cells use some of the energy and carbon of the organic substrate to make new cells, and this increasingly larger population causes increasingly rapid mineralization.

Microorganisms need nitrogen, phosphorus, and sulfur, and a variety of trace nutrients other than carbon. These requirements should be satisfied as the responsible species degrade the compound of interest. For heterotrophic microorganisms in most natural systems, usually sufficient amounts of N, P, S, and other trace nutrients are present to satisfy the microbial demand. Because carbon is limiting and because it is the element for which there is intense competition, a species with the unique ability to grow on synthetic molecules has a selective advantage.

Prior to the degradation of many organic compounds, a period is observed in which no degradation of the chemical is evident. This time interval is known as the *acclimatization period* or, sometimes, as adaptation or lag period. The length of the acclimatization period varies and may be less than 1 h or many months. The duration of acclimatization depends upon the chemical structure, subsurface biogeochemical environmental conditions, and concentration of the compound. Once the indigenous population of microorganisms has become acclimatized to the presence and degradation of a chemical and the activity becomes marked, the microbial community will retain its higher level of activity for some time. Acclimatization of a microbial population to one substrate frequently results in the simultaneous acclimatization to some, but not all, structurally related molecules.

5.2.1   Metabolism Modes

The design of bioremediation processes requires determination of the desired degradation reactions to which the target compounds will be subjected. This involves selecting the metabolism mode that will occur in the process. The metabolism modes are broadly classified as aerobic and anaerobic. Aerobic transformations occur in the presence of molecular oxygen, with molecular oxygen serving as the electron acceptor. This form of metabolism is known as *aerobic respiration*. Anaerobic reactions occur only in the absence of molecular oxygen and the reactions are subdivided into *anaerobic respiration*, *fermentation*, and *methane fermentation*.

Microorganisms have developed a wide variety of respiration systems. These can be characterized by the nature of the reductant and oxidant. In all cases of aerobic respiration, the electron acceptor is molecular oxygen. Anaerobic respiration uses an oxidized inorganic or organic compound other than oxygen as the electron acceptor. The respiration of organic substrates by bacteria is, in most cases, very similar. The substrates are oxidized to CO_2 and H_2O.

Fermentation is the simplest of the three principal modes of energy yielding metabolism. During fermentation, organic compounds serve as both electron donors and electron acceptors. Fermentation can proceed only under strictly anaerobic conditions. The process maintains a strict oxidation-reduction balance. The average oxidation level of the end products is identical to that of the substrate fermented. Thus the substrate yields a mixture of end products, some more oxidized than the substrate and others more reduced. The end products depend on the type of microorganisms but usually include a number of acids, alcohols, ketones, and gases such as CO_2 and CH_4. Table 5.1 summarizes the various microbial metabolic reactions.

Table 5.1   Summary of Metabolism Modes

Reductant electron donor	Oxidant electron acceptor	End products
Aerobic respiration		
Organic substrates (benzene, toluene, phenol)	O_2	CO_2, H_2O
NH_4	O_2	NO_2^-, NO_3^-, H_2O
Fe^{2+}	O_2	Fe^{3+}
S^{2-}	O_2	SO_4^{--}
Anaerobic respiration		
Organic substrates (benzene, toluene, phenol, trichloroethylene)	NO_3^-	N_2, CO_2, H_2O, Cl^-
Organic substrates (benzene, trichloroethylene)	SO_4^{2-}	S^{2-}, H_2O, CO_2, Cl^-
H_2	SO_4^{2-}	S_2^-, H_2O
H_2	CO_2	CH_4, H_2O
Fermentation		
Organic substrates	Organic compounds	Organic compounds CO_2, CH_4

The metabolism modes that utilize nitrate as an electron acceptor (performed by denitrifying and nitrate-reducing organisms), sulfate and thiosulfate as electron acceptors (performed by sulfate-reducing organisms), and CO_2 as an electron acceptor (performed by methanogenic organisms) can be used to biodegrade various organic contaminants. The utilization of chlorinated organic compounds as electron acceptors during anaerobic respiration is a recent observation.

Another important metabolism concept in bioremediation is cometabolism. In a true sense, cometabolism is not metabolism (energy yielding), but fortuitous transformation of a compound. As noted earlier, it was the traditional belief that microorganisms must obtain energy from an organic compound to biodegrade it. The transformation of an organic compound by a microorganism that is unable to use the substrate as a source of energy is termed cometabolism.[1] Enzymes generated by an organism growing at the expense of one substrate also can transform a different substrate that is not associated with that organism's energy production, carbon assimilation, or any other growth processes.

Contaminants that lend themselves to bioremediation by becoming a secondary substrate through cometabolism are only partially transformed. This transformation may or may not result in reducing toxicity. If all toxicity properties of a hazardous compound are removed via biotransformation, this is referred to as detoxification. Detoxification results in inactivation, with the toxicologically active substance being converted to an inactive product. Detoxification does not imply mineralization and may include several processes such as hydrolysis, hydroxylation, dechlorination, and demethylation. Fortunately, the metabolites or transformation products from cometabolism by one organism can typically be used as an energy source by another.

Since cometabolism generally leads to a slow degradation of the substrate, attention has been given to enhancing its rate. The addition of a number of organic compounds into the contaminated zone may promote the rate of cometabolism,[1] but the responses to such additions are not predictable. Addition of mineralizable compounds that are structurally analogous to the compound whose cometabolism is desired is known as *analog enrichment*.[1] The microorganism that grows on the mineralizable compound contains enzymes transforming the analogous molecule by cometabolism.

Another aspect of microbial metabolism is the recognition of preferential substrate degradation. Preferential degradation results in a sequential attack where the higher energy-yielding compounds are degraded first. In a petroleum spill, benzene will be degraded, under aerobic conditions, at a faster rate than naphthalene, and naphthalene will degrade faster than chrysene.

5.3 MICROBIAL REACTIONS AND PATHWAYS

Microbial transformations of organic compounds are frequently described by the terms, degradation, mineralization, detoxification, and activation. Degradation means that the initial substrate no longer exists. Mineralization refers to the complete conversion of the organic structure to inorganic forms such as CO_2, H_2O, and Cl^-. Detoxification is the transformation of the compound to some intermediate form that is nontoxic or less toxic. The process of forming toxic end products or intermediate products is known as activation.

Microorganisms are capable of catalyzing a variety of reactions: dechlorination, hydrolysis, cleavage, oxidation, reduction, dehydrogenation, dehydrohalogenation, and substitution.

- *Dechlorination*—the chlorinated compound becomes an electron acceptor; in this process, a chlorine atom is removed and is replaced with a hydrogen atom.
- *Hydrolysis*—frequently conducted outside the microbial cell by exoenzymes. Hydrolysis is simply a cleavage of an organic molecule with the addition of water.

- *Cleavage*—cleaving of a carbon–carbon bond is another important reaction. An organic compound is split or a terminal carbon is cleaved off an organic chain.
- *Oxidation*—breakdown of organic compounds using an electrophilic form of oxygen.
- *Reduction*—breakdown of organic compounds by a nucleophilic form of hydrogen or by direct electron delivery.
- *Dehydrogenation*—an oxidation–reduction reaction that results in the loss of two electrons and two protons, resulting in the loss of two hydrogen atoms.
- *Dehydrohalogenation*—similar to dechlorination, results in the loss of a hydrogen and chlorine atom from the organic compound.
- *Substitution*—these reactions involve replacing one atom with another.

Examples of these reactions are shown in Table 5.2.

Table 5.2 Microbially Catalyzed Reactions

Reaction	Example
• Dehalogenation	$Cl_2C = CHCl + H^+ \rightarrow ClHC = CHCl + Cl^-$
• Hydrolysis	$RCO - OR' + H_2O \rightarrow RCOOH + R'OH$
• Cleavage	$RCOOH \rightarrow RH + CO_2$
• Oxidation	$CH_3CHCl_2 + H_2O \rightarrow CH_3 CCl_2 OH + 2H^+ + 2e^-$
• Reduction	$CCl_4 + H^+ + 2e^- \rightarrow CHCl_3 + Cl^-$
• Dehydrohalogenation	$CCl_3CH_3 \rightarrow CCl_2CH_2 + HCl$
• Substitution	$CH_3CH_2Br + HS^- \rightarrow CH_3CH_2SH + Br^-$

From McCarty, P. L. and Semprini, L., Groundwater treatment for chlorinated solvents, in *Handbook of Bioremediation*, Norris, R. D., et al., Eds., Lewis Publishers, Boca Raton, FL, 1994. With permission.

5.3.1 Hydrocarbons Degradation

5.3.1.1 Aliphatic Hydrocarbons

Hydrocarbons are compounds containing carbon and hydrogen. Aliphatic hydrocarbons are straight- or branched-chain hydrocarbons of various lengths. The aliphatic hydrocarbons are divided into the families of alkanes, alkenes, alkynes, alcohols, aldehydes, ketones, and acids. There are cyclic aliphatic hydrocarbons which have diverse structures. Typical structures of aliphatic hydrocarbons are shown in Figure 5.2.

The most frequent and earliest application of *in situ* bioremediation has been to remediate hydrocarbons present in the subsurface as a result of petroleum spills. The degradation potential of alkanes is a function of the carbon chain length. Short chains are more difficult to degrade than the longer chains. Soil contains significant populations of microbes that can use select hydrocarbons as sole sources of carbon and energy. Soil populations capable of degrading hydrocarbons have been reported as high as 20% of all soil microbes.[3] Fungi and yeast are also capable of degrading aliphatic hydrocarbons in addition to bacteria.

Bioremediation of aliphatic hydrocarbons should be performed as an aerobic process. Conclusive evidence for anaerobic degradation of aliphatic hydrocarbons, although referenced, is uncertain and, at this stage, too ill-defined.

Aerobic biodegradation of aliphatic hydrocarbons involves the incorporation of molecular oxygen into the hydrocarbon structure. This is performed by oxygenase enzymes. There are two groups of oxygenases: monooxygenases and dioxygenases.[4] The most common pathway of alkane degradation is oxidation at the terminal methyl group. Oxidation proceeds as a sequence to an alcohol to the corresponding fatty acid to a ketone and eventually to carbon dioxide and water.[2] Short-chain hydrocarbons, except methane, are more difficult to degrade. Under aerobic conditions, methane is readily used as the sole carbon source by methanotrophs.

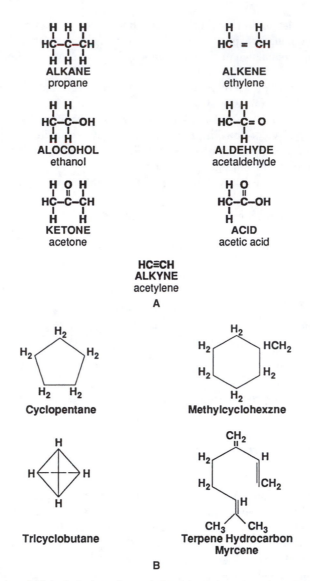

Figure 5.2 Structures of aliphatic hydrocarbons. A. Straight-chain or branched. B. Cyclic.

Alkene degradation, where a C=C double bond is involved, is more varied since microbial attack can occur at either the methyl group or the double bond.[2] Unsaturated straight-chain hydrocarbons are generally less readily degraded than saturated ones. Methyl group oxidation is considered the major degradation pathway during alkene degradation.

Hydrocarbons with branch chains are less susceptible to degradation. Even more resistant to degradation are the quaternary carbon compounds, in which one carbon atom is attached to four other carbon atoms.

Microorganisms capable of degrading cyclic aliphatic hydrocarbons are not as predominant in soils as those for the degradation of aliphatic alkane and alkene hydrocarbons.[2] Hydroxylation is vital to initiate the degradation of cycloalkanes.

5.3.1.2   Aromatic Hydrocarbons

Carbon skeletons that contain the benzene ring as the parent structure are known as aromatic hydrocarbons. They are all ring compounds and have only one free valence bond.

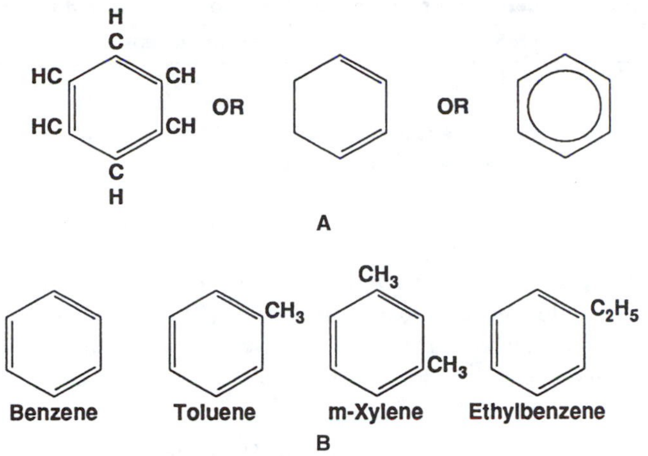

Figure 5.3   Examples of single-ring aromatic compounds. A. Benzene formula and representations. B. BTEX compounds.

The benzene ring is represented by double bonds between alternate carbon atoms. The single ring structures consist of benzene, toluene, ethyl benzene, and the three isomers of xylene (ortho, para, and meta) (Figure 5.3). These are frequently referred to as BTEX compounds and are one of the most heavily regulated group of compounds.

The hydrogens in the aromatic hydrocarbon may be substituted by a variety of different groups: OH, Cl, Br, NO_2, NO, and CN to name a few.

Aromatic compounds can be easily biodegraded, are extremely resistant, or yield undesirable intermediates. These differences depend on the number of rings in the structure, the number of substitutions, and the type and position of substituted groups. Microorganisms capable of aerobically metabolizing single-ring aromatic hydrocarbons are ubiquitous in the subsurface. The degradation is achieved by two alternate modes of oxidation.[2] The first method involves, sequentially: (1) formation of dihydrodiol, (2) formation of alkyl catechol, (3) ring fission of these oxygenated intermediates, (4) formation of either an aldehyde or an acid, and (5) eventual formation of CO_2 and H_2O. In principle, the degradation follows the dioxygenase route, which means the insertion of two oxygen groups.[5] The molecule is transformed to a smaller size, gradually "breaking off" CO_2 units. The second mechanism for degradation of aromatic compounds is oxidation of any alkyl substitutes in the ring.[2]

The stoichiometric equation of benzene degradation in the presence of O_2 is shown in equation (5.1) below. Based on this equation, the mass ratio of O_2 to benzene is 3.1:1; thus, 0.32 mg/l of benzene will be degraded per 1 mg/l of O_2 consumed by the microorganisms during aerobic biodegradation.

$$7.5O_2 + C_6H_6 \rightarrow 6CO_2 + 3H_2O. \tag{5.1}$$

Aromatic hydrocarbons can be transformed under various anaerobic conditions such as denitrifying, manganese reducing, iron reducing, sulfate reducing, and methanogenic conditions. At any given location, the benzene biodegradation sequence will depend on the availability of electron acceptors and the redox potential of the environment. This sequence is shown in Table 5.3.

Under denitrifying conditions, degradation of monoaromatic compounds has been demonstrated in a number of systems.[7] The stoichiometry of the denitrification reaction of benzene, assuming no cell growth with NO_3^- reduced completely to N_2 and benzene oxidized completely to CO_2 is

$$C_6H_6 + 6H^+ + 6NO_3^- \rightarrow 6CO_2 + 3N_2 + 6H_2O. \tag{5.2}$$

Table 5.3 Benzene Biodegradation under Various Electron Acceptor and Redox Conditions

Redox potential	Reaction type	Electron acceptors	Byproducts
>300 mv	Aerobic	O_2	CO_2, H_2O
	Denitrification	NO_3^-	NO_2^-, N_2, CO_2, H_2O
↓	Valence reduction	Mn(IV)	Mn(II), CO_2, H_2O
	Valence reduction	Fe(III)	Fe(II), CO_2, H^+
	Sulfate reduction	SO_4^{2-}	S_2^-, CO_2, H_2O
−300 mv	Methanogenesis	CO_2	CO_2, CH_4

Iron (Fe(III)) reducing conditions will also facilitate the degradation of monoaromatic hydrocarbons. Relative to other anaerobic processes, Fe(III) reduction has a very unfavorable substrate to electron acceptor ratio. The stoichiometric equation for the degradation of benzene under Fe(III) reducing conditions is

$$C_6H_6 + 30Fe^{3+} + 12H_2O \rightarrow 6CO_2 + 30H^+ + 30Fe^{2+}.$$ (5.3)

Biodegradation using sulfate as the electron acceptor involves oxidation of aromatic hydrocarbons by sulfate-reducing microorganisms coupled with reduction of sulfate to hydrogen sulfide.[7] For benzene, the stoichiometry of this reaction, assuming no cell growth is

$$C_6H_6 + 3.75SO_4^{2-} + 7.5H^+ \rightarrow 6CO_2 + 3.75H_2S + 3H_2O.$$ (5.4)

Under methanogenic (fermentative) conditions, several aromatic hydrocarbon compounds, including benzene, have been shown to transform into CO_2 and methane.[7] Assuming no cell growth, the stoichiometry for the transformation is

$$C_6H_6 + 4.5H_2O \rightarrow 2.25CO_2 + 3.75CH_4.$$ (5.5)

Higher rates of degradation are reported under denitrifying conditions than under methanogenic conditions.[2] This is expected when one considers the thermodynamics of these reactions. The amount of energy obtainable from toluene with nitrate as an electron acceptor is 20 times higher than under methanogenic conditions.[8]

Oxygenated aromatic compounds such as alcohols, aldehydes, acids, and phenols are transformed by a reductive mechanism under anaerobic conditions.[2] Reduction occurs, converting the aromatic ring to an alicyclic ring, followed by hydrolytic cleavage and mineralization. The reduction can occur, in contaminated aquifers, under denitrifying, Fe(III) reducing, sulfate-reducing and methanogenic conditions.

5.3.1.3   Polynuclear Aromatic Hydrocarbons (PAHs)

Polynuclear aromatic hydrocarbons (PAHs) are compounds that have multiple rings in their molecular structure (Figure 5.4). They include the frequently found compounds such as naphthalene and anthracene and the more complex compounds such as pyrene and benzo(a)pyrene. Biodegradation of polynuclear aromatic hydrocarbons depends on the complexity of the chemical structure and the extent of enzymatic adaptation. In general, PAHs which contain two or three rings such as naphthalene, anthracene, and phenanthrene are degraded at reasonable rates when O_2 is present. Compounds with four rings such as chrysene, pyrene, and pentacyclic compounds, in contrast, are highly persistent and are considered recalcitrant. The factors which influence the degradation of PAHs under either aerobic or anaerobic conditions are (1) solubility of the PAH, (2) number of fused rings,

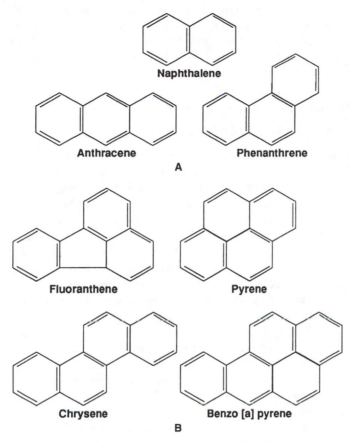

Figure 5.4 Structures of polynuclear aromatic hydrocarbons (PAHs). A. Mostly rapidly degraded PAHs. B. Slowly degraded or persistent PAHs.

(3) type of substitution, (4) number of substitution, (5) position of substitution, and (6) nature of atoms in heterocyclic compounds. The above factors are combined into a single parameter defined as structure–biodegradability relationship. Generalizations about structure–biodegradability relationships in aerobic environments do not seem to be applicable to anaerobic environments.[1]

Aerobic biodegradation of the two- and three-ring PAHs is accomplished by a number of soil bacteria. As the number of fused rings and the complexity of the substituted groups increase, the relative degree of degradation decreases. The influence of alkyl substituents is more difficult to predict.[2] One methyl addition significantly decreases the degree of degradation, and its effect varies with the substituted position. The addition of three methyl groups causes severe retardation of degradation.[9]

The importance of cometabolism for PAHs having four or more rings has been demonstrated by several investigations.[2] In fact, cometabolism may be the only metabolism mode for degradation of the heavier PAHs. Analog substrate enrichment may also be useful in enhancing the degradation of heavier PAHs. The presence of an analog substrate such as naphthalene will enhance the degradation of pyrene by many organisms.[2] Under this mode of metabolism the analog substrate is the primary substrate, and the suitable enzyme production becomes available to degrade the heavier PAH as the secondary substrate.

Many fungal species are known to degrade PAHs under aerobic conditions.[1,2] *Phanerochaete* and related fungi that have the ability to attack wood possess a powerful extracellular enzyme that acts on a broad array of PAHs. The enzyme is a peroxidase that, with H_2O_2 produced by the fungus, catalyzes a reaction that cleaves a surprising number of compounds.[1]

The fungus *Phanerochaete chrysosporium*, also known as white rot fungus, degrades many PAHs including benzo(a)pyrene, pyrene, fluorene, and phenanthrene.[2] Nitrogen-limiting conditions and lower pH (around 4.5) are favorable for this degradation.[10] The transformations by the fungus are slow, and the possibility of exploiting the catabolic activity under realistic field conditions have not been reported widely.

5.3.2 Chlorinated Organics Degradation

5.3.2.1   Chlorinated Aliphatic Hydrocarbons (CAHs)

Transformations of CAHs in the subsurface environment can occur both chemically (abiotic) and biologically (biotic). The major abiotic transformations include hydrolysis, substitution, dehydrohalogenation, coupling, and reduction reactions. Abiotic transformations generally result in only a partial transformation of a compound and may lead to the formation of an intermediate that is either more readily or less readily biodegraded by microorganisms.

Biotic transformation products are different under aerobic than anaerobic conditions. Microbial degradation of chlorinated aliphatic compounds can use one of several metabolism modes. These include oxidation of the compound for an energy source, cometabolism under aerobic conditions, and reductive dehalogenation under anaerobic conditions. However, with cometabolism, as with abiotic transformations, CAHs are generally transformed only partially by the microbial process.

With molecular oxygen as the electron acceptor, the one- to three-atom substituted chlorinated aliphatic compounds are transformed by three types of enzymes: oxygenases, dehalogenases, and hydrolytic dehalogenases.[11] With oxygenase the transformation products are alcohols, aldehydes, or epoxides. Dehalogenase transformation products are an aldehyde and glutathione. Hydrolytic dehalogenases will hydrolyze aliphatic compounds, yielding alcohols as a transformation product.

The higher chlorinated aliphatic compounds, where all available valences on carbon are substituted, such as tetrachloroethylene, have not been transformed under aerobic systems. The single-carbon saturated compound, dichloromethane, can be used as a primary substrate under both aerobic and anaerobic conditions, and completely mineralizes.[12] The two-carbon saturated CAH, 1,2-dichloroethane, can also be used as a primary energy source under aerobic conditions.[12] One unsaturated two-carbon CAH, vinyl chloride, has been shown to be available as a primary substrate for energy and growth under aerobic conditions.[12]

These observations indicate that only the less chlorinated one- and two-carbon compounds might be used as primary substrates for energy and growth, and that organisms that are capable of doing this are not widespread in the environment. The microbial transformation of most of the CAHs depends upon cometabolism.

5.3.2.1.1   *Anaerobic Cometabolic Transformation of CAHs*

Many chlorinated aliphatic compounds are transformed under anaerobic conditions. In the presence of a consortium of microorganisms, these compounds will be mineralized to CO_2, H_2O, and Cl^-. One of the predominant mechanisms for transformation of chlorinated aliphatic compounds is reductive dechlorination. The reductive process is usually through cometabolism. There are rare exceptions to the need for cometabolism, such as chloromethane serving as a primary substrate for a strictly anaerobic homoacetogenic bacterium.[2]

The pathways of anaerobic cometabolic, reductive dechlorination are shown in Figures 5.5 and 5.6. Figure 5.5 illustrates the various anaerobic biotic and abiotic pathways that chlorinated aliphatic compounds may undergo in the subsurface environment. Figure 5.6 also describes the anaerobic transformation of PCE and TCE under anaerobic conditions. During reductive dechlorination, the chlorinated compound serves as the electron acceptor.

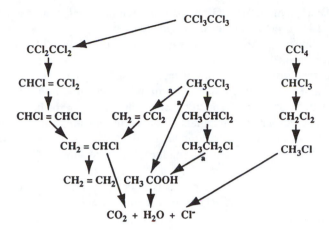

(a). Abiotic transformation

Figure 5.5 Anaerobic transformations of chlorinated aliphatic hydrocarbons.

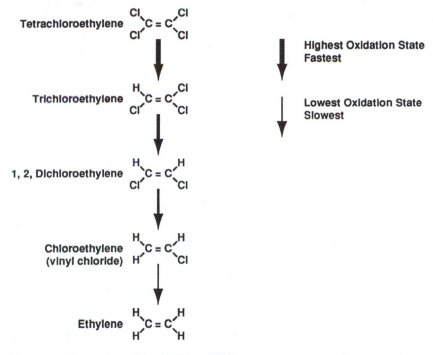

Figure 5.6 Anaerobic transformation of PCE and TCE.

The more chlorinated a compound is, the more oxidized the compound is, and the more susceptible it is to reduction (Figure 5.6). Reductive dehalogenation is carried out by electrons from the oxidation of the primary substrate.

Anaerobic transformations of tetrachloroethylene (PCE) and trichloroethylene (TCE) have been studied very intensely in the recent past.[1,2,12–14] General agreement exists that transformation of these two compounds under anaerobic conditions proceeds by sequential reductive dechlorination to dichloroethylene (DCE) and vinyl chloride (VC) (Figures 5.5 and 5.6); and in some instances, there is total dechlorination to ethene or ethane. Among the three possible DCE isomers, 1,1-DCE is the least significant intermediate, and it has been reported that *cis*-1,2-DCE predominates over *trans*-1,2-DCE.

The pathways described in Figure 5.5 indicate that any chlorinated aliphatic compound can be transformed to innocuous end products under anaerobic conditions. However, the microbial transformations generally involve cometabolism such that other primary organic substrates and suitable microbial consortium must be present. Furthermore, as noted earlier, the rates of anaerobic transformations are much greater for the highly chlorinated compounds than for less chlorinated compounds; thus, the intermediates may persist longer in the environment. Also, some of the intermediates are more hazardous than the parent compounds, examples of which are the transformation of TCE to vinyl chloride and TCA to 1,1-DCE. Hence, with anaerobic transformation, all the right conditions must be present for complete transformation to innocuous end products to occur at sufficiently high rates.

The availability of other electron acceptors in anaerobic systems affects the reductive dechlorination process by competing with the chlorinated compounds for reducing potential. For example, sulfate and nitrate can inhibit the dechlorination, since microoganisms will tend to couple half reactions that yield the greatest free energy. Introduction of nitrate and sulfate was found to decrease the dechlorination rate of PCE under field conditions.[15,16] Reductive dechlorination rates were found to be the highest under highly reducing conditions associated with methanogenic reactions rather than under less reducing conditions associated with denitrifying conditions.[16] Degradation efficiencies under various anaerobic conditions for a few selected compounds are presented in Table 5.4.

Table 5.4   Degradation Efficiencies for Chorinated Compounds under Various Anaerobic Conditions

	Removal efficiencies (percentage)		
Compound	Denitrification	Sulfate Reduction	Methanogenic
PCE	0	13	86
Chloroform	0	0	95
1,1,1-TCA	30	72	>99
Carbon tetrachloride	>99	>99	>99

5.3.2.1.2   *Aerobic Cometabolic Transformation of CAHs*

Until a few years ago, chlorinated compounds with two carbon atoms were considered nonbiodegradable under aerobic conditions. In the recent past, it was shown for the first time that TCE may be susceptible to aerobic degradation by methane utilizing bacterial communities.[17] The processes involved are illustrated by the pathways in Figure 5.7. Cometabolism of TCE is carried out by methanotrophic bacteria, which oxidize methane for energy and growth.

The responsible enzyme of methanotrophic bacteria, methane monooxygenase, catalyzes the incorporation of one oxygen atom from molecular oxygen into methane to produce methanol. The lack of substrate specificity of the monooxygenase enzyme results in its ability to oxidize a broad range of compounds, including chlorinated organic compounds. Methane monooxygenase fortuitously oxidizes TCE to form TCE epoxide, an unstable compound that chemically undergoes decomposition to yield a variety of products, including carbon monoxide, formic acid, glycoxylic acid, and a range of chlorinated acids.[12] Since these products cannot be further metabolized by methanotrophs, a community of microorganisms is necessary for mineralization to carbon dioxide, water, and chloride.

Although these oxygenase-generating microorganisms can oxidize chlorinated aliphatic compounds, engineering design of these systems are not simple. The cometabolite must always be present for sustained reactions. However, excessive methane and high oxygen concentrations will inhibit the oxidation of chlorinated compounds.[2] High methane, the

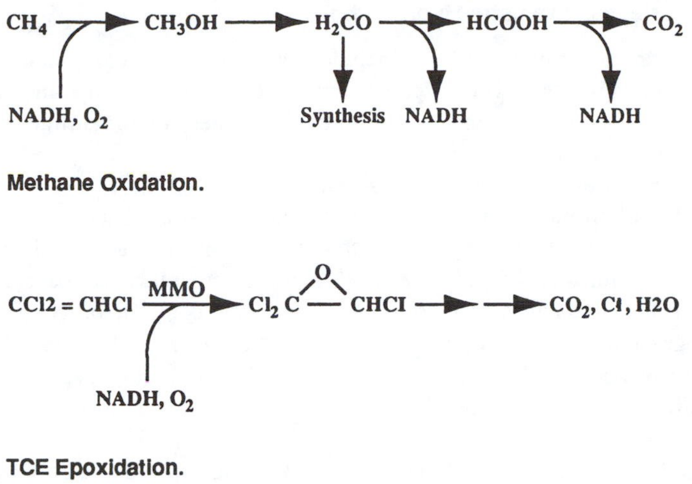

$$CH_4 \longrightarrow CH_3OH \longrightarrow H_2CO \longrightarrow HCOOH \longrightarrow CO_2$$

$$NADH, O_2 \qquad\qquad Synthesis \quad NADH \qquad\qquad NADH$$

Methane Oxidation.

$$CCl2 = CHCl \xrightarrow{MMO} Cl_2 C \overset{O}{-\!\!-\!\!-} CHCl \longrightarrow \longrightarrow CO_2, Cl, H2O$$

$$NADH, O_2$$

TCE Epoxidation.

Figure 5.7 Pathways of TCE cometabolism by methanotrophic microorganisms.

primary substrate, will hinder the reaction rate, since it will compete for the monooxygenase enzyme, making the enzyme unavailable for the target compounds. Furthermore, there is a potential for toxicity problems. It has been reported that trichloroethylene oxidation products are toxic to certain methanotrophs, and perchloroethylene (PCE) appears to inhibit trichloroethylene degradation.[2,11,18,19] Another serious limitation is that methanotrophs have not been reported to transform PCE or higher chlorinated aliphatic compounds, since the higher the degree of oxidation, the less easy it is to oxidize the compound.

Since the first report of TCE cometabolism,[17] many other groups of aerobic bacteria have been recognized as being capable of transforming TCE and other chlorinated aliphatic compounds. In addition to methane oxidizers, aerobic bacteria that are propane oxidizers, ethylene oxidizers, toluene oxidizers, phenol oxidizers, ammonia oxidizers, and vinyl chloride oxidizers also have been recognized to have the ability of cometabolizing CAHs.

Table 5.5 summarizes the discussion in the last few sections regarding the potential microbial transformations of chlorinated aliphatic hydrocarbon compounds.

Table 5.5 Potential for Biotransformation of Chlorinated Aliphatic Hydrocarbons as a Primary Substrate or through Cometabolism

Compound	Primary substrate		Cometabolism		Product
	Aerobic	Anaerobic	Aerobic	Anaerobic	
CCl_4			o	xxxx	$CHCl_3$
$CHCl_3$			x	xx	CH_2Cl_2
CH_2Cl_2	Yes	Yes	xxx		Mineralized
CH_3CCl_3			x	xxxx	CH_3CHCl_2
CH_3CHCl_2			x	xx	CH_3CH_2Cl
CH_2ClCH_2Cl	Yes		x	x	CH_3CH_2Cl
CH_3CH_2Cl	Yes		xx	a	
$CCl_2 = CCl_2$			o	xxx	$CHCl = CCl_2$
$CHCl = CCl_2$			xx	xxx	$CHCl = CHCl$
$CHCl = CHCl$			xxx	xx	$CH_2 = CHCl$
$CH_2 = CCl_2$			x	xx	$CH_2 = CHCl$
$CH_2 = CHCl$	Yes		xxxx	x	Mineralized

Note: o, very small, if any potential; x, some potential; xx, fair potential; xxx, good potential, xxxx, excellent potential; a, readily hydrolyzed abiotically.

From McCarty, P. L. and Semprini, L., Groundwater treatment for chlorinated solvents, in *Handbook of Bioremediation*, Norris, R. D., et al., Eds., Lewis Publishers, Boca Raton, FL, 1994. With permission.

5.3.2.2   Chlorinated Aromatic Hydrocarbons

Chlorinated aromatic hydrocarbons include a wide range of compounds present in the subsurface as contaminants and thus require remediation. These compounds are, to name a few, chlorophenols, chlorobenzenes, chloronitro benzenes, chloroaniline, polychorinated biphenyls (PCBs), and many pesticides.

Most chlorinated aromatic compounds that are degraded under aerobic conditions are probably acted upon through cometabolism.[2] It is also possible that a chlorinated aromatic compound is transformed to a toxic product that prevents further aerobic degradation. Complete aerobic mineralization of chlorinated aromatic compounds has been reported in the past.[2] However, the persistence of these compounds reflects the inability of many microorganisms to degrade these compounds. The nature and number of chlorine substitutions, and the substitution positions influence the extent of degradation. Degradation of chlorine-substituted aromatic compounds frequently does not follow the reaction pathways of the unsubstituted parent compounds.

Chlorinated aromatic hydrocarbons that are recalcitrant under aerobic conditions are sometimes degraded by one or more reductive dechlorinations under anaerobic conditions. Chlorinated organic compounds serve as the electron acceptors, and the primary substrate supplies the electron due to oxidation. Chlorine present at the *ortho-* and *para-*positions are more resistant to dechlorination than those at the *meta* position.[2] When the chlorine is removed, ring fission leads to methane and carbon dioxide.[20]

Methanogenic metabolism has successfully dechlorinated many aromatic organic compounds such as 3-chlorobenzoate, 2,4-dichlorophenol, and 4-chlorophenol.[2] Methanogenic cultures show preferential removal of *ortho-*chlorines, with *meta-* or *para-*chlorines removed at slower rates. Anaerobic dehalogenation and the final mineralization may require multiple species of microorganisms and reduction pathways. For example, 2,4-dichlorophenol was mineralized to CH_4 and CO_2 by as many as six species of microorganisms.[21]

Polychlorinated biphenyls (PCBs) are chlorinated aromatic compounds that are designated by numbers that represent the number of carbon atoms and the percentage of chlorine by weight. PCBs are also known under their trade name Aroclor™. For example, Aroclor 1252 contains 12 carbon atoms and has 52% chlorine by weight. PCBs are very insoluble in water and are mostly found only in soils and sediments.

No single organism is responsible for the degradation of multiple-chlorine PCBs. Both aerobic and anaerobic metabolism modes affect some biotransformation of PCBs. Analog substrate enrichments have produced varied results for PCB degradation.[2] Addition of biphenyl as an analog substrate had significant effect on the degradation of Aroclor 1242.[22] Analog enrichment, however, did not yield positive results in studies with Aroclor 1254.[23]

As noted earlier, anaerobic metabolic modes have a significant advantage over aerobic modes for PCBs. Dechlorination of the highly chlorinated Aroclor 1260 even occurs to a significant extent under anaerobic conditions.[2]

Fungi known to degrade wood, such as white rot fungi, have been documented to mineralize tri-, tetra-, and pentachlorophenol (PCP).[24] It was also reported that a consortium of microorganisms present in soil can completely mineralize PCP.[2]

5.4   BIODEGRADATION KINETICS AND RATES

Biodegradation of organic compounds, their pathways and the kinetics of defined enzymatic degradation steps, has generally been determined in well-defined, optimal laboratory conditions such as aqueous systems or shake flask experiments using water–soil/sediment suspensions or batch experiments. Mainly, these data have been used for model approaches, but they are hardly relevant for biodegradation rates *in situ*. Half-life periods as a parameter

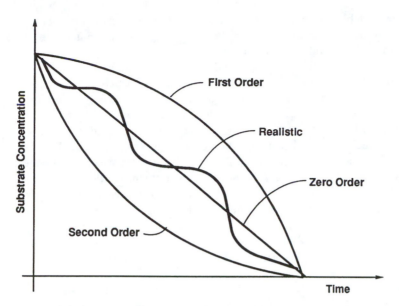

Figure 5.8   Microbial degradation kinetics order.

and first-order kinetics as a function are most commonly used for describing degradation of contaminants in the subsurface. Degradation, however, strongly depends on the site-specific environmental conditions under consideration. Measuring half-life is rather easy, since it is based on disappearance, but it does not take into account the difference between one transformation step and complete mineralization.

Kinetic models for microbial degradation are based on substrate concentration and biomass.[25] This leads to three types of kinetic order for biodegradation in natural environments, often based on empirical knowledge, and thus reflecting the rudimentary level of knowledge about microbial populations and their activity in these environments. When the substrate is completely available (i.e., its availability is not rate-limiting), the degradation only depends on the activity of the microorganisms following logarithmic growth. The degradation follows zero-order kinetics: logarithmic disappearance. A process follows first-order kinetics when the rate of biodegradation of a compound is directly proportional to its concentration. The second-order approach, in which the first-order kinetics is related to the population density, is the most realistic one. Lack of detailed stepwise degradation information may be one of the reasons why the occurrence of nonlinear reactions is presumed. This phenomena is described by equation (5.6)[1] and Figure 5.8.

$$\frac{-dC}{dt} = kC^n \tag{5.6}$$

where C = substrate concentration
      t = time
      k = rate constant for chemical disappearance
      n = a fitting parameter.

The response of the microbial community toward organic compounds does not depend on total concentration, but mainly on the water soluble concentration. Bioavailability of a compound is of extreme importance, because it frequently accounts for the persistence of compounds that are biodegradable and that might otherwise be assumed to be readily degraded.

The unavailability of a compound could result from its sorption to solids in the environment, its presence as nonaqueous phase liquid (NAPL), its entrapment within the physical matrix of the soil, and diffusional limitations.

When two or more different sequential microbial populations are required for complete degradation, longer than normal acclimatization times may be involved. For difficult to degrade compounds, this is rather a rule than an exception. Different kinetics of the various degradation steps and this acclimatization time are the reasons why the overall disappearance seems to follow a cyclic pattern (Figure 5.8). It can be assumed that nonlinear responses in reality are rather a rule than exception. Other factors that may impact the bioavailability, and thus the kinetics, of biodegradation are weathering, sequestering, and complexation of substrate.[1] *Weathering* of a contaminant results in easily biodegradable compounds being degraded early and formation of an aged residue. *Sequestering* of a compound occurs when a compound becomes less available or essentially wholly unavailable for biodegradation when it enters or is deposited in a micropore that is inaccessible to microorganisms. *Complexation of a compound* affects biodegradation when the contaminant forms insoluble complexes in association with inorganic or organic substances present in the environment.

Another factor that influences the kinetics of biodegradation is the chemical structure of the contaminant of concern. Most naturally formed compounds are biodegradable, because the relatively few catabolic pathways of microorganisms would have been exposed to these natural compounds. A synthetic chemical that is not a product of biosynthesis will be degraded only if an enzyme or an enzyme system is able to catalyze the conversion of this compound to an intermediate or a substrate which is able to participate in existing metabolic pathways. The greater the difference in structure of the xenobiotic form from the compounds produced in nature, the less is the likelihood for significant biodegradation.

Various approaches have been used to predict biodegradability in the past. These approaches include empirical, theoretical, and experimental methods.[1] The experimental methods include laboratory bench-scale experiments based on the disappearance of the compound; respirometric studies based on the oxygen uptake of aerobic microorganisms either in the laboratory or in the field; and measurement of half-lives based on the degradation of compounds.

Empirical approaches include predicting biodegradability from the properties of a compound similar to a substrate in known metabolic pathways, and predicting biodegradability based on the chemical and physical properties of a compound, such as water solubility, boiling point, melting point, molecular weight, density, partition coefficient, etc. It should be noted that empirical biodegradability predictions are qualitative at most.

Theoretical predictions of biodegradability are based on molecular topology,[1] which deals with structural features of contaminant molecules such as shape, size, presence of branching, and types of atom-to-atom connections. Of particular interest in molecular topology is molecular connectivity, which can be determined from the structural formula of the compound.

5.5 ENVIRONMENTAL FACTORS

Microbial populations capable of degrading contaminants in the subsurface are subjected to a variety of physical, chemical, and biological factors that influence their growth, their metabolic activity, and their very existence. The properties and characteristics of the environments in which the microorganisms function have a profound impact on the microbial population, the rate of microbial transformations, the pathways of products of biodegradation, and the persistence of contaminants. The impact of site-specific factors is evident from studies showing that a specific compound is biodegraded in samples from one but not another environment.[1]

A vast amount of information exists on the biochemical activities of microorganisms grown in pure or mixed cultures at various concentrations in laboratory media. This research has created a foundation for the understanding of the nutrition, population dynamics, and metabolic potential of microorganisms under controlled laboratory conditions. However, in nature, microorganisms are exposed to enormously different conditions. They may have an insufficient supply of inorganic nutrients; a paucity of essential growth factors, temperatures, and pH values at their extremes of tolerance; and contaminant levels that stress the microbial population. Contaminants at very high levels can retard the growth, inhibit the metabolic activity, and may also result in loss of viability. As a consequence, extrapolations from laboratory tests, performed under controlled conditions, to field conditions may be fraught with peril.

5.5.1 Microbial Factors

The microbial population of the soil is made up of five major groups: bacteria, actinomycetes, fungi, algae, and protozoa. Bacteria are the most abundant group, usually more numerous than the other four combined. Although transformations similar to those of the bacteria are carried out by the other groups, the bacteria stand out because of their capacity for rapid growth and degradation of a variety of contaminants. Classification of bacteria has been proposed in various forms to meet different objectives: (1) ability to grow in the presence or absence of oxygen, (2) cell morphological structure, and (3) type of energy and carbon sources.

The ability to grow in the presence or absence of oxygen is an important biochemical trait which has led to three separate and distinct categories: *aerobes*, which must have access to O_2; *anaerobes*, which grow only in the absence of O_2; and *facultative anaerobes*, which can grow either in the absence or presence of O_2.

Three morphological types are known, the *bacilli* or rod-shaped bacteria, which are the most numerous, the *cocci* or spherical-shaped cells, and the *spirilla* or spirals. The latter are not common in soils. Some of the bacilli persist in unfavorable conditions by the formation of endospores that function as part of the normal life cycle of the bacterium. These endospores often endure in adverse environments because of their great resistance to both prolonged desiccation and to high temperatures. Spore-forming genera are present among the aerobic and anaerobic bacteria. The endospore can persist in a dormant state long after the lack of substrate or water has led to the death of vegetative cells. When conditions conducive to vegetative growth return, the spore germinates and a new organism emerges.

Microorganisms are divided into two broad classes with respect to their energy and carbon sources. *Heterotrophic* forms, which require organic substrates to serve as sources of energy and carbon, dominate the soil microflora. *Autotrophic* microorganisms obtain their energy from sunlight or by the oxidation of inorganic compounds and their carbon by the assimilation of CO_2. Autotrophs are of two general types: *photoautotrophs* whose energy is derived from sunlight, and *chemoautotrophs* which obtain the energy needed for growth from the oxidation of inorganic materials.

There is frequently an initial period, also known as the acclimatization lag, during biodegradation of contaminants. During this period, no obvious biotic changes of contaminant levels take place. This period may be due to various reasons, and the causes may be in the indigenous microbial communities. The starting biomass may be so low that no appreciable degradation can happen until a critical biomass concentration is reached or the total microbial population may be abundant, but the specific degrading populations may need to be enriched. On other occasions, the contaminant must induce requisite enzyme or a new enzyme needs to be synthesized. Sometimes, the reasons for the initial lag period may lie in the contaminants themselves. The contaminants may be present in such low concentrations that they will not induce the relevant enzymes, or their chemical structure may be so unusual that they cannot interact with the active enzyme sites. The lag for the degradation of a specific contaminant can also occur due to the preferential depletion of other substrates first.

Measurement of the indigenous microbial activity is one method for evaluating potential toxic or inhibitory conditions at a site. Low bacteria counts can indicate a potential toxicity problem or a stressed microbial population. Groundwater bacterial counts range from 10^2 to 10^5 colony forming units (CFU) per milliliter of sample. Typical soil microbial counts range from 10^3 to 10^7 CFUs per gram of soil. Higher counts indicate a healthy microbial population. Counts below 10^3 organisms per gram of soil at contaminated sites may indicate a stressed microbial population.

5.5.2 Nutrients

Carbon makes up a large fraction of the total protoplasmic material of a microbial cell. Carbohydrates, proteins, amino acids, vitamins, nucleic acids, purines, pyrimidines, and other substances constitute the cell material. In addition to carbon, cell material is mainly composed of the elements hydrogen, oxygen, and nitrogen. These four chemical elements constitute about 95% by weight of living cells. Two other elements, phosphorus and calcium, contribute 70% of the remainder. The elemental composition of microbial cells on a dry weight basis is presented in Table 5.6.

Table 5.6 Elemental Composition of Microbial Cell on a Dry Weight Basis

Element	Percentage of dry weight
Carbon	50
Oxygen	20
Nitrogen	14
Hydrogen	8
Phosphorus	3
Sulfur	1
Potassium	1
Sodium	1
Calcium	0.5
Magnesium	0.5
Chlorine	0.5
Iron	0.2
All others	0.3

The microbial requirements for nutrients are approximately the same as the composition of their cells (Table 5.6). The chemical structure of bacteria is often expressed as $C_5H_7O_2N$ with only minor, but important, traces of other atoms. Carbon is usually supplied by organic substrates—organic contaminants in the case of bioremediation—for the heterotrophic microorganisms. Autotrophic microorganisms obtain their carbon supply from inorganic sources such as carbonates and bicarbonates. Hydrogen and oxygen are supplied by water. Usually, the nutrients in short supply are nitrogen, phosphorus, or both. Nearly always, the supply of potassium, sulfur, magnesium, calcium, iron, and micronutrient elements is greater than the demand. These micronutrients are present in most soil and aquifer systems.

It is widely believed that only one nutrient element is limiting at any given time, and that only when that one deficiency is overcome does another nutrient become limiting. This condition is stated by Liebig's law of the minimum: The essential constituent that is present in the smallest quantity relative to the nutritional requirement of microorganisms will become the limiting factor of growth. This law can be expanded to include the electron acceptor also.

Even in the absence of added N and P, biodegradation will continue in the subsurface, albeit at a slow rate. This phenomenon is due to the recycling of the elements as they are assimilated into microbial cells and then are converted back to the inorganic forms due to the death and lysis of microbial cells. Under such circumstances, the rate of biodegradation will be limited and will be impacted by the rate at which the limiting nutrient is recycled.

Many microorganisms also require some substances that are part of the cell structural building blocks, at trace quantities. These substances, known as the growth factors, are organic molecules such as amino acids, vitamins, or other structural units. Growth factors are not essential nutrients, but they stimulate the species of organisms that need them.

5.5.3 Physical–Chemical Factors

The activities of microorganisms are markedly affected by their physical–chemical environment. Environmental parameters such as temperature, pH, moisture content, and redox potential will determine the efficiency and extent of biodegradation.

5.5.3.1 Temperature

As temperature increases, the rates of chemical as well as biochemical reactions generally increase. This phenomenon is referred to as Arrhenius behavior (Figure 5.9A). The same phenomenon also occurs with microorganisms and the myriad of chemical and biochemical reactions that constitute "microbial activity," but only to a point. While the rates of abiotic chemical reactions might increase in an unbounded fashion with increasing temperature, this is not the case with microbial activity. Beyond some optimum temperature, the activity of any organism declines precipitously. At the lower end of the temperature range, most bacteria stop metabolic activities at temperatures just above the freezing point of water.

The decline of microbial activity at temperatures beyond the optimum is usually explained in terms of the three-dimensional shapes of enzymes and the effects of temperature on membrane integrity. Three categories of microorganisms are defined, based upon temperature optima (Figure 5.9B):

- *Psychrophiles:* Psychrophilic (or cryophilic) organisms have an optimum temperature of $15 \pm 5°C$, and a minimum temperature of $0°C$ or below. Strict psychrophiles usually die if exposed even temporarily to room temperatures. On the other hand, there are organisms with optima at 25 to $30°C$, but which can grow at $0°C$; these are sometimes called *facultative psychrophiles*. Psychrophiles usually possess membranes rich in unsaturated fatty acids, a feature which is alleged to provide a more fluid structure at low temperatures.
- *Mesophiles:* Mesophilic organisms have an optimum temperature between $25°C$ and $40°C$. Most of the microorganisms that inhabit the subsurface are mesophiles. Microorganisms commonly found effective in bioremediation perform over a temperature range of 10 to $40°C$. For many regions of the country, groundwater temperatures remain reasonably constant throughout the year at around the mean air temperature for the region.[27]
- *Thermophiles:* Thermophilic organisms have temperature optima above $45°C$. For example, there are thermophilic methanogens that prefer temperatures of 55 to $60°C$. Some are facultative thermophiles, in that their range extends into the mesophilic zone. Thermophiles have membranes rich in saturated fatty acids. The soil surface temperature around noontime during a summer day could reach 50 to $70°C$.

5.5.3.2 pH

The pH affects the microorganism's ability to conduct cellular functions, cell membrane transport, and the equilibrium of catalyzed reactions by having an impact on the three-

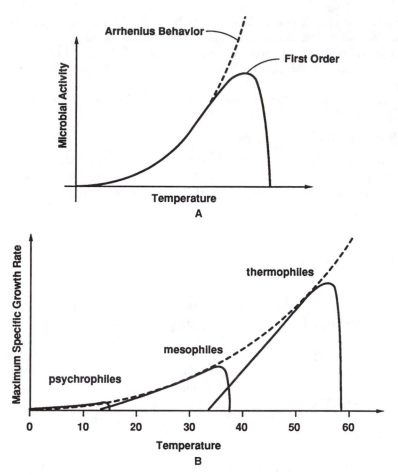

Figure 5.9 A. Microbial activity with temperature. B. Temperature dependency of growth rate of various microorganisms.

dimensional conformation of enzymes and transport proteins of microbial cells. It also affects the protonmotive forces responsible for energy production within the cell.

Most natural environments possess pH values in the range between 5 and 9. Therefore, it is not surprising to find that most microorganisms have evolved with pH tolerances within this range. Most bacteria tolerate pH 5 to 9 but prefer pH 6.5 to 7.5. There are *acidophilic* bacteria such as *Thiobacillus thioxidans*, which have pH optimum near 2.5. Also, there are *alkalophilic* bacteria that can function at pH 10 to 12.

Metabolic activities of microorganisms produce organic acids and HCl from the degradation of organic compounds (chlorinated organics). When high concentrations of organic compounds are present in the subsurface with low alkalinity, pH control may be necessary to sustain continued biodegradation.

5.5.3.3   Moisture Content

Moisture is a very important variable relative to bioremediation. Moisture content of soil affects the bioavailability of contaminants, the transfer of gases, the effective toxicity level of contaminants, the movement and growth stage of microorganisms, and species distribution.[2]

Soil moisture is frequently measured as a gravimetric percentage or reported as field capacity. Evaluating moisture by these methods provides little information on the "water availability" for microbial metabolism. *Water availability* is defined by biologists in terms of a parameter called water activity(a_w):

$$a_w = \frac{RH}{100} = \frac{P_w}{P_w^o}$$ (5.7)

where RH = relative humidity of a covered system
 P_w^o = vapor pressure of pure water at the temperature of the system
 P_w = vapor pressure at equilibrium with water in the system.

In simple terms, the water activity is the ratio of the system's vapor pressure to that of pure water (at the same temperature). Pure water has a water activity of 1.0, seawater 0.98, and dried fruit 0.7.

Microbial transport of water through the bacterial plasma membrane is a passive process, governed strictly by diffusion and the gradient in a_w across the membrane.

5.5.3.4   Oxidation-Reduction (Redox) Potential

The redox potential is a measure of how oxidizing or reducing an environment is. Redox potential is sometimes denoted by the symbol E_H. E_H greater than zero is commonly interpreted to be an oxidizing environment, and E_H less than zero, a reducing environment. The practical range of E_H in the natural environment is from +800 mV (high O_2, with no O_2-depleting processes) to about –400 mV (systems high in H_2). Redox potential is measured by use of a platinum electrode, in conjunction with some reference electrode. Unfortunately, interpretation of measured E_H values is very difficult, because natural systems are seldom at equilibrium.

Table 5.3 presents the impact of redox potential on various mechanisms of microbial transformation of contaminants during bioremediation. However, it should be noted that the concentrations of particular oxidants or reductants will affect microbial metabolic activity regardless of the redox potential. The concentration levels of oxidants or reductants will influence the enzymatic activity via effects on three-dimensional conformation.

5.6   *IN SITU* BIOREMEDIATION SYSTEMS

The most significant challenge in *in situ* bioremediation is introducing into the subsurface environment the reagents needed by microorganisms and mixing them with the contaminants to be degraded. Much of the methodology usually associated with *in situ* bioremediation can be attributed to the pioneering research and development carried out by Richard L. Raymond and Sun Tech in the 1970s. By the mid-1980s, the potential of *in situ* bioremediation was widely accepted in the remediation industry. In the last few years, there has been an explosion of activity in bioremediation which now incorporates a wide range of processes in the *in situ* environment.

A continuing source of debate among practitioners of bioremediation are the concepts of *biostimulation* and *bioaugmentation*. Biostimulation consists of adding nutrients, such as nitrogen and phosphorus, as well as oxygen and other electron acceptors, to the microbial environment to stimulate the activity of microorganisms. Bioaugmentation involves adding exogenous microbes to the subsurface where organisms able to degrade a specific contaminant are deficient. Microbes may be "seeded" from populations already present at a site and grown in an above-ground reactor, or specially cultivated strains having known capabilities for degrading a specific contaminant.

Most bioremediation systems employ some form of biostimulation. However, there is a significant resistance in the industry to use bioaugmentation. This resistance stems from the *ubiquity principle*, which states that all microorganisms are ever-present in the subsurface environment. Another argument against bioaugmentation is that indigenous organisms already present at the contaminated site would have developed the enzyme systems to degrade the target contaminants. Furthermore, the limitation of distributing the exogenous microbial cultures in the subsurface and the question of long-term survivability of these lab-grown cultures under field conditions also discourage bioaugmentation. Bioaugmentation may play a prominent role in bioremediation when the release of genetically engineered organisms is permitted.

5.6.1 Screening Criteria

Prior to designing an *in situ* bioremediation system, the feasibility of biodegradation should be carefully evaluated. This evaluation should include the ease or difficulty of degrading the target contaminants, the ability to achieve total mineralization, and the environmental conditions necessary to implement the process. There are various factors that should be incorporated into this evaluation process.

- *Biodegradability of contaminants:* Years of experience and research has established the degradation pathways of many specific contaminants. Contaminant characteristics and structure also will provide answers in terms of biodegradability. As an illustration, the following compounds have been listed with the compounds easiest to degrade at the top and the difficult ones at the bottom.

Simple hydrocarbons, C1–C15	very easy
Alcohols, phenols, amines	very easy
Acids, esters, amides	very easy
Hydrocarbons, C12–C20	moderately easy
Ethers, monochlorinated hydrocarbons	moderately easy
Hydrocarbons, greater than C20	moderately difficult
Multichlorinated hydrocarbons	moderately difficult
PAHs, PCBs, pesticides	very difficult

- *Mineralization potential of the compounds:* A review of pertinent reaction pathways will provide insight as to whether the contaminant will be utilized as a primary substrate or whether cometabolic reactions are necessary.
- *Specific microbial, substrate, and other conditions:* Of prime importance is the availability of carbon and energy in the contaminated environment. Electron acceptor availability and the redox condition should be carefully determined. In addition, the presence of microorganisms capable of degrading the contaminants, in sufficient numbers, should be evaluated. Total plate counts, specific degraders counts, and laboratory and *in situ* respiration tests can be utilized to perform this evaluation.
- *Availability of nutrients:* In general, the concentration levels of only N and P are determined.
- *Site's hydrogeologic characteristics:* Hydraulic conductivity, thickness of the saturated zone, homogeneity, and depth to the water table are parameters that should be factored into the design of the system. Distribution and transport of added nutrients and electron acceptors will be heavily influenced by the site hydrogeology.
- *Extent and distribution of contaminants:* This assessment with the site hydrogeologic parameters are the key components for developing the engineering design of the "subsurface bioreactor."

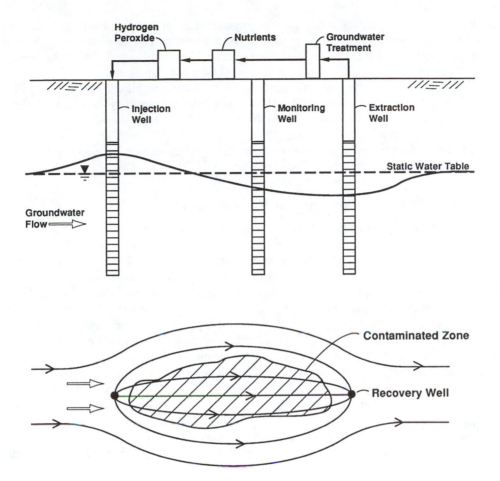

Figure 5.10 Description of the Raymond process.

- *Biogeochemical parameters:* Measurements of various biogeochemical parameters such as dissolved oxygen (DO), redox potential, CO_2, and other parameters such as NH_4^+, NO_3^-, SO_4^{2-}, S^{2-}, and Fe^{2+} will give an indication of the existing (natural or intrinsic) microbial metabolic activity at the site.

5.6.2 Raymond Process

The Raymond process shown in Figure 5.10 includes groundwater recovery wells, above-ground treatment, amendment with nutrients and possibly an electron acceptor, and reinjection of the amended groundwater.[28] This concept was developed on the premise that for most *in situ* bioremediation systems, the rate-limiting step is the rate of introduction of the electron acceptor. In the process shown in Figure 5.10, hydrogen peroxide was often used as the means of introducing oxygen to enhance the rate of aerobic biodegradation.

The perceived advantage of hydrogen peroxide is that due to its high levels of solubility compared to dissolved oxygen, a significant amount of available oxygen could be introduced into the aquifer. It was also believed that due to the high solubility, hydrogen peroxide could travel a long distance from the point of injection before being consumed. However, it was found that due to the instability of hydrogen peroxide in the presence of Fe and colloids, most of the H_2O_2 decomposed within a short distance from the point of injection.

In the Raymond process, the saturated zone of the contaminated area was manipulated to affect a "closed loop" flow system with a significant increase in groundwater flow rates.

In this manner, added O_2 and nutrients were transported faster than the natural groundwater flow velocities. Microbial populations and rate of degradation can be increased by several orders of magnitude within this "subsurface bioreactor." This configuration will also provide hydraulic containment of the plume at the downgradient edge.

One significant disadvantage of this process is the inefficient utilization of the injected H_2O_2. Less than 10 to 20% of the injected H_2O_2 only will be consumed by the microorganisms for biodegradation. The rest is lost due to the escape of the O_2 produced into the soil gas above the water table.

Injection of air into the saturated zone for the purpose of introducing O_2 to enhance biodegradation (biosparging) is described in Chapter 4.

5.6.3 Denitrification-Based *In Situ* Bioremediation

One promising alternative to the saturation limitations or high costs of the major alternative forms of oxygen involves the use of nitrate as the oxygen acceptor. In this process, the biodegradative activities of denitrifying organisms are enhanced, resulting in biodegradation of the target organic contaminants along with the transformation of NO_3^- to N_2. Nitrate feedstocks can thus be substituted for oxygen feedstocks in the groundwater manipulation system described in the previous section.

During *in situ* biosparging, the consumption of O_2 is relatively fast and the rate of O_2 transfer from the injected air to the aqueous phase is slow, due to low solubility of O_2 in water. Expansion of the aerobic zone is limited by the rate of O_2 supply to the aqueous phase. Anaerobic conditions are expected to persist within aerobically treated aquifers, especially in relatively impermeable zones and zones further away from the injection wells. The overall degradation efficiency can be increased by using nitrate, which is much more water-soluble than O_2 (9200 mg/l as $NaNO_3$ vs. 8 to 10 mg/l as O_2). The reducing equivalents that can be introduced into an aquifer using saturated sodium nitrate solution is approximately 50 times higher than with a saturated oxygen solution. However, due to regulatory and microbial toxicity considerations, the nitrate feedstock solution concentration should be significantly lower than saturated concentration.

Design of the *in situ* bioremediation system can be accomplished without downgradient groundwater extraction and upgradient injection. However, multiple injection points may be required to enhance the distribution and transport of the added reagents. Infiltration galleries can be also used to introduce the NO_3 and nutrients solution into the contaminated plume. Infiltration alone may limit the availability of the added reagents in the deeper zones of the contaminated plume.

Possible injection scenarios are shown in Figures 5.11 and 5.12. Based on Figure 5.11, if only injection wells are used, distribution and transport of reagents will be less effective than using both injection and extraction wells. When injection and extraction wells are used, lateral and vertical dispersion of the injected reagents will be increased, and thus the effectiveness of the "subsurface" bioreactor will be enhanced.

Denitrification-based *in situ* bioremediation has been field tested, and limited information is available in the literature. In one field study, a gasoline-contaminated plume was bioremediated by the injection of nitrate-spiked water.[29] In another study, nitrate addition into treatment cells within a JP-4 jet fuel contaminated plume resulted in degradation of specific contaminants.[30]

5.6.4 Pure Oxygen Injection

Providing an electron acceptor such as oxygen for enhanced bioremediation often becomes the critical limiting factor during system design. Continuous or intermittent oxygen delivery into the saturated zone is a challenging task, with field options primarily limited to

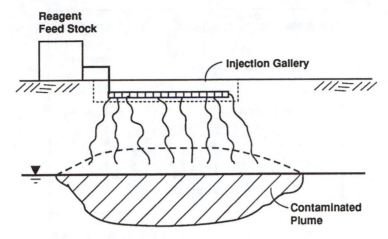

Figure 5.11   Injection of reagents via injection gallery.

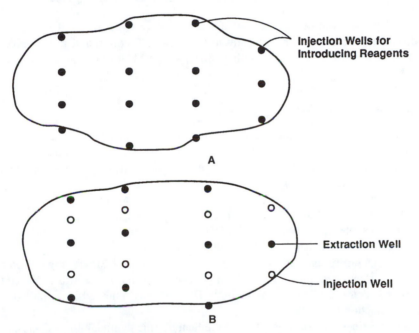

Figure 5.12   Injection well configurations for introducing reagents for bioremediation. A. Injection wells alone. B. Injection and extraction wells.

sparging air and adding hydrogen peroxide. These two options have been discussed in detail in Chapters 4 and 8, respectively.

An innovative technique to inject pure oxygen in the form of microbubbles has been reported in the literature.[31] A coarse soil matrix in the saturated zone is preferred for this technique to provide both a high permeability for flowing groundwater and a suitable saturated matrix for adhering and retaining microbubbles. The use of oxygen microbubbles for *in situ* bioremediation has the advantages of increased oxygen transfer rate to flowing groundwater (DO pickup), and increased oxygen utilization (percent of O_2 injected) compared to air injection.

The microbubbles are typically made from a low-surface-tension water containing 100 ppm or more of an appropriate surfactant. The microbubbles upon generation resemble a thick cream with much of the volume made up of microbubbles. The bubbles are generated by a colloidal gas apron.

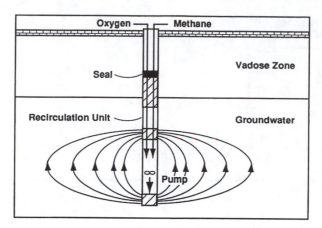

Figure 5.13 Subsurface recirculation system for methane and O$_2$ injection. (From McCarty, P. L. and Semprini, L., Groundwater treatment for chlorinated solvents, in *Handbook of Bioremediation*, Norris, R. D., et al., Eds., Lewis Publishers, Boca Raton, FL, 1994. With permission.)

In a field study, it was demonstrated that approximately 15 to 20% of the oxygen injected can be dissolved into the flowing groundwater.[31] With 10% committed to biodegrading the surfactant, a minimum of 5 to 10% net utilization was available for biodegrading contaminated groundwater.

5.6.5 Methanotrophic Biodegradation

Injection of methane and other required nutrients can enhance the cometabolic degradation of TCE and some other chlorinated aliphatic hydrocarbons. The methane provides the necessary material substrate for the indigenous microorganisms to produce the enzyme methane monooxygenase which, in turn, will degrade the TCE.

Typical injection rates of methane lie in the range of 1 to 4% in the methane–air mixture. Since methane is injected as a gaseous reagent, it is prudent to select the nutrients also in the form of gases. Nitrogen in the form of NH$_3$ gas or nitrous oxide and phosphorus in the form of triethyl phosphate can be used as nutrient sources.

In a field demonstration test using a horizontal injection well, 300 ft long and 35 ft below the water table, it was determined that 40% of the contaminant removal was achieved through methanotrophic cometabolic biodegradation.[32] The rest was removed by volatilization as a result of air injection.

In another field demonstration study of methanotrophic degradation of CAHs, the stimulation of indigenous methanotrophs was accomplished through methane and oxygen addition.[12] In this case methane and oxygen were added to the extracted groundwater and injected in the dissolved form. Concentrations of methane and oxygen were in the range of 16 to 20 mg/l and 33 to 38 mg/l, respectively. The conceptual application of this process can be implemented the same way as that shown in Figure 5.10. Another possible system for delivering the needed chemicals is subsurface groundwater circulation (Figure 5.13). This eliminates the need to pump contaminated groundwater to the surface treatment and reinjection. Methane and oxygen would be introduced directly into the well, which has a pump to induce flow from the bottom of the well and release at the top screen intersecting the water table. Instead of a pump, air injection to effect air lifting of the water will serve the dual purpose of pumping the water and introducing O$_2$. Multiple recirculation wells installed in a line across the direction of the groundwater flow will serve as a biologically reactive zone.

The study showed that the rates and extents of transformation were compound-specific and also that the cometabolic transformation was strongly tied to methane utilization; upon stopping methane addition, transformation rapidly ceased.[12] The percentage of transforma-

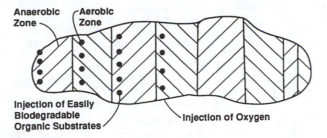

Figure 5.14   Anaerobic–aerobic sequential biodegradation.

tions achieved were TCE, 20%; *cis*-1,2-DCE, 50%; *trans*-1,2-DCE, 90%; and vinyl chloride, 95%. The difference in rates for *cis*-DCE and *trans*-DCE shows that a small change in chemical structure can have a large effect on the cometabolic transformation rate.

At the same site, higher rates of TCE degradation were accomplished by aerobic cometabolic stimulation of phenol-utilizers through phenol and oxygen addition.[12] The cometabolic transformation was strongly tied to the amount of phenol utilized and transformations achieved were TCE, 85%; and *cis*-DCE, over 90%.

5.6.6   Enhanced Anaerobic Biodegradation

Addition of easily biodegradable organic substrates will enhance the reductive dechlorination of many of the chlorinated aliphatic hydrocarbons. Many organic substrates such as acetate, butyric acid, lactic acid, methanol, ethanol, vitamin B_{12}, and sucrose have been shown to be effective in acting as the primary substrate to enhance the anaerobic cometabolic transformations. However, anaerobic dechlorination reaction rates are slower compared to the possible aerobic transformations of some of the intermediates. Hence, an anaerobic–aerobic sequential transformation will be able to achieve mineralization at a much faster rate than completely anaerobic pathways. If the contamination plume to be remediated is large, multiple anaerobic–aerobic sequencing segments can be implemented to achieve faster cleanup times (Figure 5.14).

5.6.7   Oxygen Release Compounds

Formulations of very fine, insoluble magnesium peroxide (MgO_2) release oxygen at a slow, controlled rate when hydrated. Their use has been demonstrated to increase the dissolved oxygen concentrations within contaminated plumes, and thus enhance the rate of aerobic biodegradation.[33,34] Magnesium peroxide releases oxygen when it comes in contact with water as shown by the following equation:

$$MgO_2 + H_2O \rightarrow \tfrac{1}{2}O_2 + Mg(OH)_2 .\tag{5.8}$$

The byproducts of the reaction are oxygen and magnesium hydroxide, which will also help in maintaining moderate pH levels within the contaminated plume.

MgO_2 is normally placed in an inert matrix and is available in easily installable socks of various diameters. These socks can be stacked in wells screened across the entire thickness of the contaminated zone.

5.6.8   Natural Intrinsic Bioremediation

The basic concept behind "natural intrinsic bioremediation" is to allow the natural indigenous microorganisms to biodegrade the contaminant present in the groundwater. While natural

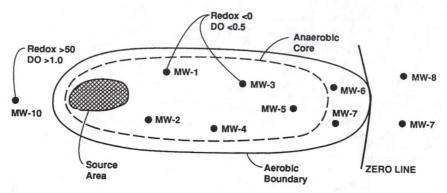

Figure 5.15 Concept of zero line.

attenuation processes include biodegradation, abiotic oxidation, hydrolysis, dispersion, dilution, sorption, and volatilization, intrinsic bioremediation is the primary mechanism for the attenuation of biodegradable contaminants. Intrinsic bioremediation, abiotic oxidation, and hydrolysis are the only attenuation mechanisms that destroy the contaminants to innocuous end products.

The use of intrinsic bioremediation as part of the site remediation strategy can significantly reduce cleanup costs. Intrinsic bioremediation is not a "no action" alternative. Implementation of a natural bioremediation system differs from conventional techniques, such that the contaminated plume is allowed to remain contaminated. Acceptance of natural bioremediation as a remediation alternative will be greatly enhanced if a *zero line* can be established (Figure 5.15).

The definition of the zero line (in plan view) is the location of a vertical plane in which the rate of natural degradation of contaminants exceeds the mass flux rate of contaminants. In the absence of any new release of contaminants, the zero line will not be a fixed, stationary line. Due to the dynamic natural attenuation processes, this line will be shifting toward the source area, thus resulting in a gradual shrinkage of the contaminant plume.

Existence of a zero line can be inferred by evaluating groundwater quality data over a period of time. Three to four rounds of sampling data collected over a period of time may indicate the existence of the zero line. However, the credibility of the argument will be greatly enhanced by providing supporting biogeochemical evidence collected from within and outside the plume. The data collected will indicate, in most cases where petroleum contamination is present, four distinct zones of biogeochemical dynamics: (1) the heart of the plume, (2) an anaerobic zone, (3) an aerobic zone, and (4) a remediated zone (Figure 5.16).

As noted in previous sections, various biodegradation pathways will take place in these four zones. Almost all dissolved petroleum hydrocarbons are biodegradable under aerobic conditions, where microorganisms utilize O_2 as the electron acceptor and the contaminants as the substrate for their growth and energy. When oxygen supply is depleted and nitrate is present, facultative anaerobic microorganisms will utilize NO_3^- as the electron acceptor. Once the available oxygen and nitrate are depleted, microorganisms may use oxidized ferric ion (Fe(III)) as an electron acceptor. When the redox conditions are further reduced (near the source area due to the abundance of contaminant mass), sulfate may act as the electron acceptor. Under significantly lower redox conditions (within the heart of the plume), methanogenic conditions will exist and the microorganisms can degrade the petroleum contaminants using water as the electron acceptor.

5.6.8.1 Concept of Bio-Buffering

Intrinsic bioremediation can use a wide range of electron acceptors under varying redox conditions. The biochemical reactions facilitated by these electron acceptors fall into two categories:

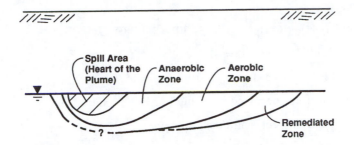

Figure 5.16 Four different zones depicting the natural intrinsic bioremedation process.

1. Relatively fast transformations that involve the use of O_2 and NO_3^-.
2. Relatively slow transformations that involve the reduction of Fe(III), SO_4^{2-}, and methanogenesis using H_2O.

The first reactions to occur are nearly instantaneous. Once the O_2 and NO_3^- are depleted and the environment turns more anaerobic, the slower reactions will begin. It is worth noting that even in a reducing environment, multiple reactions occur simultaneously, including the continuing reduction of O_2 owing to the replenishment of all electron acceptors by inflowing groundwater.

The concept of *bio-buffering* is based on the premise that the degradative capacity of the aquifer is a lot more than the available DO in the system. Bio-buffering also could be defined as the stability of the assimilative capacity of the natural system in response to the introduction of the contaminant mass flux into the aquifer. Among all the electron acceptors, O_2 and CO_2 are the most readily available, due to natural recharge processes and aquifer geochemistry. Sulfate, iron, and manganese also occur naturally, but are dependent on site minerology. Sulfate may be also introduced by manmade activities. The predominant sources of nitrate are anthropogenic activities such as agricultural fertilization.

Estimation of the assimilative capacity of benzene in an intrinsic bioremediation system is presented in the next few steps.

Aerobic Oxidation:

$$7.5O_2 + C_6H_6 \rightarrow 6CO_2 + 3H_2O$$

Mass ratio of O_2 to $C_6H_6 = 3.1:1$

0.32 mg/l benzene degraded per 1 mg/l of O_2 consumed

If the background DO concentration is 4.0 mg/l,

$$\text{assimilative capacity (aerobic biodegradation)} = \frac{0.32}{1} \times 4$$
$$= 1.28 \text{ mg/l}$$

Denitrification:

$$6NO_3^- + 6H^+ + C_6H_6 \rightarrow 6CO_2 + 6H_2O + N_2$$

Mass ratio of NO_3^- to $C_6H_6 = 4.8:1$

0.21 mg/l benzene degraded per 1 mg/l of NO_3^- consumed

If the background NO_3^- concentration is 12 mg/l,

$$\text{assimilative capacity (denitrification)} = \frac{0.21}{1} \times 12$$
$$= 2.52 \text{ mg/l}$$

Iron Reduction:

$$60H^+ + 30Fe(OH)_3 + C_6H_6 \rightarrow 6CO_2 + 30Fe^{2+} + 78H_2O$$

Mass ratio of $Fe(OH)_3$ to C_6H_6 = 41:1

Mass ratio of Fe^{2+} produced to C_6H_6 degraded = 15.7:1

0.045 mg/l of benzene degraded per 1 mg/l of Fe^{2+} produced

If the background Fe^{2+} concentration is 25 mg/l,

$$\text{assimilative capacity (Iron)} = \frac{0.045}{1} \times 25$$
$$= 1.125 \text{ mg/l}$$

Sulfate Reduction:

$$7.5H^+ + 3.75SO_4{}^{2-} + C_6H_6 \rightarrow 6CO_2 + 3.75H_2S + 3H_2O$$

Mass ratio of $SO_4{}^{2-}$ to C_6H_6 = 4.6:1

0.22 mg/l benzene degraded per 1 mg/l of sulfate consumed

If background $SO_4{}^{2-}$ concentration is 60 mg/l,

$$\text{assimilative capacity (sulfate reduction)} = \frac{0.22}{1} \times 60$$
$$= 13.2 \text{ mg/l}$$

Methanogenesis:

$$4.5H_2O + C_6H_6 \rightarrow 2.25CO_2 + 3.75CH_4$$

Mass ratio of CH_4 produced to C_6H_6 = 0.8:1

1.3 mg/l benzene degraded per 1 mg/l of CH_4 produced

If background methane concentration is 0 mg/l and

measured methane concentration is 4.0 mg/l,

$$\text{assimilative capacity (methanogenesis)} = \frac{1.3}{1} \times 4.0$$
$$= 5.2 \text{ mg/l}$$

5.6.8.2   Evaluation of Natural Intrinsic Bioremediation

Evaluation of natural intrinsic bioremediation can be performed by collecting and analyzing site-wide groundwater quality data with the following objectives:

- Documented loss of contaminant mass at the field scale.
- Biogeochemical indicator trends.
- Laboratory confirmation of microbial activity.

Collection of an adequate database during the site characterization process, over a period of time, is an important step in documenting intrinsic bioremediation. At a minimum, the site characterization phase should provide data on the location and extent of contaminant sources; the location, extent, and concentration of dissolved-phase contamination; geologic information on the type of soil distribution; hydrogeologic parameters such as hydraulic conductivity and hydraulic gradients; and groundwater biogeochemical data.[35–36]

Biogeochemical trends can be established by collecting groundwater samples and analyzing for the following parameters: dissolved oxygen (DO), redox potential, pH, temperature, conductivity, alkalinity, nitrate, sulfate, sulfide, ferrous iron, carbon dioxide, methane, and chloride, in addition to the contaminants.[36] The extent and distribution (vertical and horizontal) of contamination and electron acceptor and metabolic by-product concentrations are important in documenting the occurrence of intrinsic bioremediation.

If the dissolved oxygen concentration levels within the contamination plume are below background levels, it is an indication of aerobic biodegration at those locations. Similarly, nitrate and sulfate concentrations below background levels in the plume are indications of anaerobic biodegradation through denitrification and sulfate reduction. Presence of nitrite and H_2S in the plume will further enhance the evidence of denitrification and sulfate reduction. Furthermore, elevated concentrations of metabolic by-products such as ferrous ion and methane will indicate the occurrence of Fe(III) reduction and methanogenesis inside the plume. Contour maps should be developed to provide clearly visible trends of these processes inside the contamination plume.

Significant quantitative differences in the concentration levels of the various electron acceptors and metabolic by-products will be more than sufficient, in more cases, to claim the occurrence of natural intrinsic bioremediation. However, estimating the rate constants of contaminant degradation will further support the selection of this alternative as the preferred remediation method.

First-order rate constants can be calculated by the following equation:

$$Rate = \frac{\ln \text{(highest concentration downgradient/highest concentration upgradient)}}{\text{distance traveled/plume velocity}}$$

It should be noted that the concentration at the downgradient location should be corrected for dilution. When estimating the plume velocity, the retardation factor of the contaminants considered should be taken into account.

Laboratory confirmation of microbial activity can vary from enumerating the microbial population to performing full-blown microcosm studies. Respirometric studies of the groundwater samples also can be performed in the laboratory.

Intrinsic bioremediation of CAHs should be evaluated differently than that of petroleum hydrocarbons. Since most of the metabolic pathways are induced by cometabolic mechanisms, presence of primary organic substrates and acclimatization of indigenous microorganisms will play significant role in natural bioremediation. Mineralization and nonmineralization pathways and accumulation of metabolic intermediates should be taken into serious consideration.

Starting concentration of the target contaminants, availability and concentration of electron acceptors, and presence of native organic compounds will all play a significant role in intrinsic bioremediation of chlorinated compounds.

Under optimal conditions, natural intrinsic bioremediation should be capable of completely containing a dissolved hydrocarbon plume. While there are an increasing number of well-documented cases where this has occurred, there is a great deal of anecdotal evidence that suggests this is possible.

One of the interesting questions that is currently being investigated by researchers is whether it may be possible to complete a mass balance on the supply of electron donors and electron acceptors at a given site. This question is complicated because of sampling and field data collection limitations. Other complicating factors involve the temporal nature of the distributions of electron acceptors and donors.

5.7 BIOMODELING

Groundwater transport and fate models have traditionally focused on modeling advection, dispersion, and sorption as three main attenuation mechanisms in groundwater. A fourth key variable that impacts the fate of contaminants is biodegradation. Development of biodegradation models is not simple, because of the complex nature of microbial kinetics, the lack of accurate field data, and the lack of robust numerical schemes that can simulate the physical, chemical, and biological processes accurately. Several researchers have developed groundwater biodegradation models.[37] The main approaches used for modeling biodegradation kinetics are

- first-order degradation models
- biofilm models (including kinetic expressions)
- instantaneous reaction models, and
- dual-substrate Monod models.

One of the most popular models used in biomodeling of petroleum hydrocarbons degradation is the BIOPLUME model. This model incorporates a system of equations to simulate the simultaneous growth, decay, and transport of microorganisms combined with the transport and removal of hydrocarbons and oxygen.[37] This model was later expanded and extended and released as BIOPLUME II.

BIOPLUME II model included two expressions for simulation of biodegradation: (1) first-order decay of the contaminants, and (2) aerobic decay based on the dissolved oxygen concentrations present in the groundwater. BIOPLUME II relied on the same concept used in BIOPLUME I, which showed that when biodegradation occurs rapidly relative to groundwater velocities, the process can be assumed to occur instantaneously. In other words, the rate of reaction can be neglected, and the biodegradation of contaminants using oxygen as an electron acceptor is based solely on the stoichiometry of the chemical reaction.[35,37]

BIOPLUME II incorporated three different sources of oxygen: (1) background levels of oxygen prior to contamination, (2) oxygen supply from external sources (H_2O_2 or air injection), and (3) dissolved oxygen replenished by the moving groundwater. Despite its popularity, BIOPLUME II has two main limitations: (1) it does not account for slowly degrading compounds, and (2) it does not allow for simulating anaerobic processes.

BIOPLUME III, in contrast, simulates the transport and fate of six components in groundwater: contaminant, DO, NO_3^-, Fe, SO_4^{2-}, and CO_2. The biodegradation model assumed in BIOPLUME III is the sequential utilization of electron acceptors.[38]

$$O_2 \rightarrow NO_3^- \rightarrow Fe^{3+},\ SO_4^{2-} \rightarrow CO_2$$

For each of the electron acceptors, a number of biodegradation kinetic expressions such as first-order decay and instantaneous or Monod kinetics can be selected. Except for the aerobic and denitrification pathways, instantaneous reaction kinetics should be avoided.

5.8 PRIMARY KNOWLEDGE GAPS

Knowledge gaps include both those items that are not understood well and the myriad of information known by practitioners that has not been disseminated to the general audience.[39] As the published reports are getting more and more detailed, increasingly specific questions are being asked.

- Identification of the cause and effect of some unexpected results obtained in some field demonstration studies.
- Better scale-up of laboratory results to field performance.
- Evaluation of suitable habitats, nutritional requirements, lag times, and degradation rates (in the field) for various contaminants.
- Metabolic pathways of many contaminants of concern which are still unknown.
- Optimization of environmental conditions, and stimulation of favorable growth conditions under site-specific variations.
- Effect of NAPLs.
- *In situ* methods for monitoring process efficiency.
- Mass balance of electron donors and acceptors within a given system.
- Impact on aquifer permeability due to enhanced bioremediation.
- Bioavailability of higher molecular weight hydrocarbons. Enhancing bioremediation in low permeable environments.

REFERENCES

1. Alexander, M., *Biodegradation and Bioremediation*, Academic Press, New York, 1994.
2. Cookson, J. T., *Bioremediation Engineering: Design and Application*, McGraw-Hill, New York, 1994.
3. Bitton, L. N., Microbial degradation of aliphatic hydrocarbons, in *Microbial Degradation of Organic Compounds*, Gibson, D. T., Ed., Marcel Dekker, New York, 1984.
4. Wackeff, L. D., Brusseau, G. A., Householder, S. R., and Hansen, R. S., Survey of microbial oxygynases: trichloroethylene degradation by propane-oxidizing bacteria, *Appl. Environ. Microbiol.*, 55, 2960, 1989.
5. Doelman, P., Microbiology of soil and sediments, in *Biogeodynamics of Pollutants in Soils and Sediments*, Salomons, W. and Stigliani, W. M., Eds., Springer, Berlin, 1995.
6. Salomons, W., Long term strategies for handling contaminated sites and large scale areas, in *Biogeodynamics of Pollutants in Soils and Sediments*, Salomons, W. and Stigliani, W. M., Eds., Springer, Berlin, 1995.
7. Reinhard, M., In situ bioremediation technologies for petroleum derived hydrocarbons based on alternate electron acceptors (other than molecular oxygen), in *Handbook of Bioremediation*, Norris, R. D., et al., Lewis Publishers, Boca Raton, FL, 1994.
8. Grbić-Galić, D., Microbial Degradation of Homocyclic and Heterocyclic Aromatic Hydrocarbons under Anaerobic Conditions, unpublished report, Department of Civil Engineering, Environmental Engineering and Science, Stanford University, Palo Alto, CA, 1990.
9. McKenna, E., Biodegradation of Polynuclear Aromatic Hydrocarbon Pollutants by Soil and Water Microorganisms, Final Report, Project No. A-073-ILL, University of Illinois, Water Resources Center, Urbana, IL, 1979.
10. Dhawale, S. W., Dhawale, S. S., and Dean-Ross, D., Degradation of phenanthrene by *Phanarochaete chrysosporium* occurs under ligninolytic as well as nonligninolytic conditions, *Appl. Environ. Microbiol.*, 53, 3000, 1992.
11. Semprini, L., Grbić-Galić, D., McCarty, P. L., and Roberts, P. V., Methodologies for Evaluating In Situ Bioremediation of Chlorinated Solvents, U.S. Environmental Protection Agency, EPA/600/R-92.042, 1992.
12. McCarty, P. L. and Semprini, L., Groundwater treatment for chlorinated solvents, in *Handbook of Bioremediation*, Norris, R. D., et al., Eds., Lewis Publishers, Boca Raton, FL, 1994.
13. Vogel, T. M., Criddle, C. S., and McCarty, P. L., Transformations of halogenated aliphatic compounds, *Environ. Sci. Technol.*, 21(8), 722, 1987.
14. Bouwer, E. J., Bioremediation of chlorinated solvents using alternate electron acceptors, in *Handbook of Bioremediation*, Norris, R. D., et al., Lewis Publishers, Boca Raton, FL, 1994.
15. Bouwer, E. J., Rithmann, B. E., and McCarty, P. L., Anaerobic degradation of halogenated 1- and 2-carbon organic compounds, *Environ. Sci. Technol.*, 15, 596, 1981.

16. Bouwer, E. J. and Wright, J. P., Transformations of trace halogenated aliphatics in anoxic biofilm columns, *J. Contam. Hydro.*, 2, 155, 1988.
17. Wilson, J. T. and Wilson, B. H., Biotransformation of trichloroethylene in soil, *Appl. Environ. Microbiol.*, 49(1), 242, 1988.
18. Tsien, H. C., Bousseau, A., Hanson, R. S., and Wackeff, L. P., Biodegradation of trichloroethylene by *Methylosinus trichosporium, Appl. Environ. Microbiol.*, 55, 3155, 1989.
19. Palumbo, A. V., Eng, W., Boerman, P. A., Strandberg, G. W., Donaldson, T. L., and Herber, S. E., Effects of diverse organic contaminants on trichloroethylene degradation by methanotrophic bacteria and methane-utilizing consortia, in *On Site Bioreclamation Processes for Xenobiotic and Hydrocarbon Treatment*, Hinchee, R. E. and Offenbuttel, R. F., Eds., Butterworth-Heinemann, Stoneham, MA, 1991.
20. Young, J. C., Anaerobic degradation of aromatic compounds, in *Microbial Degradation of Organic Compounds*, Gibson, D. T., Ed., Marcel Dekker, NY, 1984.
21. Zhang, X. and Wiegel, J., Sequential anaerobic degradation of 2,4-dichlorophenol in freshwater sediments, *Appl. Environ. Microbiol.*, 58, 2993, 1990.
22. Brunner, W., Sutherland, F. H., and Focht, D.-D., Enhanced biodegradation of polychlorinated biphenyls in soil by analog enrichment and bacterial innoculation, *J. Environ. Qual.*, 14, 324, 1985.
23. Rhee, G. Y., Sokul, R. C., Bush, B., and Bethoney, C. M., Long term study of the anaerobic dechlorination of Aroclor 1254 with and without biphenyl enrichment, *Environ. Sci. Technol.*, 27, 714, 1993.
24. Luey, J., Brouns, T. M., and Elliott, M. L., Biodegradation of Hazardous Waste Using White Pot Fungus: Project Planning and Concept Development Document, report prepared by Pacific Northwest Laboratory to U.S. Department of Energy, Richland, Washington, 1990.
25. Simkins, S. M. and Alexander, M., Models for mineralization kinetics with the variables of substrate concentration and population density, *Appl. Environ. Microbiol.*, 47, 1299, 1984.
26. Stanier, R. Y., Ingraham, J. L., Wheelis, M. L., and Painter, P. R., *The Microbial World*, 5th ed., Prentice-Hall, Englewood Cliffs, NJ, 1986.
27. Lee, M. D., Thomas, J. M., Border, R. C., Bedient, P. B., Ward, C. H., and Wilson, J. T., *CRC Critical Reviews in Environmental Control — Biorestoration of Aquifers Contaminated with Organic Compounds*, Report, Rice University, vol. 18, 1988.
28. Jamison, V. W., Raymond, R. L., and Hudson, J. O., Biodegradation of high octane gasoline in groundwater, *Devel. Indust. Microbiol.*, 16, 305, 1976.
29. Berry-Spark, K. and Barker, J. F., Nitrate remediation of gasoline contaminated groundwaters: results of a controlled field experiment, in *Proceedings of the NWWA/API Conference on Petroleum Hydrocarbons and Organic Chemicals in Groundwater: Prevention, Detection and Restoration*, Houston, TX, Nov. 1987.
30. Hutchins, S. R., Miller, D. E., Beck, F. P., Thomas, A., Williams, S. E., and Willis, G. D., Nitrate based bioremediation of JP-4 jet fuel: Pilot scale demonstration, in *Applied Bioremediation of Petroleum Hydrocarbons*, Hinchee, R. E., Kittel, J. A., and Reisinger, H. J., Eds., Battelle Press, Columbus, OH, 1995.
31. Michelsen, D. L. and Lofti, M., Oxygen microbubble injection for in situ bioremediation: Possible field scenario, in *Biological Processes: Innovative Waste Treatment Technology Series*, vol. 3., Freeman, H. M. and Sferra, P. R., Eds., Technomic Publishing, Lancaster, PA, 1993.
32. Saaty, R. P., Showalter, E. W., and Booth, S. R., Cost effectiveness of in situ bioremediation at Savannah River, in *Bioremediation of Chlorinated Solvents*, Hinchee, R. E., Leeson, A., and Semprini, L., Eds., Battelle Press, Columbus, OH, 1995.
33. Norris, R. D., personal communication, 1995.
34. Ochs, L. D., personal communication, 1995.
35. Wilson, B. H., Wilson, J. T., Kampbell, D. H., and Bledsoe, B. E., *Traverse City: Geochemistry and Intrinsic Bioremediation of BTEX Compounds*, U.S. Environmental Protection Agency, USEPA/540/R-94/515, 1994.
36. Weidemeier, T. H., Wilson, J. T., Miller, R. N., and Kampbell, D. H., United States Air Force guidelines for successfully supporting intrinsic remediation with an example from Hill Air Force Base, *National Water Well Association/American Petroleum Institute Outdoor Action Conference*, Las Vegas, NV, 1994.
37. Bedient, P. B. and Rifai, H. S., Modeling in situ bioremediation, in *In Situ Bioremediation: When Does it Work*, National Research Council, National Academy Press, Washington, DC, 1993.

38. Rifai, H. S., Newell, C. J., Miller, R. N., Tiffinder, S., and Rounsaville, M., Simulation of natural attenuation with multiple electron acceptors, in *Intrinsic Bioremediation*, Hinchee, R. E., Wilson, J. T., and Downey, D. C., Eds., Battelle Press, Columbus, OH, 1995.
39. Norris, R. D., In situ bioremediation of soils and groundwater contaminated with petroleum hydrocarbons, in *Handbook of Bioremediation*, Norris, R. D., et al., Eds., Lewis Publishers, Boca Raton, FL, 1994.

6

VACUUM-ENHANCED RECOVERY

6.1 INTRODUCTION

Vacuum-enhanced recovery (VER) is a technique of applying a high vacuum or negative pressure on a recovery well and the formation in order to enhance the liquid recovery of that well by increasing the net effective drawdown. VER also increases the mass removal of the volatile and semivolatile contaminants, by maximizing dewatering and facilitating volatilization from previously saturated sediments via the increased air movement. In addition, mass removal of aerobically biodegradable contaminants will be enhanced by the resulting increase of subsurface O_2 levels. VER is a cost-effective technology that has been successfully applied to

- enhance the overall recovery of contaminants, especially under low permeability conditions;
- remove both dissolved and free-phase (nonaqueous liquids (NAPL)) contamination present in groundwater; and
- dewater contaminated zones and then use the vacuum to move air through the dewatered zone to volatilize and/or biodegrade the residual contamination in soil.

Use of VER systems in environmental remediation projects involves a modification to the approach used in classic dewatering systems. For decades, the VER technique has been used as a standard approach for dewatering and stabilizing low permeability sediments, and increasing the rate of dewatering in more permeable sediments.[1] In the remediation industry, this technique is also known as *dual-phase extraction* due to its inherent ability to extract liquids and vapors at the same time.

The use of vacuum-enhanced recovery systems has significant advantages over conventional recovery systems in certain hydrogeologic settings. The advantages of vacuum-enhanced systems include

- increased capture zone,
- reduced number of recovery wells required to achieve the same remedial objectives,
- reduced time for remediation due to the accelerated rate of removal of both liquid and residual contaminants, and
- effective source removal at low permeability sites; in many instances the only other viable remedial option may be excavation.

6.2 PROCESS DESCRIPTION AND BASIC PRINCIPLES

Groundwater containment and/or liquid hydrocarbon (NAPL) recovery is often accomplished through the use of some form of pumping systems. By removing groundwater at a controlled rate, a gradient is created toward the extraction point. The area within which the groundwater or NAPL moves toward the extraction point is defined as the *capture zone*. The capture zone of a particular extraction point is limited by the natural hydrogeological properties of the site and the rate at which groundwater is extracted. The transmissivity of the formation and the existing natural gradient will both affect the capture zone. In general, the lower the transmissivity and steeper the natural gradient are, the smaller the capture zone will be of a particular extraction system. The capture zone of an extraction point can be increased by increasing the pumping rate, assuming all other parameters remain unchanged. A limiting factor is that drawdown resulting from withdrawal cannot exceed the total saturated thickness of a water table aquifer. The hydraulic gradient and the yield cannot be increased by increasing the pumping rate if drawdowns have reached their limiting value.[2,3]

The application of a vacuum to the extraction point provides a method to further enhance the hydraulic gradients. By definition, the hydraulic gradient between two points is the difference in hydraulic head divided by the distance along the flowpath. The flow rate through the aquifer varies directly with the hydraulic gradient. If drawdown is at a maximum, then the head difference cannot be increased by lowering the water level. However, the effective head difference and hence the hydraulic gradient can be increased by applying a vacuum (negative pressures) at the point of extraction. This results in a corresponding increase in the rate of groundwater extraction (yield). This is the fundamental principle behind vacuum-enhanced recovery systems (Figure 6.1).

For example, the drawdown in a pumping well operating without a vacuum will be equal to the difference between the static water level and dynamic water level in the pumping well. The effective drawdown in the same pumping well operating with a vacuum is equivalent to the difference in static and dynamic level plus the amount of vacuum that is applied (Figure 6.1).

The application of a vacuum to an extraction point has several benefits: it increases the gradient and thus increases the capture zone and the rate of recovery or formation dewatering. In areas of extremely limited saturated thickness, vacuum systems provide one of the very few alternatives for containment when cut-off walls or trenches are not feasible and/or cost-prohibitive. It also provides an alternative for cost-effective remediation in low permeability formations.

6.3 MASS REMOVAL MECHANISMS

Application of high vacuums on the extraction well in a low permeability formation can result in three major mass removal mechanisms:

1. The application of high subsurface vacuums, as indicated earlier, can increase groundwater recovery rates, thus enhancing the rate of removal of dissolved contaminants. Figure 6.2 provides representation of this phenomenon. In this figure, a recovery well is being pumped (without vacuum application) at a rate of Q_1, which is the maximum flow rate achievable from the well due to the limitations on the amount of drawdown. Application of vacuum on the well increases the net hydraulic head differential, allowing the groundwater flow rate to be increased to a higher rate (Q_2). At this pumping rate (with vacuum application), the physical level of drawdown observed near the well (in the zone of vacuum influence) is actually the same as that of conventional pumping (Q_1). However, the vacuum application creates a greater effective drawdown (shown in Figure 6.2 by the broken line) which will allow for increased pumping rate (Q_2).

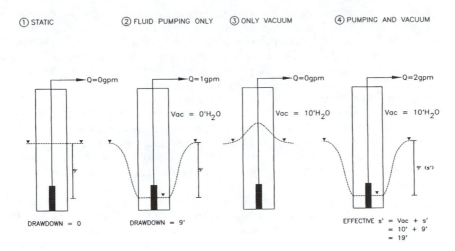

Figure 6.1   Effect of a vacuum on pumping level.

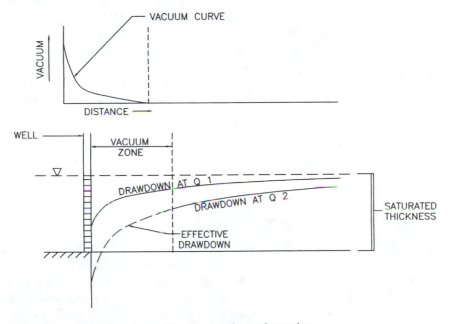

Figure 6.2   Schematic of drawdown with vacuum-enhanced pumping.

Although the increase in groundwater yield from VER systems allows for more efficient mass removal of the soluble groundwater constituents, other benefits also are achieved. Fine-grained sediments can trap water in soil pores under high capillary pressure. Through application of high vacuum, these capillary pressures can be overcome, forcing the release of retained water. Once these sediments are dewatered, they are then open to airflow created by the high-vacuum system, which will allow for conventional vapor extraction type removal of the adsorbed phase constituents that were previously trapped beneath the water table. This provides the greatest increase in mass removal rates.

2. In the same manner that application of high vacuum to recovery systems can increase groundwater yields in low permeability formations, they can also increase the recoverability of LNAPLs (those which float on the water table such as separate phase petroleum products). Recovery of LNAPLs is often the first step in remediating the aquifer, since they usually constitute the bulk of the contaminant mass and are a continuing source for soluble and absorbed phase constituents. Increased recovery of the free product by high-vacuum appli-

cations is accomplished in several ways. The increased hydraulic gradient and effective drawdown will allow for more free product to be drawn to the well and be recovered as a liquid. In addition, if the product contains a significant volatile fraction (e.g., gasoline), the airflow created by the vacuum-enhanced pumping system along the free product/vadose zone interface will cause increased partitioning from the free phase to the vapor phase, allowing for the product to be recovered in the vapor phase.

3. Application of high vacuums (24 in. of mercury in comparison to 1 to 5 in. for a conventional system) creates a large driving force for airflow in the vadose and the dewatered zones. This increase in pressure differential allows for increased airflow in the vadose zone, further allowing enhanced removal of contaminants from both the vapor and adsorbed phases. In most cases, the higher pressure differential also increases the air velocity at a distance away from the extraction well, which in turn enhances the contaminant mass removal rates.

   The ability of a high-vacuum system to create airflow in low permeability formations also makes it a valuable and viable technology for remediation by enhancing the biodegradation of aerobically biodegradable constituents. High-vacuum bioventing can be an important complementary process to the physical removal mechanisms in achieving remedial objectives.

The application of high vacuum offers additional benefits for NAPL recovery by overcoming the capillary displacement pressure of water against the NAPL and by increasing the relative permeability of the NAPL (compared to water).

When VER is used to recover free product without creating excessive drawdown in a well, the extraction rate can be controlled to induce drawdown equal to upwelling due to vacuum application. This method will allow the water level to be maintained at static levels, which reduces smearing of the free product. In contrast, conventional free-product recovery methods draw down the water and free-product level in the extraction well. Gravity is the only driving force that causes the free product to migrate downgradient into the recovery well. Due to the drawdown, free product can get trapped in void spaces of the soil matrix in the cone of depression. Due to this smearing the amount of recoverable free product may be slightly reduced.

6.4   APPLICABILITY OF THE TECHNOLOGY

The use of high-vacuum systems can be very beneficial to the overall remedial program at a particular site provided the system is applied in the proper hydrogeologic setting. These systems are applicable only within a limited range of conditions. If they are applied outside of this range they will be ineffective in remediating the problem or they will not be cost effective. Based on the data collected from numerous applications of VER systems under different hydrogeologic conditions, some basic guidelines have been developed.[2] VER systems are normally considered as a remedial option for

- low transmissivity formations—normally less than 500 gpd/ft (gpd: gallons per day)
- low hydraulic conductivities—normally in the range of 10^{-3} and 10^{-5} cm/s; application of the systems at sites with hydraulic conductivities less than 10^{-6} cm/s may be possible if some secondary permeability exists
- perched NAPL or groundwater layers
- total fluids recovery in low permeability formations
- formations consisting of interbedded sand and clay stringers
- formations with limited saturated thickness
- low permeability fractured systems

6.5   PILOT TEST PROCEDURES

A pilot test should be conducted at the site to determine site-specific engineering parameters. The pilot test will be utilized to determine the feasibility of the VER technology and obtain engineering design parameters such as radius of influence, groundwater pumping rate and airflow rate, and applied vacuum. Furthermore, contaminant levels in the extracted air and water can be measured. The collection and analysis of site-specific data should be utilized for accurate evaluation and final design of the remediation system.

Typically, a pilot test plan will include the following:

- installation of test and monitoring wells
- equipment needs
- test method and monitoring requirements
- mass removal estimation

6.5.1   Test and Monitoring Wells

A properly designed network of extraction and monitoring wells is critical for a successful pilot test. Existing wells should be carefully evaluated for use in the pilot test. The test-recovery well should be installed in the vicinity of the impacted area of groundwater and soils. The test-recovery well should be screened both in the saturated and unsaturated zone. Depending upon site hydrogeologic conditions, it is recommended that at least four monitoring wells be installed at 10, 30, 50, and 100 ft away from the test-recovery well. The extraction wells should be constructed properly using a continuously wrapped screen if appropriate and a fine-grained sand pack.

6.5.2   Equipment Needs

The following is a general list of equipment that may be needed to conduct the pilot test. The list is provided for guidance and should be evaluated on a site-specific basis.

- Liquid ring or high vacuum blower system
- Generator
- Generator fuel supply
- Submersible pump (if required)
- Discharge hose from pump to tank
- Plastic groundwater collection tank (approximately 1000 to 5000 gal)
- Vapor-phase granular activated carbon (if required)
- Fittings and pipes to connect liquid ring pump to wells
- Air sampling equipment
- Water sampling equipment
- Airflow meter
- Water flow meter and totalizer
- Magnehelic vacuum gauges
- Drop tubes and well head attachments for wells
- Water level indicator/interface probe
- VER well head vacuum gauge ($\geq$24 in. Hg)
- Oxygen/carbon dioxide meter
- Explosimeter

6.5.3   Test Method and Monitoring

In general, the vacuum-enhanced recovery test should be performed after completion of a short-term, conventional, non-vacuum-enhanced pumping test, unless the formation is of such low transmissivity that any sustainable groundwater recovery is not possible. In this short-term pumping test, time and distance drawdown data can be collected to evaluate the hydrogeologic parameters for the formation. Non-vacuum-enhanced pumping is usually performed initially, followed by application of high vacuums for the balance of the test.

During the high vacuum test, a vacuum pressure is applied to the test-recovery well to evacuate the well and surrounding soils of liquids and air containing volatile organic compounds. The test well can be equipped with a drop tube that extends below the static liquid level to near the bottom of the casing. The well casing head is sealed to withstand the applied vacuum pressure, and the vacuum is applied to the drop tube to evacuate the well.

At least three to four monitoring wells are required to monitor the vacuum and groundwater drawdown. These monitoring wells are equipped with drop tubes that extend below the well static liquid level to near the bottom of the casing, and the well casing heads are sealed. Well liquid levels are measured in the drop tube (with adjustments for casing side vacuum pressure). A pressure gauge is provided on the casing to measure the induced soil vacuum pressure.

The preferred portable vacuum pump is a centrifugal, liquid ring type and is capable of producing and sustaining vacuum pressures from 0 to 24 in. of mercury (from 0 to 27 ft of water). The liquid ring pump is especially suited for this application because of its high vacuum pressure capability and because of the minimum risk for internal source of ignition while compressing potentially explosive mixtures of volatile organic compounds.

Prior to beginning the design test, the fluid levels in all of the on-site monitoring wells are measured, using an electronic water level indicator, or electronic product/water interface probe. The recovery well should be surveyed and tied into the elevation network for the existing monitoring wells. During the course of the design test, the following parameters are measured and recorded:

- test-well applied vacuum pressure using calibrated vacuum pressure gauge or mercury manometer every hour
- monitoring-well fluid levels using electronic fluid level indicators
- monitoring-well vacuum pressure using calibrated vacuum pressure gauge or well manometer every hour
- test-well liquid recovery rate, monitored continually with a totalizing-type flow meter
- test-well airflow rate and volatile organic compounds concentration, continuously measured and recorded using an orifice meter and an OVA/PID/explosimeter, respectively

The process flow diagram for the pilot-test plan is shown in Figure 6.3. Figure 6.4 and Figure 6.5 include sketches of recovery well head and monitoring well head details.

6.5.4   Estimation of Mass Removal

The vacuum-enhanced recovery technique removes contaminant mass by recovering NAPLs and contamination present in the dissolved and vapor phases. The NAPL volume and the organic concentrations present in extracted groundwater and vapors indicate the mass removed. The NAPL mass is generally measured by calculating the amount removed from the oil–water separator. The dissolved portion in groundwater is calculated by multiplying the total volume of groundwater pumped by the weighted average concentrations of dissolved

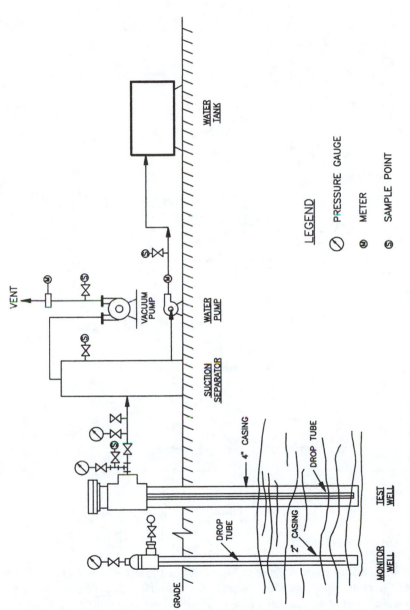

Figure 6.3 Pilot test process flow diagram.

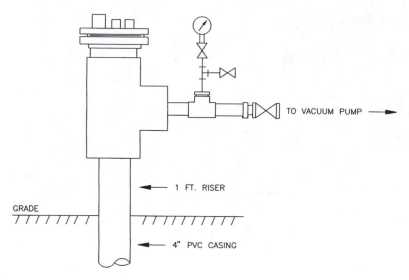

Figure 6.4 Recovery well head.

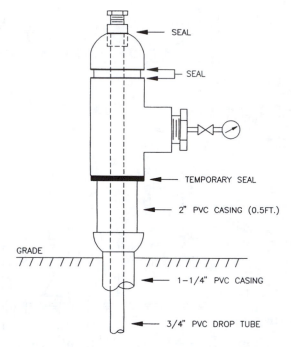

Figure 6.5 Monitoring well head.

constituents measured in water samples collected during the test. The mass in the vapor phase is calculated by multiplying the total volume of air discharged throughout the test by the weighed average concentration measure in air samples collected during the test.

In addition, soil vapor should be monitored for parameters such as carbon dioxide and oxygen to assess the potential biological activity during the pilot test.

6.6 SYSTEM DESIGN

After the pilot test is completed, there are a number of engineering design parameters that must be determined to design a full-scale system, as follows:

- groundwater influence
- well spacing based on groundwater influence
- design flow rate
- water treatment options (from groundwater concentrations)
- required vacuum pressure
- vapor extraction influence
- airflow rate
- off-gas treatment (from vapor concentrations)
- potential to enhance biological activity
- equipment specifications
- estimated cleanup time

In this chapter emphasis will be placed on well design, well spacing and groundwater influence, and pumping system. Information regarding airflow rate, vapor extraction influence, and selection of off-gas treatment technologies is provided in Chapter 3.

6.6.1   Well Design

In many ways, the well is constructed like a typical groundwater extraction well. Variations are as follows: Filter pack should be selected as if the well was a normal groundwater extraction well. Soil vapor extraction wells commonly use a very large grain size for the filter pack; however, under VER operation this may allow too many fines to be drawn into the well. As in any groundwater extraction well, the screen slot is sized to suit the filter pack. Both the screen and casing are typically of the same material; PVC is commonly used. Screen and casing diameters of 4 in. are common.

The use of high vacuums coupled with the low permeability formations can result in rapid well plugging and/or silting. The reduced pressures can result in more rapid precipitation of dissolved inorganic constituents on the well screen, gravel pack, or within the formation. The increased gradients can result in fines plugging the screens or silting of the wells in poorly designed wells. Thus, proper well design and wrapped screens are normally employed and periodic redevelopment of the wells may be necessary.

6.6.1.1   Drop Tube

The diameter of the drop tube should be selected to provide a vertical air velocity of 3000 standard feet per minute (sfm) for sufficient lift. The value of 3000 sfm for the uphole velocity is empirically based; air drillers have generally found that 3000 sfm is usually necessary to efficiently remove water and cuttings during air drilling, which is quite similar to lifting an air–water mix up a drop pipe. The base of the draw pipe is commonly cut at an angle to prevent cutting off flow if the draw pipe is inadvertently placed flush to the base of the well.

6.6.1.2   Valves

All systems should have a throttle valve on each well for equalizing the airflow rate. Some wells are likely to be more or less efficient to airflow than others, thus the valve is necessary for equalization. Some systems are designed with a vacuum release valve that allows atmospheric air into the well to reduce vacuum within the well.

6.6.2   Well Spacing and Groundwater Influence

This section contains recommended modifications to the way we calculate capture zones for low permeability formations.[4,5] First of all, it would be constructive to look at an example of an "over-predicted" capture zone in a low permeability formation. Let us examine a hypothetical 20-ft-thick silt formation having a hydraulic conductivity of 2.1 gpd/ft^2 (10^{-4} cm/s), a storage coefficient of 0.05, and a hydraulic gradient of 0.01 ft/ft. Assume a fully penetrating, 100% efficient extraction well operates for 30 consecutive days, resulting in a drawdown of 10 ft. Assume further that the borehole radius is 0.5 ft (12 in. borehole).

The discharge rate of the well may be calculated using the Cooper–Jacob equation, but the observed drawdown must be corrected for dewatering first. The dewatering correction is as follows:

$$s_t = s_a - \frac{s_a^2}{2b} \tag{6.1}$$

where s_t = theoretical drawdown corrected for dewatering
         s_a = actual drawdown
         b = aquifer thickness.

With an observed drawdown of 10 ft,

$$s_t = 10 - \frac{10^2}{2 \cdot 20}$$

$$= 7.5 \text{ ft}.$$

The Cooper–Jacob equation permits calculating discharge as follows:

$$Q = \frac{s_t T}{264 \log\left(\dfrac{0.3Tt}{r^2 S}\right)} \tag{6.2}$$

where Q = discharge, in gpm
         s_t = drawdown corrected for dewatering, in ft (7.5)
         T = transmissivity, in gpd/ft (42 [$T = Kb$])
         t = pumping time, in days (30)
         r = borehole radius, in ft (0.5)
         S = storage coefficient (0.05).

The resulting discharge rate is 0.266 gpm (383 gpd).

The standard capture zone equations calculate the distance to the stagnation point (x_0), the capture width at the well (w_0), and the upgradient capture width (w) as follows (assuming consistent units):

$$x_0 = \frac{Q}{2\pi TI} \tag{6.3}$$

$$w_0 = \frac{Q}{2TI} \tag{6.4}$$

$$w = \frac{Q}{TI} \tag{6.5}$$

where I = gradient, ft/ft. For the example presented here, a flow rate of 383 gpd (0.266 gpm) gives the following results:

$$x_0 = 145 \text{ ft}$$

$$w_0 = 456 \text{ ft}$$

$$w = 912 \text{ ft}.$$

Using vacuum-enhanced recovery techniques, it might be possible to double or triple the yield of this well. Assuming we could double the discharge rate to 766 gpd (0.532 gpm), the capture zone dimensions compute to the following:

$$x_0 = 290 \text{ ft}$$

$$w_0 = 912 \text{ ft}$$

$$w = 1824 \text{ ft}.$$

The key question then is whether design engineers can really rely on a 0.5 gpm well to provide over one third of a mile of capture width? In practice, intermittent recharge events probably "swamp out" the cone of depression periodically, because an ordinary recharge event can overwhelm the small discharge rates generally associated with tight formations. With the cone of depression being flooded out periodically, its lateral extent may be limited and the well may not be able to influence gradients at the great distances predicted by conventional capture theory.

One could argue that the hydraulic gradient is an expression of site recharge and, because it is incorporated into the capture zone equations, the equations still should be trustworthy. This argument breaks down, however, since it is possible to hypothesize heterogeneous situations in which the operation of extraction wells induces greater recharge than that which would occur with no wells operating.

All of this leaves us without a rigorous procedure for establishing a capture system design in tight sediments (unless we want to rely on a single 0.5 gpm well to capture over one third of a mile). To fill the breach, we must propose an algorithm that should lead to a conservative design.

The essence of the procedure is to base capture on the configuration of the cone of depression after a fixed, limited pumping time—say, 30, 60, or 90 days. In other words, after the fixed time has passed, it is assumed that there is no further growth in the cone of depression. After selecting the arbitrary pumping time (perhaps 30 days for humid climates and 90 days for desert climates), drawdowns and gradients are calculated based upon the Theis equation (not the log equation), and capture analysis is based upon the resulting drawdown configuration. A description of this procedure follows, along with several required graphs and a few examples.

After the arbitrary pumping time has been chosen, the first step is to compute the so-called log-extrapolated radius of influence of the well, R. This is done primarily for mathematical convenience, because R consolidates several other parameters. (R is commonly called the radius of influence because it is the distance to zero drawdown on an extrapolated semilog distance-drawdown graph. However, it is not a true radius of influence because the Theis equation predicts some additional drawdown beyond this point.) R may be computed from the following equation:

$$R = \sqrt{\frac{0.3Tt}{S}} \tag{6.6}$$

where R = log-extrapolated radius of influence, in feet
       T = transmissivity, in gpd/ft
       t = pumping time, in days
       S = storage coefficient.

If consistent units are used, i.e., transmissivity in ft²/day, the equation is as follows:

$$R = \sqrt{\frac{2.246Tt}{S}} \; . \tag{6.7}$$

The log-extrapolated radius of influence is significant in that we will want to express capture zone dimensions in terms of R.

6.6.2.1   Distance to Stagnation Point

Differentiating and manipulating the Theis equation gives rise to the following expression relating Q, discharge, to x_0, distance to downgradient stagnation point.

$$Q = 2\pi TIx_0 \, e^{\frac{x_0^2}{1.781R^2}} \; . \tag{6.8}$$

This is the same as the conventional capture-zone equation except for the exponential term. When the exponent is small (x_0 is small in relation to R), the exponential term is close to 1 and the standard capture zone equations work just fine. As x_0 increases, however, to a significant fraction of R or beyond, the exponential term is substantially greater than 1 and the extraction well must produce more water than would be determined by conventional analysis. In essence, the exponential term represents a multiplier that must be applied to the Q computed from the standard equations.

Figure 6.6 shows the magnitude of the exponential term as a function of the ratio x_0/R. For example, if x_0 is half the radius of influence, Q will be 1.15 times the conventional calculation. If x_0 equals R, the discharge must be 1.75 times that calculated from conventional analysis. And if x_0 equals $2R$, nearly a tenfold increase in discharge rate is required over conventional theory.

6.6.2.2   Width of Capture Zone at Extraction Well

Conventional theory predicts that the width of the capture zone at the extraction well will be times the distance to the stagnation point. That is,

$$w_0 = \pi \, x_0 \; . \tag{6.9}$$

For tight formations (time-limited cones of depression), however, the ratio w_0/x_0 decreases as discharge rate and capture zone size increase. Figure 6.7 shows how this ratio decreases with increasing x_0/R. Figure 6.7 was developed empirically by using analytic element modeling and particle tracking to assess the relationship between capture width and distance to stagnation point.

Figure 6.6 was based upon an exact equation, whereas Figure 6.7 was determined empirically. Combining these two graphs produces Figure 6.8, which shows the magnitude of the discharge increase required for capture compared to conventional theory, all as a function of desired capture width. For example, reading from the graph, when the capture width is twice the radius of influence, the discharge must be 1.4 times the value calculated

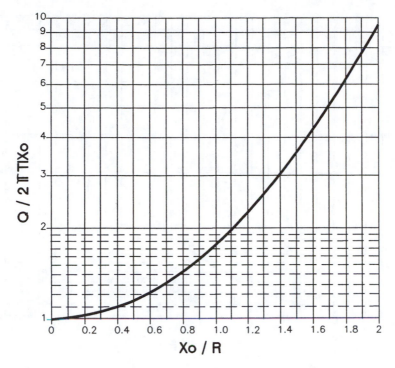

Figure 6.6 Ratio of required Q to conventional Q as as a function of distance to stagnation point.

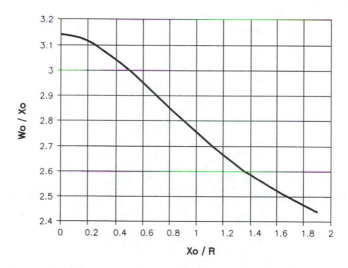

Figure 6.7 Decline in w_0/x_0 with increasing x_0 (R = log-extrapolated radius of influence).

using the conventional equations. For a capture width equal to the radius of influence, the graph shows that the required discharge is 1.09 times the conventional discharge.

Figures 6.6 and 6.8 show that in tight formations we pay a penalty in terms of discharge rate, and that the penalty increases as the capture zone dimensions approach and exceed the log-extrapolated radius of influence. If the remediation design includes a large number of wells, each with a small capture zone, the total required discharge is minimized. As the number of wells is reduced and the individual capture zone size is increased, the required discharge rate increases by the multiplication factor shown on the vertical scale of the figures. An optimum remediation design must weigh the costs of drilling more wells versus pumping more water.

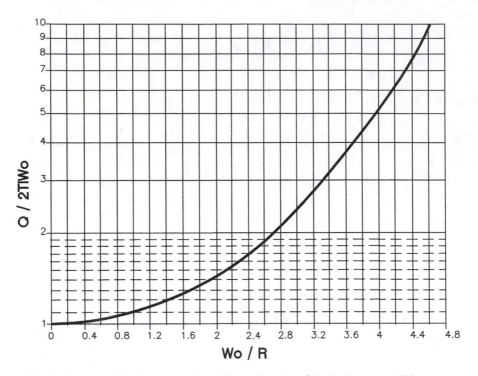

Figure 6.8 Ratio of required Q to conventional Q as a function of desired capture width.

6.6.2.3   Example Calculations

An example will illustrate these procedures. Calculations will be made based upon capture width, w_0, because this is the parameter of greater importance.

Using the 20-ft silt described above and a 30-day pumping time, the radius of influence is calculated as follows:

$$R = \sqrt{\frac{0.3 \cdot 42 \cdot 30}{0.05}}$$

$$= 87 \text{ ft}.$$

The required flow rate for a particular capture width may be computed as follows:

1. Select the desired capture width. For example, a w_0 of 200 ft.
2. Compute discharge required using conventional equation.

$$Q = 2TIw_0$$

$$= 2 \cdot 42 \cdot 0.01 \cdot 200$$

$$= 168 \text{ gpd } (0.12 \text{ gpm}).$$

3. Calculate w_0/R and use Figure 6.8 to determine the multiplier to apply to the conventional rate. The ratio w_0/R equals 2.3 and, reading from the graph, the multiplier is 1.6. Thus, the required Q is

$$Q = 168 \cdot 1.6$$

$$= 269 \text{ gpd } (0.187 \text{ gpm}).$$

Repeating this process for another assumed value of w_0, for example, 300 ft, yields the following:

$$\text{conventional equation}: \quad Q = 2.42 \cdot 0.01 \cdot 300$$

$$= 252 \text{ gpd } (0.175 \text{ gpm})$$

$$w_0/R = 3.45$$

$$\text{Multiplier} = 3.2 \text{ (from Figure 6.8)}$$

$$Q \text{ actually required} = 252 \cdot 3.2$$

$$= 806 \text{ gpd } (0.56 \text{ gpm}).$$

These same calculations were performed for capture widths ranging from 50 ft to 400 ft in 50-ft increments. Table 6.1 shows the results, including the discharge associated with the conventional approach and the actual required discharge associated with this procedure. Recall that earlier analysis showed a required discharge rate of 0.266 to achieve a w_0 of 456 ft. According to Table 6.1, this discharge rate would result in a capture width between 200 and 250 ft. Similarly, earlier calculations showed that a discharge rate a little over 0.5 gpm resulted in more than 900 ft of capture, whereas Table 6.1 predicts less than 300 ft of capture.

Table 6.1 Comparison of Conventionally Calculated Discharge Rates and Actual Required Rates for 30 Days of Uninterrupted Pumping

w_0 (ft)	w_0/R	Q from conventional equation ft³/day	gpm	Multiplier from Figure 6.8	Q actually required ft³/day	gpm
50	0.58	42	0.029	1.02	43	0.030
100	1.15	84	0.058	1.1	92	0.064
150	1.73	126	0.088	1.29	163	0.113
200	2.30	168	0.117	1.6	269	0.187
250	2.88	210	0.146	2.2	462	0.321
300	3.45	252	0.175	3.2	806	0.560
350	4.03	294	0.204	5.1	1499	1.041
400	4.60	336	0.233	9.5	3192	2.217

The calculations described here can be made more conservative by choosing a shorter pumping time. For example, if a pumping time of 10 days is used in the above example, the radius of influence, R, is 50 ft. Recalculating the capture zone information, then, produces the results shown in Table 6.2. Note that in this case a discharge rate of over 0.5 gpm would produce less than 200 ft of capture width and a discharge rate about 0.25 gpm would produce a little over 150 ft of capture width.

This methodology should be applied any time the expected capture width, w_0, for an individual well exceeds the log-extrapolated radius of influence, R.

6.6.3 Pumping System Design

There are two basic options for vacuum-enhanced pumping. One method uses a single pump to remove the fluid and to apply a vacuum to the well(s) and formation (Figure 6.9). The second method uses a submersible pump to remove the produced fluids and a separate vacuum pump to apply a suction to the well(s) and formation (Figure 6.10).

Table 6.2 Comparison of Conventionally Calculated Discharge Rates and Actual Required Rates for 10 Days of Uninterrupted Pumping

w_0 (ft)	w_0/R	Q from conventional equation		Multiplier from Figure 6.8	Q actually required	
		ft³/day	gpm		ft³/day	gpm
50	1.00	42	0.029	1.09	46	0.032
75	1.49	63	0.044	1.2	76	0.053
100	1.99	84	0.058	1.4	118	0.082
125	2.49	105	0.073	1.75	184	0.128
150	2.99	126	0.088	2.35	296	0.206
175	3.49	147	0.102	3.35	492	0.342
200	3.98	168	0.117	5	840	0.583
225	4.48	189	0.131	8.2	1550	1.076

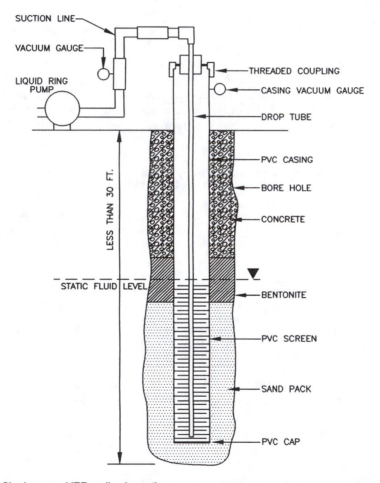

Figure 6.9 Single pump VER well schematic.

At pumping depths of less than 25 ft, a single pump can be used to recover both the air and fluids. This type of system may employ a simple surface-mounted diaphragm pump or a much more complex liquid ring system. In low permeability conditions, surface-mounted diaphragm pumps are capable of evacuating the well and producing a vacuum in the well and formation. Yields of liquid hydrocarbons and the influence of individual wells have been significantly increased by applying this suction lift technique. However, the primary goal of such a system must be liquid recovery or gradient control, and the amount of air entering

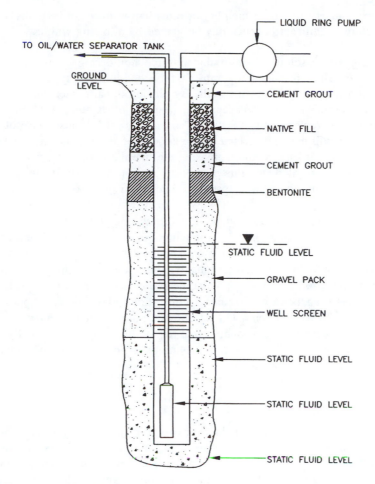

LIQUID RING PUMP

TO OIL/WATER SEPARATOR TANK

GROUND
LEVEL

CEMENT GROUT

NATIVE FILL

CEMENT GROUT

BENTONITE

STATIC FLUID LEVEL

GRAVEL PACK

WELL SCREEN

STATIC FLUID LEVEL

STATIC FLUID LEVEL

STATIC FLUID LEVEL

Figure 6.10 Two pump VER schematic.

the system must be kept at a minimum. If a single pump is to be used, the type of pump selected must match the remedial goals and the hydrogeologic conditions present.

In situations where the pumping levels exceed about 25 ft, the use of the two pump type system is required. In these cases, suction lift restrictions limit the use of a single surface-mounted pump. The vacuum pump is simply used to remove vapors and induce a vacuum on the well and formation, and a downhole submersible pump is used to extract liquids.

The separation of vapor and liquid extraction can allow for the utilization of one liquid ring pump for multiple extraction wells (no risk of losing groundwater suction lift from some wells). If it is used for dual extraction (liquid and vapors) from multiple wells, "balancing" the wells and the drop tube depth becomes critical. Balancing of multiple dual extraction wells can be very difficult if the subsurface is even slightly heterogeneous. Once some wells become dry, and if the liquid ring pump induces all of its airflow from these dry wells, the remaining wells will lose suction and the system will be only partially operable.

In order to alleviate the well balancing problems, three alternatives may be considered. First, one may cycle the vacuum pump between the various wells using solenoid valves operated by a timer control. Although this is a low-cost solution, it is imperfect, since the wells will only operate part of the time and will not assist in significantly reducing remedial time-frames. This solution may be acceptable when mass removal rates are diffusion-limited and constant liquid recovery and airflow are not critical to maximize remediation. In most cases, the early stages (first to 12 months) of remediation may not be entirely diffusion-limited; therefore, significant benefits may be gained by allowing withdrawal of constant flow.

A second possible solution is utilizing multiple liquid ring pumps and a common knock-out tank for all of the vacuum pumps. In this configuration, typically each pump is utilized for extraction from two to six wells. By placing the air/water knock-out after the pumps (the vacuum of each pump is separated from the other pumps), all the pumps can be connected to it, thereby reducing the capital costs. This solution also may not be perfect, because the pumps will not be protected by the knock-out system, thus necessitating more frequent pump seal maintenance and increasing the likelihood of pump screen filter clogging. In addition, some liquid ring pumps will not operate properly if the recovery wells produce water exceeding the pump capacity; a knock-out prior to the pump will prevent this problem.

A third possible solution is utilizing multiple liquid ring pumps and multiple knock-out systems (one knock-out per pump). In this configuration, each pump is utilized for extraction from two to six wells. By placing the air/water knock-out tank before the pump, the pumps will be protected from fouling and water inundation, thus minimizing pump seal replacements. Although this is likely to be the most efficient method of operation, it involves higher costs.

In selecting between the various methods of operation and seal water supplies, the design engineer is often faced with a dilemma of selection of a high capital expenditure option with high reliability or lower capital cost options that require higher maintenance. System downtime that will lengthen remediation time frames needs to be considered when choosing the most appropriate system configuration.

6.6.3.1   Liquid Ring Pump

Liquid ring pumps can be used for single or two pump recovery systems. Typically these pumps are used to obtain a vacuum of 24 to 25 in. of Hg. As with most standard vacuum pumps, a liquid ring pump consists of several blades attached to a drive shaft that is enclosed in a casing. The shaft is attached to the motor or another driver and the casing contains both a suction and a discharge port. A cross section of the liquid ring pump is shown in Figure 6.11. As the impeller begins to form a seal, liquid is forced to the outside of the casing via centrifugal forces. The seal liquid forms a "liquid ring" along the casing interior. As the impeller moves around the casing, the space between the cell and the seal liquid increases, thereby creating a vacuum. As each impeller cell passes, a vacuum is created at the suction port of the pump. As the impeller cell continues to rotate, the space then begins to be reduced, thereby causing pressure to build in the cell. The pressure created then forces the fluids (extracted vapors/liquids) out of the discharge port. The typical seal fluids used include water or oil. If water is used as the seal liquid, the pump should be provided with complete fresh water makeup in which 100% of the fluid is discharged with each pass through the pump. If water treatment becomes difficult due to increased cost, the seal fluid can be recirculated using a transfer pump and a heat exchanger.

Liquid ring pumps have the option of using oil as a pump seal fluid. Using oil as the pump seal eliminates the need for continuous flow of water as the seal fluid and thus decreases the cost of operation and maintenance cost of the system. The oil-cooled system requires the additional temperature control valve as a standard to maintain the temperature between 150 to 170°F.

6.6.3.1.1   Sizing of Liquid Ring Pump

The information needed to accurately size a liquid ring vacuum pump includes

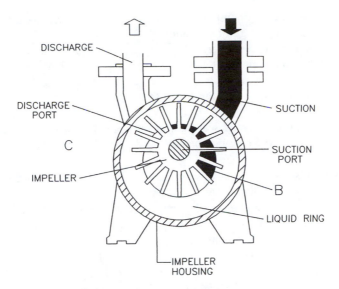

Figure 6.11 Cross section of liquid ring pump.

- inlet vacuum, usually expressed in inches of mercury
- inlet temperature
- mass flow rate, usually expressed in pounds per hour
- vapor pressure data for each fluid component
- seal fluid data, if other than water: specific gravity, specific heat, viscosity, thermal conductivity, molecular weight, and vapor pressure data
- temperature of the seal fluid or cooling water
- discharge pressure, usually expressed in pounds per square inch (PSIG)

Each of these factors influences the sizing of the liquid ring vacuum pump.

6.6.3.1.2   Cavitation

Cavitation can be detected by a rumbling noise that sounds like marbles rolling around inside the pump. This sound should not be confused with that occurring as a result of "water hammer." If the seal fluid vapor pressure is near the operating inlet pressure of the vacuum pump, then the sound heard is most likely from cavitation.

Liquid ring vacuum pump manufacturers are able to determine if cavitation can result from a particular application. Cavitation occurs as a result of rapid boiling of the seal liquid. During boiling, bubbles form in the liquid and seek to escape. As the bubbles rise in the liquid, they are subjected to higher pressure zones and can collapse. The void of space that is left after the bubbles collapse is instantly filled with liquid. This phenomenon leads to extensive erosion or pitting of the pump internals. If the liquid ring pump is disassembled and the impellers and port plates are pitted, cavitation is evidently occurring.

Possible ways to eliminate cavitation include the following:

- Use of a colder seal fluid. This will lower the vapor pressure of the seal fluid and keep it from flashing.
- Use of a seal fluid with a lower vapor pressure. This will prevent it from flashing and causing cavitation. The seal fluid should be compatible with the process gas mixture and the materials of construction. Certain accessories may need to be redesigned to operate with different seal fluid.

- Increasing the inlet operating pressure (if this pressure will be tolerated by the process) beyond the range of cavitation.
- Checking the loading and operating pressure. If the load is less than design, the pump may be operating at a lower absolute pressure, which may lead to boiling of the seal liquid and cause cavitation. A vacuum relief valve or air bleed valve can be used to introduce additional load so the pump will be able to operate closer to the design pressure.
- Increasing the throughput of seal fluid, if possible. This will decrease the temperature rise of the liquid ring pump and lower the vapor pressure of the seal fluid.

The discharge pressure of the pump is needed to determine any reduction of pump capacity or any increase in operating horsepower. Most liquid ring vacuum pumps are designed to operate from a vacuum suction to an atmospheric discharge pressure. If the discharge pressure is increased, then more energy or brake horsepower is required to operate at the increased compression range. By changing the discharge pressure, the efficiency of the pump is changed and the capacity of the pump may be reduced to compensate for this greater compression range.

Visual inspection should be performed once the liquid ring pump is up and running. The vacuum level, seal fluid temperature, and seal fluid pressure should be checked and compared with the design conditions. The bearing housing should be warm to the touch. A hot bearing housing can indicate a bad or worn bearing, or misalignment.

A liquid ring vacuum pump is a very versatile machine. It can handle "wet" loads and has no metal-to-metal contact. It also acts like a direct contact condenser; it can absorb the heat generated by the compression, friction, and condensation of the incoming gas; and it will absorb and wash out any contaminants entrained in the gas. In spite of this versatility, however, sizing and selection of the most economical system still requires full information concerning not only what fluids it will handle, but how it will be operated.

6.6.3.2   Jet Pumps (Eductor-Type Pumps)

In operation, eductor-type systems generally use a circulation tank and centrifugal pump to force pressurized water through the pressure nozzle of an eductor or group of eductors, producing a high-velocity jet.[6] This jet action creates a vacuum in the line, which causes the suction fluid (recovered groundwater and/or vapors) to flow up into the body of the eductor where it is entrained by the pressure liquid (water supply). Both liquids are thoroughly mixed in the throat of the eductor and are discharged against back pressure. The streamlined body with no pockets permits the pressure liquid to move straight through the eductor and reduces the possibility of solids collecting in and clogging the suction material. In addition, pressure drop in the suction chamber is held to a minimum. Figure 6.12 is a diagram of a typical eductor, which is usually manufactured of ductile steel, bronze, or stainless steel. The primary advantages of eductor-type systems compared to liquid-ring pump systems are the relatively low cost, along with easier operation and maintenance.[6,7] The primary disadvantage is the low airflow capability of eductors.

6.7   LIMITATIONS

The limitations of the technology include the following:

- applicable only to a limited range of hydrogeology settings
- normally have higher operation and maintenance costs, and hence rapid remediation must be achieved to make them cost effective

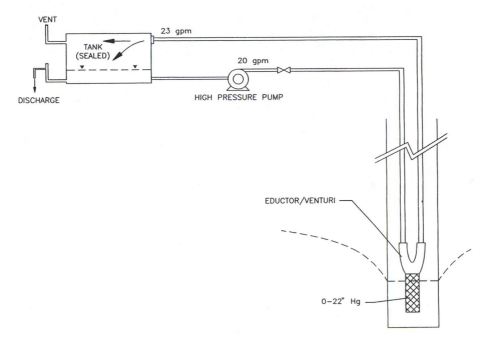

Figure 6.12 Schematic of jet pump.

- produce off-gases with high concentrations that will require treatment in many locations, and
- due to the high vapor concentrations produced by the systems in many instances, provisions for handling explosive vapors must be incorporated in the system design.

6.8 CASE STUDY

6.8.1 Background

In March 1989, three 4000-gal underground storage tanks (USTs) were removed from the former site of a maintenance garage located in the northeastern U.S. One of the USTs was used for gasoline storage and two were used for diesel fuel storage. Soil sampling conducted at the time of tank closure indicated that a discharge had occurred. Six monitoring wells, identified as MW-1 through MW-6 on Figure 6.13, were installed to evaluate ground-water quality.

The geology of the site consists of low permeable formation (fill, silt, and clay) and a depth to water of 20 ft. A second investigation performed in 1992 included the installation of eight additional monitoring wells (identified as MW-7 through MW-14 on Figure 6.13), soil borings, and piezometers (identified as P-1 through P-4 on Figure 6.13); collection of groundwater and soil samples; and performance of aquifer tests. As part of the investigation, several tests were performed for collection of data for treatment system design: a percolation test, aquifer pumping tests, and a vacuum-enhanced recovery (VER) pilot test.

Based on the pilot test results, it was determined that a VER system could remediate the source area and contain the migration of contaminated groundwater. A remedial system was designed after the pilot test was conducted. Construction of the system began in June 1993 after procurement of the necessary permits. The remediation system began operation in August 1993.

The treatment system consisted of both vapor- and liquid-phase treatment components. The groundwater was recovered using submersible pumps from four recovery wells (RW-1 through RW-4) under the influence of high-level vacuum, and discharged to an oil and water separator,

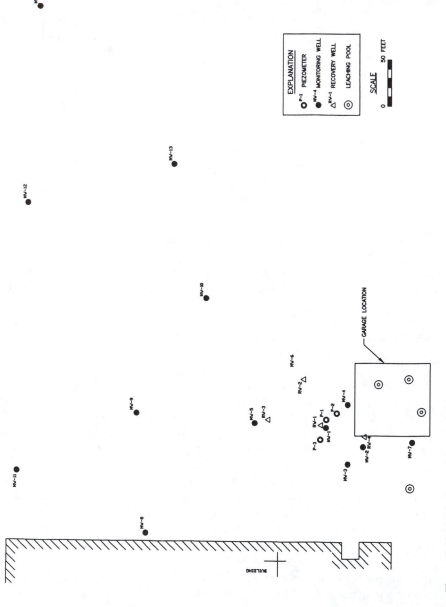

Figure 6.13 Monitoring well locations.

a diffused air stripper, a particulate filter, a 30-μm bag filter, and a liquid-phase granular activated carbon (GAC) unit for treatment. The vapor recovery system consisted of a liquid ring vacuum pump and cooling system, liquid knock-out tank and transfer pump, and a thermal oxidizer. Vapors were recovered from four recovery wells, identified as RW-1 through RW-4.

The recovery wells were 6 in. in diameter, constructed of stainless steel, and equipped with a submersible pump. The oil–water separator removed the recovered LNAPL by separating it from the groundwater before the groundwater was treated. The diffused air stripper was designed to treat dissolved VOCs present in the groundwater. The particulate filter and bag filter were used to remove iron and fine colloidal sediment entrained in the groundwater. Final polishing of groundwater was achieved by passing it through two 55-gal GAC drums situated in series. Components of both the groundwater and vapor treatment system were housed within an intrinsically safe trailer. The process flow diagram for the remediation system is shown on Figure 6.14.

The remediation system was designed in such a way that all vapors from both the vapor recovery stems and the air stripper were directed through the thermal oxidizer for treatment. The system also employed ultraviolet/infrared (UV/IR) and lower explosive limit (LEL) detectors for fire and explosion protection. An auto-dialer alarm system was provided to automatically inform the operator via telephone of a system failure.

Treated effluent was discharged to injection wells located upgradient of the recovery wells. The injection wells consisted of four leaching pools 10 ft in diameter and 8 ft deep.

6.8.2 Operating Parameters

The dual-phase vacuum enhanced remediation system was designed to extract and treat up to 5 gallons per minute (gpm) of groundwater and a total of 260 cubic feet per minute (cfm) of vapor (60 cfm from the vapor extraction wells and 200 cfm from the air diffuser). A liquid ring vacuum pump was utilized to apply a high vacuum to the wells (approximately 20 in. of mercury) to enhance groundwater and product recovery and to extract vapors from the impacted subsurface. From system start-up in August 1993 through May 1995, the remediation system treated approximately 800,000 gal of impacted groundwater.

6.8.3 Influent Quality

Influent groundwater concentrations of target VOCs at the start-up of the remediation system in August 1993 were 46 mg/l and included 9.6 mg/l of benzene, 16 mg/l of toluene, 1.3 mg/l of ethylbenzene, 11 mg/l of total xylenes, 2.2 mg/l of MTBE, and 5.9 mg/l of TBA. Total influent VOC concentrations were reduced from a total of 46 mg/l in August 1993 to 8.4 mg/l in May 1995.

Groundwater quality in the area of concern showed significant improvement after the start-up of the remediation system in August 1993. A comparison of the total VOC concentrations in groundwater samples collected from the monitoring wells on April 15, 1992 (prior to the installation of the remediation system) to the total VOC concentrations in groundwater samples collected from the monitoring wells on April 21, 1995 (after approximately 21 months of remediation system operation) indicated that concentrations of VOCs in groundwater were reduced and LNAPL was no longer present. The total VOC concentrations in groundwater on April 15, 1992 are shown on Figure 6.15.

Results of April 1995 sampling of the network wells indicate that concentrations were reduced to a range of nondetect to 10.1 mg/l. The areal extent of the total targeted VOC plume was considerably reduced. Based on the 1.0 mg/l isopleth (Figures 6.15 and 6.16), the plume was reduced in size by more than one fourth from April 1992 to April 1995.

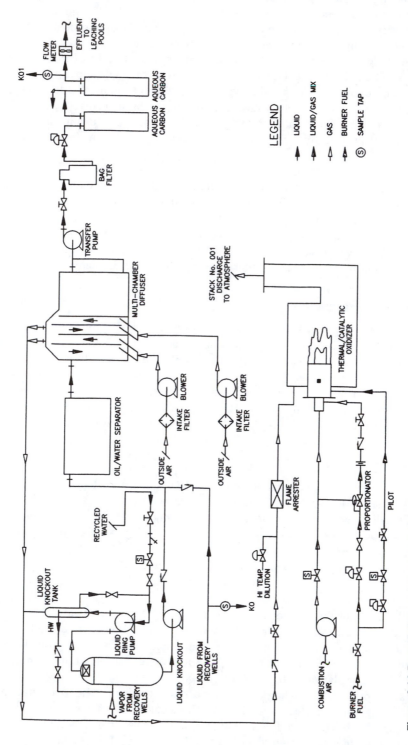

Figure 6.14 Remediation system process flow diagram.

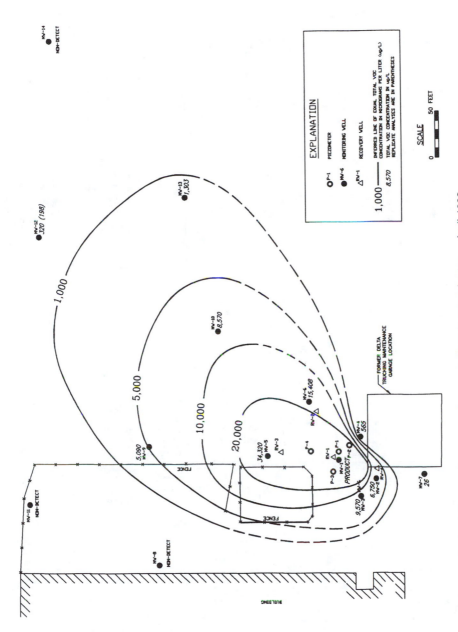

Figure 6.15 Total volatile organic compound concentrations in groundwater — April 1992.

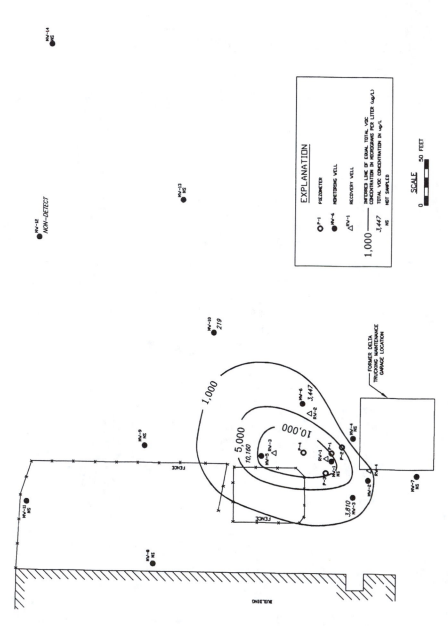

Figure 6.16 Total volatile organic compound concentrations in groundwater — April 1995.

6.8.4 Summary

- Over the 21 months of system operation from August 1993 to May 1995, approximately 800,000 gal of groundwater was extracted and treated and over 350 gal of LNAPL product were removed from the subsurface.
- Currently there is no LNAPL present at the site.
- Total VOC concentrations, which ranged from 3.2 to 34.3 mg/l in the monitoring network wells in April 1992, were reduced to a range of nondetect to 10 mg/l based on April 1995 sampling results.
- The areal extent of the total target VOC plume based on a concentration of 1 mg/l was considerably reduced. A reduction of more than one fourth in plume area from April 1992 to April 1995 was evident and showed that VER reduced LNAPL, vapor-, and dissolved-phase contamination.
- Based on data and the continuous decline in the LNAPL, vapor-, and dissolved-phase contamination at the site, it was proposed to shut down the remediation system in December 1995. This recommendation was approved by the regulatory agency based on the fact that natural bioremediation will remove the remaining constituents at the site with time.

REFERENCES

1. Blake, S. B. and Gates, M., Vacuum enhanced recovery: A case study, in *Proc. 1986 Conf. Petrol. Hydrocarbons Org. Chem. Groundwater: Prevention, Detection Restoration*, Las Vegas, NV, 1986.
2. Blake, S. B., Hockman, B., and Martin, M., Applications of vacuum dewatering techniques to hydrocarbon remediation, in *Proc. 1989 Conf. Petrol. Hydrocarbons Org. Chem. Groundwater: Prevention, Detection Restoration*, Las Vegas, NV, 1989.
3. Ayyaswami, A., Vacuum enhanced recovery: Theory and applications, in *Proc. 1994 Conf. Georgia Water Pollution Control Assoc.*, Atlanta, GA, 1994.
4. Schafer, D., Internal Memorandum on Capture Zones in Low-Permeability Formations, Geraghty & Miller, Inc., 1995.
5. Schafer, D., personal communications, 1995.
6. Hansen, M. A., Flavin, M. D., and Fam, S. A., Vapor extraction/vacuum enhanced groundwater recovery. A high-vacuum approach, presented at *Purdue Indust. Waste Conf.*, West Lafayette, IN, 1994.
7. Johnson, R. L., Bagby, W., Matthey, P., and Chien, C, T., Experimental examination of integrated soil vapor extraction techniques, in *Proc. Org. Chem. Groundwater: Prevention, Detection Restoration*, Las Vegas, NV, 1994.

7

IN SITU REACTIVE WALLS

7.1 INTRODUCTION

In situ reactive walls are an emerging technology that have been evaluated, developed, and implemented only within the last few years. This technology is gaining widespread attention due to the increasing recognition of the limitations of pump and treat systems, and the ability to implement various treatment processes that have historically only been used in above-ground systems in an *in situ* environment. This technology is also known in the remediation industry as "funnel and gate systems" or "treatment walls" and will be referred to as *in situ* reactive walls in this chapter.

The concept of *in situ* reactive walls involves the installation of impermeable barriers downgradient of the contaminated groundwater plume and hydraulic manipulation of impacted groundwater to be directed through porous reactive gates installed within the impermeable barrier. Treatment processes designed specifically to treat the target contaminants can be implemented in these reactive or treatment gates. Treated groundwater follows its natural course after exiting the treatment gates. The flow through the treatment gates is driven by natural groundwater gradients, and hence these systems are often referred to as passive treatment walls. If a groundwater plume is relatively narrow, a permeable reactive trench can be installed across the full width of the plume, and thus preclude the necessity for installation of impermeable barriers.

In situ reactive walls eliminate or at least minimize the need for mechanical systems, thereby reducing the long-term operation and maintenance costs that so often drive up the life cycle costs of many remediation projects. In addition, groundwater monitoring and system compliance issues can be streamlined for even greater cost savings.

Most of the developmental work on *in situ* reactive walls was performed at the Waterloo Center for Groundwater Research, University of Waterloo, Ontario, Canada.[1,2] Since this is an innovative and emerging technology, as of yet there are not many reported case studies in the literature.

7.2 DESCRIPTION OF THE PROCESS

Application of *in situ* reactive walls should be considered as an alternative to pump and treat systems. Reactive walls can be installed at the downgradient edge of the plume as a containment system and/or immediately downgradient of the source area to prevent further migration of elevated levels of contaminant mass. Physical, chemical, or microbial processes can be implemented at the porous reactive gates.

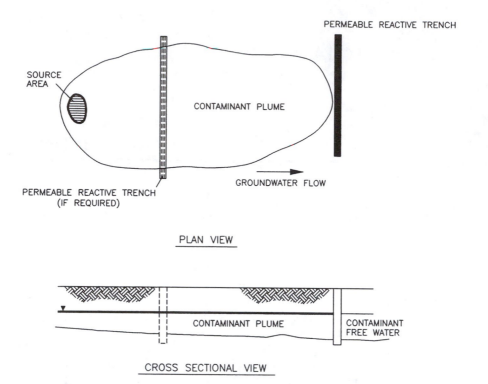

Figure 7.1 Plan and cross-sectional view of permeable reactive trench.

Several configurations of *in situ* reactive walls systems are feasible, and the applicability of this technology will depend on the geologic and hydrogeologic conditions, as well as contaminant distribution in the vertical and horizontal dimensions at a given site.

7.2.1 Permeable Reactive Trench

The simplest configuration is a permeable reactive trench that extends across the entire width of the plume (Figure 7.1). This system can be installed by digging a trench and filling it with permeable material to create an artificially permeable environment. The permeable material selected will depend upon the required porosity and permeability in the trench. As the contaminant plume moves through the wall, contaminants can be removed by various mass transfer processes such as air stripping, biodegradation, adsorption, and metal-enhanced dechlorination. Suitability of this configuration for a specific site will depend upon the contaminant type and distribution, preferred mass removal process, geologic and hydrogeo- logic conditions, and ease of implementation of the selected process in a cost-effective manner.

Permeable reactive trenches can be considered as a system that incorporates an *in situ* reactor to achieve the same mass transfer reactions that are utilized in an above-ground treatment system during a pump and treat operation. It should be noted that the *in situ* system has a distinct advantage over above-ground systems due to the significantly higher residence times available within the *in situ* reactor. Typical groundwater flow velocities will provide residence times of days, if not weeks, within the *in situ* reactor in comparison to minutes or hours available in above-ground reactors. As a result, mass removal efficiencies could be expected to reach close to 100% in a properly designed system. However, other more cost-effective configurations may be available to achieve the same mass transfer process depending on the site geologic conditions. For example, if the geology is homogeneous and permeable and the contamination is deep, conventional air sparging (with air injection wells) may be more cost-effective in comparison to permeable reactive walls with air injection points. If

Table 7.1 Potential Mass Transfer Reactions for Various Contaminants

Contaminant(s)	Air stripping volatilization	Aerobic biodegradation	Absorption; (carbon, ion exchange)	Abiotic dechlorination	Anaerobic biodegradation
TCE, PCE, DCE	•		•	•	•
Benzene, toluene, ethyl benzene, xylenes	•	•	•		•
Alcohols, acetone, MEK, ketones, phenol		•			
Dissolved heavy metals (Pb, Zn, Ni)			•		
NO_3^-					•

the geology is less permeable and contamination is shallow, permeable reactive walls may be the preferred treatment technique.

Table 7.1 describes the potential mass transfer reactions suitable for implementation within the reaction trench for various contaminants.

7.2.2 Funnel and Gate Systems

Funnel and gate systems can be implemented using various configurations, depending upon the plume width and depth and the type of mass transfer reaction required. Options for the funnel (cut-off walls) include sheet pile walls or slurry walls and options for the gate include permeable nonreactive materials such as pea gravel (for air sparging) or oyster shells (for biodegradation) and reactive materials such as activated carbon or zero-valence iron.

7.2.2.1   Single Gate System

The simplest configuration is a single gate with cut-off walls extending on both sides of the gate (Figure 7.2). This configuration is more suitable for a narrow, elongated plume. Funnel and gate systems can be constructed through the entire thickness of an aquifer, if the contamination extends across the full depth of an aquifer, as seen in situations where DNAPLs are present. When the contaminant plume is shallow, as may be the case under LNAPL conditions, penetration of the system may be required only to the depth of contamination. The above two configurations are known as a *fully penetrating gate system* and a *hanging gate system*,[1] respectively (Figure 7.3).

The main advantage of a funnel and gate system over a permeable reactive trench, across the full plume width, is that a smaller-scale reactor, at the gate, is used for treating a given plume at a lower cost. If a reactor requires periodic media replacement (such as a carbon or ion exchange bed) or flushing out the precipitated metals it will be easier to accomplish the change out using a "small" gate.

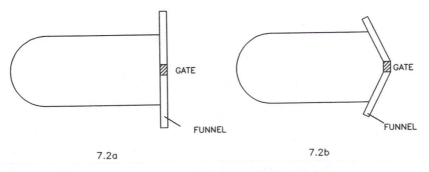

7.2a   7.2b

Figure 7.2 Possible configurations of single gate funnel and gate system.

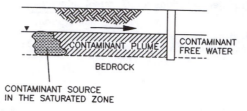

7.3a FULLY PENETRATING GATE

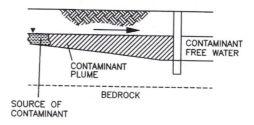

7.3b HANGING GATE

Figure 7.3 Cross-sectional views of fully penetrating gate and hanging gate systems. a. Fully penetrating gate. b. Hanging gate.

The type of treatment processes potentially applicable at the gate include

- air stripping, volatilization
- microbial degradation
- adsorption (carbon or ion exchange)
- chemical oxidation
- metal enhanced dechlorination
- metals precipitation

As can be seen from the above list, most treatment processes used in an above-ground pump and treat system can be implemented in an *in situ* reactor by carefully designing a funnel and gate system. In addition, physical recovery techniques for LNAPL and DNAPL recovery can also be implemented within the gate.

The funnel and gate geometry required to direct all the contaminated groundwater through the gates and the ease of installation of the selected configuration will determine the applicability in most cases. Selected mass removal reactions to address the specific contaminants and the ease of implementation will also influence the applicability of funnel and gate systems. Table 7.2 describes the potential mass removal reactions for various contaminants that could be implemented within the reactor gate.

When dealing with a groundwater plume that contains multiple contaminants, two or more gates in series may be required to implement different mass removal reactions. For example, a plume containing trichloroethylene (TCE) and acetone may require a gate with an air injection and extraction system (for removing TCE) to be followed by a fixed film bioreactor gate (for degrading acetone). If the same plume has pentachlorophenol (PCP) in it, a third gate with activated carbon may be required for PCP removal. Gates in series are shown in Figure 7.4.

Table 7.2 Potential Mass Removal Reactions for Various Contaminants in the Gate

Contaminant	Air stripping volatilization	Microbial degradation	Adsorption (carbon, ion exchange)	Chemical oxidation (O_3, H_2O_2)	Metal enhanced dechlorination
Chlorinated aliphatics (TCE, PCE, DCE)	•	•	•	•	•
Benzene, toluene, ethyl benzene, xylenes	•	•	•	•	
Chlorinated aromatics (pentachlorophenol)		•	•	•	
Alcohols, ketones, phenols (acetone, MEK, phenol)		•		•	
Dissolved heavy metals (Pb, Ni, Zn, Cd)			•		
1,4-Dioxane				•	

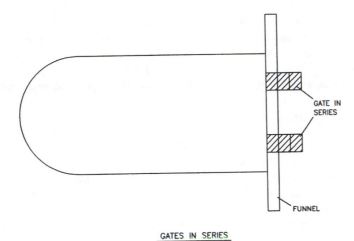

GATES IN SERIES

Figure 7.4 Illustration of reactive gates in series.

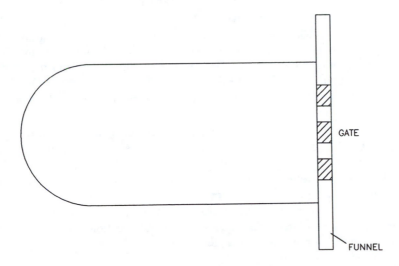

Figure 7.5 Multiple gate configuration of a funnel and gate system.

7.2.2.2   Multiple Gate System

When the contaminant plume is relatively wide, multiple gates may be required to direct contaminated groundwater to flow through the gates (Figure 7.5). Multiple gate systems can be installed as fully penetrating or hanging gates as dictated by the depth of contamination. When designing a multiple gate system, a balance between maximizing the size of the capture zone for a gate and maximizing the retention time in the gate should be achieved. In general, capture zone size and retention time are inversely related.[1]

Mass removal reactions discussed in Table 7.2 can also be implemented in multiple gate systems. The width and depth of the funnel (cut-off wall), the required number of gates, and the total width of the gates play a significant role in deciding whether this technique is the most cost-effective alternative to address a given contaminated site.

7.3   DESIGN APPROACHES

As noted earlier, design of *in situ* reactive walls is influenced by site geologic and hydrogeologic conditions and the type, concentrations, and vertical and lateral distribution of the contaminants. Optimal system geometry will be influenced by the geologic and hydrogeologic conditions as well as by the contaminant distribution. Selection and design of the reaction processes in the gate will be influenced by the contaminant type and concentrations.

7.3.1   System Geometry

System geometry is a simplified term describing the dimensions of the cut-off walls, gates, and the trench and the number of gates in parallel or in series. The orientation of the system in relation to the contaminated groundwater flow direction will also influence the system geometry considerations. The designer has to balance several conflicting criteria when designing an *in situ* reactive wall system. For example, the designer has to balance the need for more gates to better control groundwater flow with the need to minimize the number of gates to be cost-effective in implementing the required reaction processes. In most cases construction of the gates is far more expensive than the cut-off walls. Hence the optimum design will minimize the number and width of gates while still accommodating flow from the entire contaminated plume and providing adequate residence time within the gate.

As a result, the designer should rely upon groundwater flow and transport models to evaluate the most effective configuration from both technical and economic viewpoints. Iterative modeling simulations and particle tracking should be performed to arrive at the optimum configuration.

The factors that should be evaluated during the modeling exercise include the following:

- Total width of the wall and the minimum number of required of gates
- Particle tracking for the selected configuration to ensure complete capture of the plume
- Residence times through the gates
- Back-pressure development when the number of gates are inadequate and the resulting "dam" effect
- Potential for fouling with time and hence increased resistance to flow through the gates
- Impacts due to water table fluctuations and natural variation in flow directions
- Any geologic/hydrogeologic anomalies that may lead to preferential flow within the contaminated plume
- If the proposed system is a hanging gate system (Figure 7.3B), the potential for any underflow of the contaminated water
- Model calibration and sensitivity to the site-specific hydrogeologic parameters such as hydraulic conductivity and transmissivity

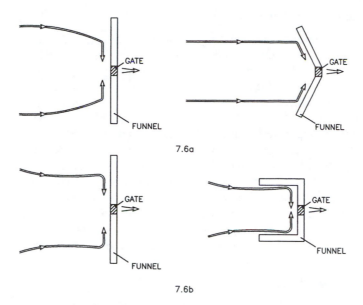

7.6a

7.6b

Figure 7.6   Illustration of effective width of a funnel and impact on capture zone by the funnel configuration. a) Comparison between straight- and apex-angled funnels. b) Comparison between straight- and U-shaped funnels.

Selection of optimal system geometry for the permeable reactive trench system is much simpler than for the funnel and gate system. Since the trench is constructed with a much more permeable media than native soils, there will be less hydrogeologic concerns for achieving complete capture of the plume. Constructibility of the trench in a cost-effective manner will determine the depths to which this system can be installed.

7.3.1.1   Funnel Width and Angle

Increased funnel (cut-off wall) widths provide a larger capture zone, and thus the required flow capacity through the gates will have to be increased accordingly. Beyond the localized influence around the cut-off walls, the width of the capture zone is generally proportional to the flow through the gate for a given site. The required capture zone will determine the optimum funnel width and the number of gates.

It has been shown that the maximum discharge and capture occurs when the funnel is perpendicular to the groundwater flow direction,[1] Figure 7.6A. The effective width of the funnel decreases when the funnel is not straight and has an apex angle. U-shaped funnels provide a smaller capture zone in comparison to a straight funnel of the same length (Figure 7.6B).

7.3.1.2   Gate Width

As the total width of the gate(s) increases, via a single gate or multiple gates, relative to the total funnel width, the absolute and relative flow and the width of the capture zone increase. The *absolute flow* is defined as the flow through the gate(s) at a given time. The *relative flow* is the fraction of flow intersected by the funnel width which flows through the gate(s). It is calculated by dividing the flow through the gate(s) by the flow through a section of the aquifer equal to the width and depth of the funnel at an upgradient location in the absence of the funnel and gate system.

A gate that is as wide as possible is always desirable. However, economic considerations may limit this desirable objective.

7.3.1.3   Gate Permeability

It is easy to understand that when the permeability of the gate is higher, the flow through the gate also will be higher. However, it has been reported that there is relatively little increase in flow through the gate when the gate permeability is greater than 10 times that of native soils of the aquifer.[1] This is due to the fact that flow is limited by the transmissivity of the aquifer upgradient and downgradient of the gate. While there is a general tendency among designers of funnel and gate systems to make the gate permeability as high as possible, very high values are not required for proper functioning of the system and may not be desirable due to the potentially shortened residence times. However, consideration should be given to the potential fouling of the gate media due to microbial growth and/or inorganic precipitation, and the eventual decrease in gate permeability. In addition, mass removal reactions such as adsorption or biodegradation via immobilized biomass may require increased surface area in the reactor media. Under these circumstances, the porosity of the media that will provide a higher permeability will have to be balanced with the need to have a higher available surface area. Media with higher available surface area will be of finer particles and thus will have a lower porosity.

7.3.2   System Installation

In addition to being cost-effective, *in situ* reactive wall systems should be able to last for years with little or no maintenance. System installation objectives should also be cost-effective in order to enable the *in situ* reactive wall technology as the preferred remediation alternative at a given site. *In situ* reactive walls may require a high initial capital investment in some cases, but long-term life cycle cost savings will be realized from greatly reduced operating, maintenance, and monitoring costs.

7.3.2.1   Permeable Reactive Trenches

Construction of a permeable reactive trench will be very much influenced by the depth to which the trench has to be excavated. The biggest drawback for choosing this technique is the cost of disposing the contaminated soil removed during excavation. The deeper the trench has to be, the wider it may need to be to facilitate installation. In certain situations, a portion of the contaminated soil can be backfilled on top of the more permeable media such as pea gravel (Figure 7.7) and soil vapor extraction (SVE) can be installed to remediate VOC-contaminated soil. If *in situ* air stripping or aerobic biodegradation is the selected technology to be implemented in the trench, an SVE system may have to be installed anyway to collect the stripped vapors.

There are various methods to install a trench. Backhoes, or trenchers (also known as ditchers), can be used to excavate the trenches. Depending on the depths of excavation and location of the trenches, shoring techniques may have to be implemented to prevent trench collapse. If the trench has to be excavated near a building, shoring may be required to ensure the safety of the building foundations. Application of a biodegradable slurry has been used in the recent past to minimize the cost of shoring. The slurry helps to stabilize the slopes until the backfill material is placed in the trench. The slurry will be completely biodegraded within a short time frame and the trench will, thus, function as a permeable trench.

Various backfill materials, depending on the reactive processes taking place in the trench, can be used as the porous media in the trench. Table 7.3 describes the various porous media that can be used as the backfill material.

7.3.2.2   Types of Funnel Walls

There are various types of impermeable or less permeable subsurface barriers that can be used as cut-off walls in a funnel and gate system. The purpose of the cut-off walls is to

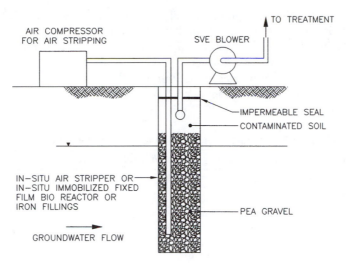

Figure 7.7 Cross-sectional view of a permeable reactive trench.

Table 7.3 Porous Media That Can Be Used as Potential Backfill Material in Permeable Reactive Trenches

Porous media	Applicability to potential mass removal reactions
Pea gravel	*In situ* air stripping *In situ* aerobic biodegradation
Iron filings	Reductive dehalogenation of chlorinated aliphatics
Crushed sea shells	*In situ* immobilized, fixed film bioreactor (aerobic and anaerobic)
Limestone	Changing of pH and redox in the trench for metal precipitation and pH neutralization
Granular activated carbon	Adsorption of organics
Ion exchange—zeolites	Removal of heavy metals and ionic contaminants

erect a continuous impermeable barrier to direct the contaminated groundwater to flow through the gates to be remediated. Slurry walls, sheet pile walls, and geomembranes are commonly used techniques to construct the cut-off walls.

7.3.2.2.1   *Slurry Walls*

Slurry walls are a means of placing a low-permeability, subsurface cut-off wall. They consist of vertically excavated trenches that are initially filled with low permeability materials. These walls are described by the material used to backfill the slurry trench. Soil–bentonite walls are composed of soil materials (often the trench spoils) mixed with bentonite slurry. Cement–bentonite walls are constructed using a slurry of Portland cement and bentonite and set to form a permanent, low-permeability wall. Diaphragm walls are installed of precast or cast in place reinforced concrete panels (diaphragms) installed during slurry trenching. Each of these, as well as hybrids of the three, have different characteristics and applications.

Deep soil mixing can also be used in the construction of slurry walls. Deep soil mixing walls are installed using large overlapping augers equipped with mixing paddles. As the group of augers is advanced into the ground, additives such as bentonite or cement are injected and mixed with the soil. Construction costs for deep soil mixing are compatible with slurry walls. In general, soil–bentonite slurry walls are less permeable and more resistant to chemical degradation than cement–bentonite walls. Slurry walls can be constructed to significant depths (150 to 200 ft).

Due to the subsurface nature of construction, there could be some construction and postconstruction defects encountered during slurry wall construction. These defects include zones of the wall that may not provide the same resistance to groundwater flow as the good, intact portions of the same wall; poorly mixed backfill; trapped pockets of slurry; and loss of the filter cake in portions of the trench wall. Postconstruction defects include cracking due to change in moisture, temperature, consolidation and stress as well as increase in permeability due to chemically aggressive contaminants.

Table 7.4 presents the estimated permeability of various slurry walls. As expected, laboratory results performed under controlled conditions exhibit much lower permeabilities than field conditions.

Table 7.4 Permeability of Barrier Materials with Nonaggressive Contaminants

Material	Permeability (lab) cm/s $\times 10^{-8}$	Permeability (field) cm/s $\times 10^{-8}$
Bentonite slurry	5	50
Cement—bentonite	3	100
Compacted soil—bentonite	0.1	10
Soil–cement	0.5	50
Cement–grout	0.1	1,000
Concrete	0.1	—
Silicate grout	47	10,000

From Boscardin, M. D. and Ostendorf, D. W., Cutoff walls to contain petroleum contaminated soils, in *Petroleum Contaminated Soils*, Calabrese, E. J. and Kostecki, P. T., Eds., Lewis Publishers, Boca Raton, FL, 1989. With permission.

It is known that some contaminants can react with barrier materials and potentially cause significant increases in permeability over time. Bentonite-based slurry walls tend to fare poorly with some organic contaminants.[4] However, proprietary mixes of treated bentonite may provide satisfactory performance under appropriate conditions.

A structural cap may have to be installed to maintain the strength and integrity of the slurry walls. At a site with significant vehicular traffic, a structural cap on the wall will be required.

7.3.2.2.2   Sheet Pile Walls

An impermeable or low-permeable cut-off wall can be installed using closely spaced steel sheet piles in a longitudinal direction. However, the space between adjoining sheet piles may allow leakage of contaminated groundwater without being deflected toward the gate for treatment. The concept of driving sheet piles and sealing the joints between adjacent piles was recently introduced, specifically for the purpose of installing funnel and gate systems.[5] This proprietary technique, known as the Waterloo Barrier™, incorporates a sealable cavity into the pile interlocks (Figure 7.8). Extensive field-scale tests conducted indicate that bulk hydraulic conductivity values of less than 10^{-8} cm/s can be achieved with this arrangement.

The sheet pile cut-off wall can be installed using conventional pile driving techniques. Vibratory or impact pile drivers can be used depending upon the soil conditions. Vibratory equipment is suitable for most soil conditions, but better results can be achieved with impact equipment in certain cohesive soils. A foot plate at the toe of each larger interlock prevents most of the soil from entering the sealable cavity during driving. After the pile driving is done, the cavities are water jetted to remove the loose soil caught up in the interlocks. A number of clay-based, cementitious polymer and mechanical sealants are available to meet a variety of site conditions.

Potential leak paths through the barrier are limited to the sealed joints, and therefore the joints should be inspected before the sealing operation to confirm that the complete length of the cavity is open and can be sealed. Each joint should be sealed from bottom to top, facilitating the emplacement of sealant into the entire length of the joint.

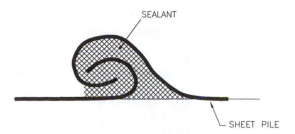

SEALANT

SHEET PILE

Figure 7.8 Sealable, interlocking joints for sheet pile walls.

7.3.3 Applicable Reactive Processes

Construction and operation of the *in situ* reactor is the key to successful implementation of any *in situ* reactive wall system. Several types of *in situ* reactors can be installed.

7.3.3.1 Air Stripping

Application of air stripping in below-ground, *in situ* systems is often called air sparging. The application of this technology in specially designed subsurface reactors or trenches should be considered as equivalent to operating an *in situ* air stripper.

Volatile organic compounds (VOCs) can be easily stripped from contaminated groundwater by injecting air into specially designed *in situ* reactors. Strippability of dissolved contaminants is governed by the Henry's law constant (Appendix C). In general, the higher the Henry's law constant, the more readily a compound would partition to the vapor phase.

In situ air strippers can be installed in the form of sparging trenches (Figure 7.9) or sparging gates/wells. The most economical configuration of a sparging gate is shown in Figure 7.10 and Figure 7.11 for shallow and deeper depths, respectively. The system shown in Figure 7.10 was installed at a site where the depth to the confining bedrock layer was less than 15 ft. A concrete vault with two chambers was constructed as the reactor gate. Clean sand and pea gravel were filled on both sides of the reactor gate to facilitate the flow of water through the gate. The chamber with packing material was provided as a means to develop a bacterial culture to enhance the microbial degradation of the contaminant at this site.

Figure 7.11 shows the configuration of an air stripping well, where the contaminated groundwater water is directed to flow through the well into which compressed air is injected. Injected airflow rates can be optimized to achieve the required stripping efficiencies. Due to the groundwater velocities encountered at most of the sites, the residence time in the *in situ* air strippers will be more than sufficient to achieve the required stripping efficiencies. The stripped contaminants can be easily collected, via a vapor extraction system, and treated prior to discharge to the atmosphere.

7.3.3.2 *In Situ* Bioreactors

When the contaminants present in the groundwater are aerobically biodegradable, implementation of an *in situ* bioreactor will minimize the overall system cost by minimizing the need for treatment of stripped contaminants or the replacement of adsorption media.

The configurations shown in Figures 7.9 and 7.10 can be easily converted into a bioreactor when the conditions are shallow. Air injection rates required for the operation of the bioreactor will be significantly lower than those required for air stripping. At steady state, most or all of the contaminant mass will be biodegraded, and hence the need for capturing the injected air can be eliminated. If the *in situ* reactive wall system is to function as a bioreactor, the

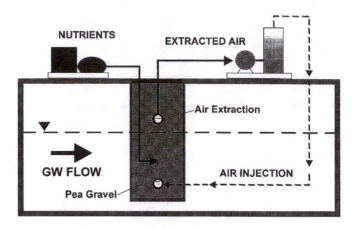

Figure 7.9 Horizontal trench sparging (section view).

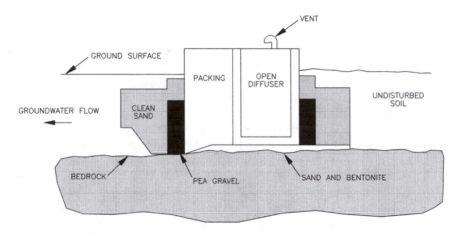

Figure 7.10 Air stripping gate configuration for shallow water table.

porous media used in the trenches or the gate should be able to support the growth of the biomass. The rough surface of oyster shells is an example, where the enhanced attachment will increase the biomass per unit volume of the reactor.

A single well as shown in Figure 7.11 will not be able to achieve complete biodegradation of the contaminants due to insufficient reactor volume. Growth of biomass in a small reactor volume will also increase the resistance to flow through the gate. Multiple wells placed parallel to each other (Figure 7.12) will be adequate to maintain an *in situ* bioreactor, especially under deep situations.

Inoculation or seeding of the biomass into the *in situ* reactor should be accomplished with a microbial population acclimated to the contaminants present in the groundwater. A small above-ground tank filled with contaminated groundwater, collected from the source area of the plume, should be aerated with increasing levels of contamination for a few weeks. The water in the tank should be emptied into the *in situ* reactor when it becomes cloudy with a significant microbial population.

Contaminants that are readily degraded under anaerobic or anoxic conditions can also be treated in an *in situ* bioreactor. The same configurations shown in Figures 7.9, 7.10, and 7.11 can be used for the implementation. However, the dissolved oxygen or other election acceptors, such as NO_3^-, and SO_4^{2-} present in the incoming groundwater may have to be removed prior to the entrance into the bioreactor. Injection of an innocuous readily biodegradable (labile) compound such as sugar into the surrounding trenches or in an upgradient location will maintain anaerobic or anoxic conditions in the reactor.

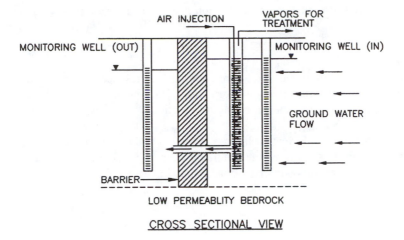

CROSS SECTIONAL VIEW

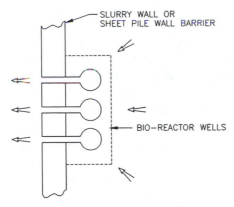

PLAN VIEW

Figure 7.11 Air stripping well for deeper water table conditions.

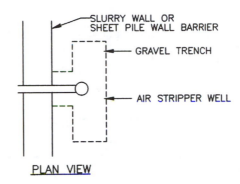

Figure 7.12 Configuration of bioreactor wells.

7.3.3.3 Metal-Enhanced Abiotic Dechlorination

Metal-enhanced abiotic dechlorination is a new twist on the age-old corrosion process. The use of zero-valence metals in the degradation of chlorinated aliphatic compounds and the potential application in an *in situ* environment have been studied at the University of Waterloo during the last few years.[6,7] Reductive dechlorination of pesticides was reported as early as 1972.[8]

The mechanism of reductive dechlorination is explained by the following equation.[9]

$$Fe^\circ + RCl + H^+ \rightarrow Fe^{2+} + RH + Cl^-.$$ (7.1)

Alkyl chlorides, RCl, can be reduced by iron and, in the presence of a proton donor such as water, they will undergo reductive dechlorination. The reaction represented by the above equation is a well-known member of a class of reactions known as dissolving metal reductions.[9]

The net reductive dechlorination by iron (equation (7.1)) is equivalent to iron corrosion with the alkyl chloride serving as the oxidizing agent. The characteristic reaction of iron corrosion, shown by the following equation, results in oxidative dissolution of the metal at or neutral pH.[10]

$$Fe^\circ - 2e \Leftrightarrow Fe^{2+}.$$ (7.2)

In the absence of strongly oxidizing solutes, dissolved oxygen, when present, is the preferred oxidant, resulting in rapid corrosion.

$$2Fe^\circ + O_2 + 2H_2O \Leftrightarrow 2Fe^{2+} + 4OH^-.$$ (7.3)

Further oxidation of Fe^{2+} by O_2 leads to the formation of rust (ferric hydroxide). Water alone can serve as the oxidant under anaerobic conditions according to the following equation.

$$Fe^\circ + 2H_2O \Leftrightarrow Fe^{2+} + H_2 + 2OH^-.$$ (7.4)

Both reactions (7.3) and (7.4) result in increased pH, and this effect is more pronounced under aerobic conditions due to the rapid rates of corrosion. The pH increase favors the formation of iron hydroxide precipitates, which may form a surface layer on the metal, thus inhibiting its further dissolution.

The above discussion suggests[9] three general pathways leading to dechlorination of alkyl chlorides. The first pathway (Figure 7.13) involves the metal directly and implies that reduction occurs by election transfer from the Fe° surface to the adsorbed alkyl chloride[9] according to equation (7.1). The second pathway involves further oxidation of Fe^{2+} that is an immediate product of corrosion in aqueous systems (Figure 7.13).

$$2Fe^{2+} + RX + H^+ \Leftrightarrow 2Fe^{3+} + RH + X^-.$$ (7.5)

The third pathway for reductive dechlorination by iron involves the hydrogen produced during corrosion (Figure 7.13).

$$H_2 + RX \rightarrow RH + H^+ + X^-.$$ (7.6)

Determining the relative importance of these three dechlorination pathways will be essential to predicting field performance of iron-based remediation technologies.

Since dechlorination apparently occurs at the iron–water interface, the following transport and reaction mechanisms influence the dechlorination process:[9] (1) mass transport of the contaminant to the Fe° surface; (2) adsorption of the chlorinated aliphatic contaminant to the surface; (3) chemical reaction at the surface; (4) desorption of the by-products; (5) mass transport of the by-products into bulk solution. Any one or a combination of these steps may

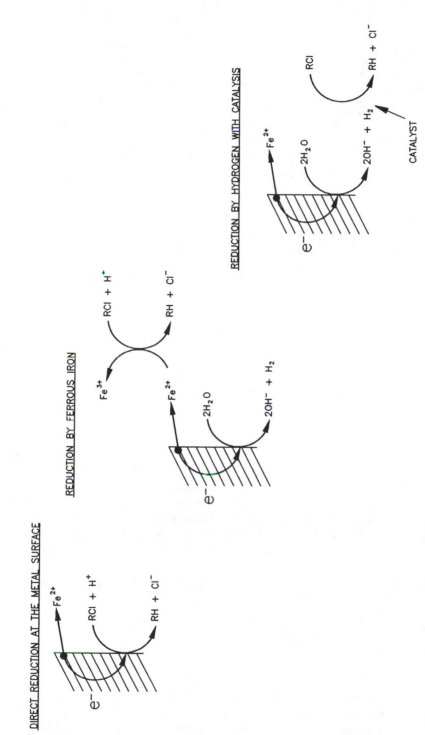

Figure 7.13 Three general pathways of dechlorination of alkyl chlorides.

Table 7.5 Rates of Reductive Dechlorination for Various Chlorinated Aliphatic Compounds

Compound	Half life (days)	Chlorinated by-products
Trichloromethane	6.5	Dichloromethane (low concentration)
Tetrachloromethane	5.4	Trichloromethane, dichloromethane
Monochloroethane	14.9	None detected
1,1-Dichloroethene	55	None detected
cis-1,2-Dichlorethane	37	None detected
1,1,1-Trichloroethane	5.5	1,1-Dichloroethane
1,1,2-Trichloroethane	7.8	Trace of monochloroethane
Trichloroethene	7.1	cis-1-2-Dichloroethene
1,1,2,2-Tetrachloroethane	10.2	cis-1-2-Dichloroethene
		trans-1-2-Dichloroethene
Tetrachloroethene	13.9	Trichloroethene
		cis-1,2-Dichloroethene

From Wilson, E., K., Zero-valent metals provide possible solution to groundwater problems, *Chem. Eng. News*, July 3, 1995.

be rate-limiting and, hence, careful consideration should be given to controlling those steps during system design.

Table 7.5 presents degradation rates, in the form of half lives, for various chlorinated aliphatic compounds.[8] By-products formed as a result of incomplete dechlorination are also shown in Table 7.5. The rates presented in Table 7.5 seem to be rather slow compared to other observations reported in the literature.[7]

As seen from Table 7.5, residence times required for complete dechlorination are very high even under slow-moving groundwater flow conditions. It should also be noted that due to the potential formation of anaerobic conditions in the in-site reactor, microbial (biotic) dechlorination also could take place in addition to the abiotic dechlorination. In addition, mixing of pH neutralizing reagents with the iron filings will also enhance the rate of dechlorination.

Configuration of the *in situ* reactors to implement metal-enhanced abiotic dechlorination can be accomplished as shown in Figures 7.9 through 7.12. The specific configuration will be very much influenced by the site-specific conditions and the required residence times to achieve complete reduction.

Another possible use of placement of iron filings in an *in situ* reactor is the conversion of the highly toxic and soluble hexavalent chromium (Cr(VI)) to much less toxic and less soluble trivalent chromium (Cr(III)). The reaction is quite rapid, occurring on the order of less than a minute.[7] The reaction is probably driven by several processes, including reduction at the surface of metallic iron and in solution from the production of ferrous iron. The resultant Cr(III) precipitates out of solution as a chromic hydroxide precipitate. Hence design of the system should incorporate options to backwash or flush out the precipitates accumulated in the reactor after a certain period of operation.

Another variation of this technique includes adding palladium to iron. Contaminants such as *cis*-1-2-dichloroethene can be resistant to dechlorination by iron within reasonable residence times. Palletized iron has been reported to dechlorinate this compound within a few hours.[7]

7.3.3.4   Adsorption

Adsorption mechanisms can be employed in the *in situ* reactors to remove a variety of contaminants. Liquid-phase granular activated carbon (GAC) can be used to remove many organics, especially those not easily removable by air stripping or biodegradation (e.g., pentachlorophenol and tetrachlorophenol). Ion exchange resins can be used to remove dissolved heavy metals present in the groundwater.

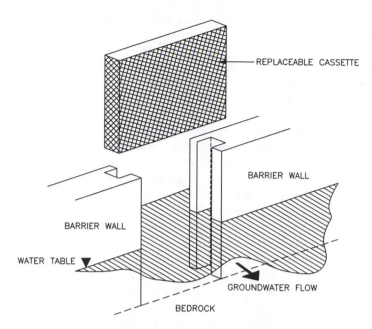

Figure 7.14 Easily replaceable, porous reactive cassettes at reactor gates.

7.3.3.4.1   *Liquid-Phase Granular Activated Carbon (GAC)*

During the adsorption process from aqueous solution, dissolved organics are transported into the solid sorbent (GAC) grain by means of diffusion and are then adsorbed onto the extensive inner surface of the activated carbon granules. Traditionally, the adsorption phenomenon has been categorized as physical adsorption and chemisorption. *Physical adsorption* is a rapid process, caused by nonspecific secondary binding mechanisms and is reversible. The adsorbate will desorb into the solution in response to a decrease in solution concentration. *Chemisorption* is more specific, because it entails the transfer of electrons between adsorbent and adsorbate and may or may not be reversible.

Due to the cost of liquid-phase GAC, only the configurations similar to those shown in Figures 7.10 through 7.12 will be preferred. The virgin carbon can be slurried into the reactor and the spent carbon can be vacuum slurried out of the reactor. Easily replaceable porous cassettes filled with carbon also can be retrofitted into the reactor (Figure 7.14).

Frequency of changeouts of the carbon bed can be estimated by performing a column isotherm study in the laboratory. Published information in the literature can also be used for this purpose. However, it should be noted that the estimated adsorption capacity of the *in situ* bed can be compromised by the potential fouling caused by growth of microorganisms and deposition of inorganic precipates within the bed. In addition, contaminant adsorptive capacity may be reduced due to the adsorption of natural organic matter such as humic substances and other micropollutants. Fouling of the carbon bed may also increase the resistance to flow through the gate. Design enhancements that provide the flexibility for periodic backwashing of the bed may overcome this problem. Design of an upflow configuration in the *in situ* reactor will also alleviate some of the potential problems caused by inorganic fouling.

7.3.3.4.2   *Ion Exchange Resins*

An *ion exchange reaction* may be defined as the reversible interchange of ions between a solid phase (the ion exchange bed) and solution phase, the ion exchange bed being insoluble in the medium in which the transfer is carried out. The important features characterizing ion exchange media are

- a hydrophylic structure of regular and reproducible form
- controlled and effective ion exchange capacity
- rapid rate of exchange
- chemical stability
- physical stability in terms of mechanical strength
- consistent particle size and effective surface area

Ion exchange reactions take place due to the preference of a specific ion to the media in comparison to the existing ion attached to the media. Most of the dissolved heavy metals present in groundwater are in the divalent or trivalent state cations, with the exception of hexavalent chromium. Depending on the heavy metal(s) to be removed, a strong acid or a weak acid ion exchange media may have to be used.

The most cost-effective use of the ion exchange method will be when the dissolved heavy metal concentrations in the groundwater are at low levels. At high concentrations, media replacement will be more frequent and the cost of disposal or regeneration of the spent media will be prohibitive.

The placement of the fresh ion exchange media in the *in situ* reactor can be carried out in the form of a wet slurry and the spent media can be vacuumed out as a slurry also.

Presence of other ions and nonoptimum pH conditions will significantly impact the loading capacity of ion exchange beds. The use of ion exchange beds in *in situ* reactive wall systems will be very specific and under very optimum conditions. The configurations similar to those shown in Figures 7.10 through 7.12 can be implemented for the use of ion exchange as the mass removal reaction.

7.3.3.5   Precipitation

Precipitation of dissolved heavy metals can be achieved by manipulating the pH and the Eh (redox) conditions of the contaminated groundwater as it flows through the *in situ* reactor. Heavy metals can be precipitated as hydroxide, carbonate, or sulfide precipitates. Precipitation of heavy metals has been successfully achieved in wastewater treatment processes. Hence, implementation of this process in an *in situ* reactor only needs creative design considerations. The biggest advantage in an *in situ* reactor is the availability of significantly higher residence times in comparison to wastewater treatment systems. Table 7.6 presents the various metals that can be precipitated as hydroxide, carbonate, and sulfide.

Table 7.6   Metals That Can Be Removed as Hydroxide, Carbonate, and Sulfate

Metal	Type of precipitate
Ba	Hydroxide
Cd	Hydroxide, carbonate, sulfide
Cr^{3+}	Hydroxide
Cu	Hydroxide, carbonate, sulfide
Pb	Hydroxide, carbonate, sulfide
Zn	Hydroxide, carbonate, sulfide
Hg	Hydroxide, sulfide
Ag	Hydroxide, sulfide
Ni	Hydroxide, carbonate, sulfide
As	Sulfide

Proper mixing of the added reagents and maintenance of optimum pH and Eh conditions are important to achieve complete precipitation of the heavy metals. Hence, the *in situ* reactor for precipitation has to be constructed like a reaction tank or rapid mixer in an underground vault (similar to Figure 7.10). Another gate in series has to be provided downstream to filter out the precipitates. This filter has to be designed in such a way to remove the filtered solids by backwashing. Due to the constructibility considerations, this reactor can be installed only under shallow water table conditions.

Metals precipitation can be achieved in configurations such as permeable reactive trenches too. Limestone can be used as the backfill materials to adjust the pH of the incoming groundwater and precipitate metals as metallic hydroxides. For instance, at an abandoned mine site, the pH of the groundwater was less than 3.0, and it had elevated concentrations of copper and zinc. A reactive wall providing a bed of limestone was able to precipitate the metals at a residence time of 1 h.[11] In such situations, the potential problem of plugging by the precipitates will have to be overcome.

7.3.3.6 Chemical Oxidation

In the oxidative degradation of organic compounds, the compound is converted by means of an oxidizing agent into new harmless compounds typically having either a higher oxygen or lower hydrogen content than the original compound.

Use of hydrogen peroxide or ozone or a combination of both are common oxidative processes. Implementing an *in situ* oxidation system can be very cumbersome and can also raise safety concerns. Hence, the use of *in situ* oxidation systems should be chosen only when all other reactive processes are considered not effective to treat a specific contaminant. A compound that falls in this category is 1,4-dioxane. *In situ* oxidation reactors have to be built like an underground vault (or like a parshall flume), and hence will be applicable only under shallow water table conditions.

7.3.4 Residence Time

An important factor in designing an *in situ* reactive wall system is the relationship between the residence time of contaminated groundwater in the gate or the trench and the rate of contaminant degradation reactions. The average residence time in the reactor can be calculated by dividing the empty bed (or void) volume of the reactor and dividing it by the flow through the gate.

In situ reactive wall systems have the distinct advantage of providing significantly higher residence times in comparison to similar above-ground reactors. Increased residence times are possible due to the relatively slow flow rates through the gates and will be in the order of days, if not weeks, depending on the site-specific groundwater flow velocities.

Concentration of the contaminants in the incoming groundwater also has a significant impact on the required residence times. For degradation processes that are first-order reactions, the retention time necessary is given by the following formula:[1]

$$N_{\frac{1}{2}} = \frac{\left[\ln\left(C_{eff}/C_{inf}\right)\right]}{\ln \frac{1}{2}} \tag{7.7}$$

where $N_{\frac{1}{2}}$ = number of half lives required
C_{eff} = desired concentration in the effluent
C_{inf} = concentration in the influent.

Faster and greater mass removal can be achieved by either faster reaction rates or longer residence times. Increased residence times require increased reactor volume and thus higher costs. Depending on the type of mass removal reaction employed in the reactor, both of these factors may have to be optimized. Processes such as air stripping and adsorption take place at such a fast rate that available residence times will be more than sufficient in most cases. If not, optimizing the air-to-water ratio or providing packing material to enhance the mass transfer efficiencies may be warranted for *in situ* air stripping.

Biodegradation processes are governed by the rate of biodegradation of specific contaminants. An empirical parameter used to estimate the allowable mass loading in biological treatment systems is the food-to-microorganisms ratio, the unit of which is mass of contaminants/mass of viable biomass. From this expression it can be noted that the higher the available biomass, the faster the reaction will be. Increased available surface area in the reactor for the attachment of biomass will increase the biomass concentration per unit volume of the reactor. In some cases increased retention times may be required and can be achieved by increasing the reactor volume or by having multiple gates to reduce the flow through each gate or by providing gates in series to achieve complete reaction.

Required residence time is a key factor for the design of metal-enhanced dechlorination systems. As seen from Table 7.5, half lives required for various contaminants are significantly higher in comparison to any other process. Residence times required for the precipitation and chemical oxidation processes are relatively small and will not be a concern in most situations.

7.3.4.1   Downgradient Pumping

In situ reactive wall systems are often considered as passive treatment systems, because contaminants have to follow the natural groundwater flow gradients to reach the gate for remediation. In some cases where accelerated cleanup times are preferred, pumping of clean groundwater downgradient of the gates will increase the flow through the gates (Figure 7.15). Disposal of clean water will not pose any regulatory limitation, and reinjection of this clean water at an up-gradient location may further enhance the gradient available for the flow in the area of concern.

7.4   CASE STUDY

Pentachlorophenol (PCP) and tetrachlorophenol (TCP) were detected in on-site groundwater monitoring wells at a former wood treating facility in the western U.S.[12,13] (Figure 7.16). The groundwater system consists of a shallow aquifer containing a heterogeneous mixture of marine deposits and artificial fill which is underlain by low-permeability siltstones and mudstone. The shallow aquifer ranges in thickness from 10 to 20 ft and averages 15 ft. Based on the results of the site investigation, it was determined that impacted groundwater had the potential to move off site and adversely affect downstream domestic water supply wells.

A number of remediation alternatives were evaluated, and an *in situ* reactive wall incorporating a slurry wall as the barrier and liquid-phase activated carbon (GAC) as the gate reactor was chosen as the preferred alternative. A number of studies were performed to evaluate the applicability of this technique at the given site. These studies focused on the potential for underflow beneath the barrier wall, the spatial relationship between gates(s) and funnel(s), mass loadings at the gate, and interferences with gate performance.

7.4.1   Groundwater Flow Patterns

A key site condition which that the funnel and gate system feasible was the high contrast between permeabilities in the water-bearing shallow aquifer and in the bedrock. To evaluate

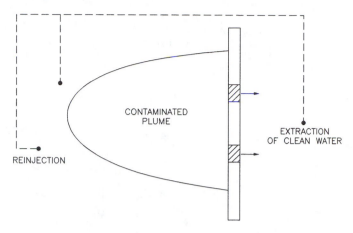

Figure 7.15 Downgradient pumping of clean water to enhance the flow through the gates.

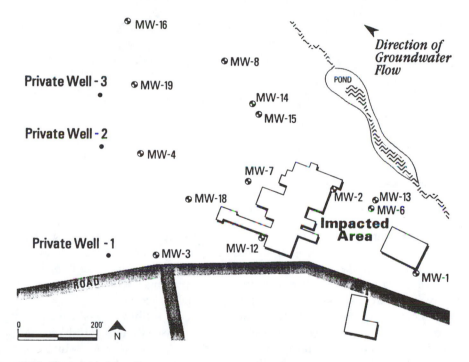

Figure 7.16 Plan view of the site.

the effect of a funnel-and-gate system on the horizontal distribution of water, a computer simulation of the hydrogeologic conditions in the impacted upper aquifer was developed. Using this model, various configurations of the funnel and gate were tried to optimize the configuration (Figures 7.17 through 7.20). The model provided for a design layout that routed the water from underneath the mill through the gates and avoided flow of contaminated water around the ends of the barrier.

With respect to minimizing the disruption of natural groundwater flow patterns, there were three key concerns that needed to be addressed. The first was the effect on water levels both up- and downgradient of the wall. The modeling showed that pressure required to move water through the gate was slightly greater than the natural gradient and was a function of the gate configuration and permeability of granular activated carbon. The selection of a moderate carbon grain size and the design of the treatment gates shown in Figure 7.21 minimized the pressure

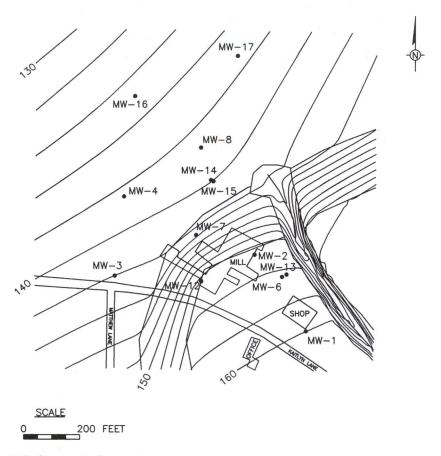

Figure 7.17 Simulated 2-ft groundwater contours.

loss across the treatment gate by presenting a large cross-section flow area through the carbon. This prevented water from backing up behind the gate and increasing the hydraulic head on the up-gradient side of the barrier. The pressure required to move the water through the gates corresponds to an increase in water differential of approximately 2 in. The treatment gates themselves therefore had a negligible effect on groundwater elevation.

The second concern was ensuring that water could readily access a treatment gate. Because of the nonuniform distribution of hydraulic transmissivities across the site, water could be impeded in its lateral movement toward and away from the gates, which could have resulted in undesirable hydraulic mounding behind the portion of the wall midway between gates. To minimize this effect, gravel-filled collection and distribution galleries were installed at each gate to collect water from the up-gradient side of the gate, guide it through the gate, and then redistribute it uniformly after treatment.

7.4.2 Underflow of Barrier

The potential for flow under the barrier wall was examined mathematically, using data collected during aquifer tests conducted at the site. Several test trenches were also excavated to evaluate the character of the bedrock along the alignment of the slurry wall. A comparison of the transmissivity of the shallow drinking-water aquifer with that of the underlying bedrock indicated that the hydraulic conductivity of the bedrock is approximately 1/1000 that of the overlying shallow aquifer. This resulted in a conservatively calculated underflow that is less than 1% of the total flow through the combined aquifers.

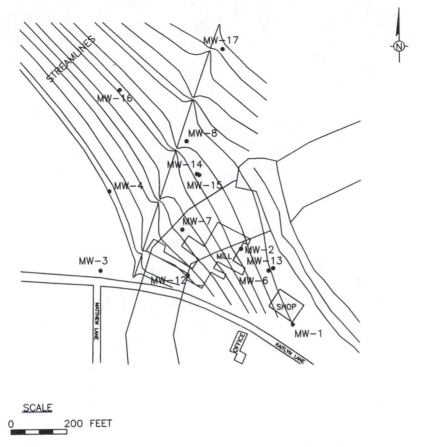

Figure 7.18 Streamlines showing capture zone for five 10-ft-wide reactive walls.

7.4.3 Number and Location of Gates

To determine the volume of water each gate is expected to treat, the total flux of water beneath the site was calculated using previously obtained aquifer data.

Because of the nonlinear configuration of the funnel-and-gate system, the flux across the barrier itself cannot be easily calculated. This is due to the influence of the funnel-and-gate arrangement on the flow lines across the site. However, the flow can be approximated by taking the cross-sectional flow of a region where the particle traces are relatively straight. This length is approximately 400 ft of potentially impacted aquifer.

The combined flow rate through all four treatment gates is the product of the transmissivity, the length of the cross-sectional area, and the hydraulic gradient. This results in a total flow of approximately 20 gpm. The flow through each gate is estimated as one fourth of the total flow, or 5 gpm per gate.

7.4.4 Gradient Control

Figure 7.22 shows the plan view of the treatment gate with collection and distribution galleries installed to guide water through the gates. Both galleries are downcut into the shallow aquifer and backfilled with gravel. Because the aquifer will tend to have higher horizontal permeability than vertical permeability, the collection and distribution galleries are downcut into the aquifer to expose a large cross-sectional area to groundwater flow. This minimizes the pressure required to move water from the aquifer to the carbon treatment gate. This is

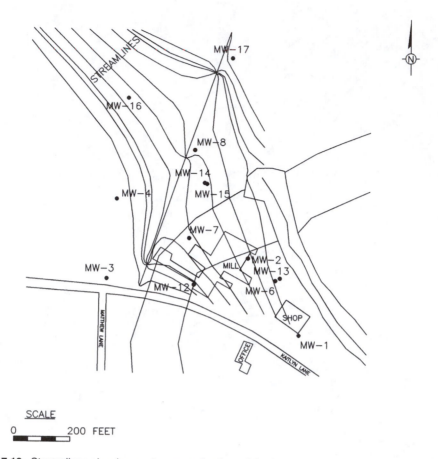

Figure 7.19 Streamlines showing capture zone for three 3-ft-wide reactive gates.

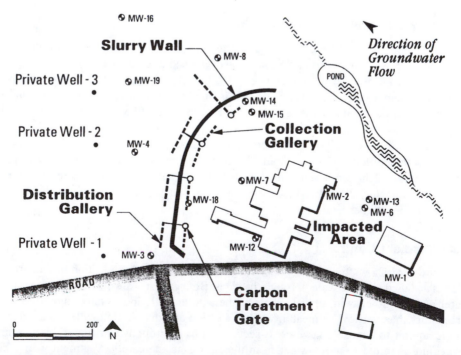

Figure 7.20 Configuration of funnel and gate system.

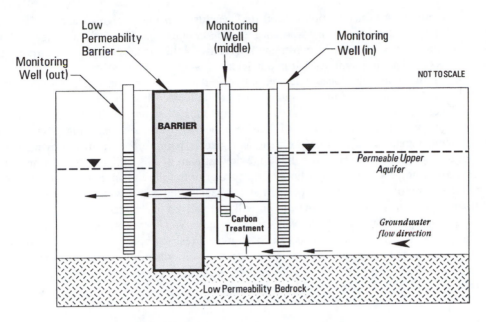

Figure 7.21   Cross section of treatment gate.

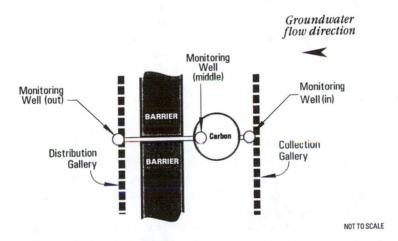

Figure 7.22   Plan view of treatment gate with galleries.

especially important on the downgradient side of the wall, where infiltration of water will be limited by the infiltration rate of the gallery. The installation of these collection and distribution galleries further ensures that the pressure drops across the wall will minimally affect the natural groundwater gradients and flow patterns.

7.4.5   Gate Design

For the purpose of designing the carbon gates, a very large margin of safety was factored in to the mass of carbon installed in each gate. Factors of design, e.g., contaminant concentration, flow rate, and the time between carbon changeouts, were selected very conservatively to ensure that there is more treatment capacity than required. The design has inherent flexibility, should the actual conditions change during the life of the project. The estimated gate flow rate is 5 gpm. The time between changeouts has been selected to be 4 years, although more frequent changeouts could easily be accommodated. In the vicinity where the gates will be installed, concentrations of compounds of concern have ranged from nondetect to the low parts per billion. A concentration of 200 μg/l was used in this design to account for the loading due to natural occurring wood-degradation compounds. Using the average carbon loading efficiency of 1% and the constraint of a relatively thin aquifer results in a carbon bed 4 ft tall placed within a cylinder 4 ft in diameter.

Monitoring points are located before, within, and after the treatment gate. Groundwater samples will be collected from the monitoring points located before each of the treatment gates. These samples will be analyzed for compounds of concern. Should any of these compounds be detected, a sample will then be collected from the monitoring point within the treatment gate to verify that the treatment gate is being effective in removing these compounds. There is a treatment buffer zone downstream of this midgate monitoring point to ensure that impacted water does not exit the gate prior to full removal of these compounds. Upon detection of a compound of concern at a concentration above water-quality objectives at the midgate measuring point, the carbon will be removed and replaced.

Carbon replenishment is a relatively easy procedure. Wet spent carbon will be vacuumed out as a slurry, using above-ground slurry pumps. Fresh carbon will be emplaced after dewatering the gate using the upgradient monitoring well, which is completed in the gravel packing adjacent to the gate. While water is being evacuated and the gate is dry, fresh carbon will be poured into the gate to the desired thickness.

7.5   LITERATURE REPORTINGS

Many research demonstrations of *in situ* reactive walls have been summarized by the U.S. Environmental Protection Agency.[14] Some of these case studies have been excerpted below.

Demonstration Study 1

- Description of Demonstration: A series of large-diameter augered holes in a staggered three-row array were located within an aquifer to intercept a contaminant plume of chromate and chlorinated organics. A mixture (by volume) of 50% iron filings (two types), 25% clean coarse sand, and 25% aquifer material was poured down hollow-stem augers from 22 ft to 10 ft below ground surface. Each iron column was approximately 8 in. in diameter and a total of 21 columns were installed in a 60 ft^2 area. The mixed waste contaminant plume is between 14 and 20 ft below ground surface and the water table ranges from 5 to 6 ft below ground surface. One iron type was shown to be an effective reductant for chromate in a 2-year laboratory study, while other iron has been shown to be more effective in the reductive dechlorination of the organics. This field experiment is evaluating the effectiveness of this method of treatment wall emplacement and is providing additional *in situ* field data for full field-scale implementation.
- Wastes Treated: TCE, DCE, vinyl chloride, and Cr^{+6}.
- Status: The demonstration has been in operation since September 1994.
- Preliminary Results: Preliminary results show complete reduction of Cr^{+6} to below detection (<0.01 mg/l) limits and greater than 75% reduction in initial TCE concentrations and reduction of vinyl chloride concentrations to less than 1 μg/l. These results are very promising, especially for the chlorinated organics, because the experiment was primarily designed to optimize chromate remediation, not the chlorinated organics.

Demonstration Study 2

- Description of Demonstration: An above-ground field test reactor containing 50% iron was installed at a semiconductor manufacturing facility to test the feasibility of installing an *in situ* permeable reactive zone. Groundwater was pumped through the reactor at a rate of 4 ft/day for 9 months. Initial concentrations of contaminants were 50 to 200 ppb trichloroethene (TCE), 450 to 1000 ppb *cis*-1,2-dichloroethene (*cis*-DCE), 100 to 500 ppb vinyl chloride (VC), and 20 to 60 ppb freon-113. The time required to degrade one half of the contaminant mass in the above-ground field test was less than 1.7 h for TCE, 1 to 4 h for *cis*-DCE, 2 to 4 h for VC, and less than 1.6 h for freon-113. Mineral precipitation, hydrogen gas production, and microbial effects also were evaluated.
- Wastes Treated: TCE, DCE, vinyl chloride, freon-113.
- Status: Based on the feasibility tests described above, a full-scale *in situ* permeable treatment wall was approved by the state regulatory agency and was installed in December 1994. The permeable reactive zone, which will be 4 ft thick, 40 ft long, 10 ft high, and about 25 to 30 ft deep, will contain 100% reactive iron. It will be installed using a trench box design. The cost is $175,000 to $200,000.

Demonstration Study 3

- Description of Demonstration: Researchers installed a permeable reactive wall containing an iron-based catalyst about 16.5 ft downgradient from the source of a contaminant plume to pilot test the capability of the wall to degrade halogenated organic compounds. The plume was about 6.5 ft wide and 3.3 ft thick with maximum concentrations along the axis of about 250,000 µg/l trichloroethene (TCE) and 43,000 µg/l tetrachloroethene (PCE). The source of the plume was located about 13 ft below ground surface and 3.3 ft below the water table.

  Using sealable joint sheet piling, a rectangular cell was constructed on the surface and driven to a depth of 32 ft, the cell was sealed and dewatered, and the native sand was replaced by the reactive material consisting of 22% by weight granular iron and 78% by weight course sand. The sand, which was coarser than the native materials, ensured that the wall would be more permeable than the surrounding sand. After installing the reactive material, the sheet piling was completely removed and natural flow conditions were achieved. The wall dimensions were 18 ft long, 5.2 ft thick, and 7.2 ft deep and it was positioned 3.3 feet below the water table.
- Wastes Treated: VOCs.
- Status: The pilot test was completed in 1993.
- Demonstration Results: Multilevel monitoring wells were located 1.6 ft up-gradient of the wall, in the wall at distances of 1.6 and 3.3 ft, and 1.6 ft downgradient of the wall for a total of 348 sampling points. Concentration distributions through the wall were determined on 13 occasions over 474 days, during which there was no decline in the effectiveness of the barrier. Most of the mass loss occurred within the first 50 cm of the wall; at greater distances into the wall, performance was below that expected. However, the reactive wall reduced the TCE concentration by 95% and the PCE concentration by 91%. No vinyl chloride was detected in the samples. Increased chloride concentrations downstream of the wall were consistent with the quantity of TCE and PCE that had been degraded. Only trace amounts of dichloroethene (DCE) were detected downstream of the wall.

REFERENCES

1. Starr, R. C. and Cherry, J. A., In situ remediation of contaminated groundwater, *Groundwater*, 32, 465, 1994.
2. Burris, D. R. and Cherry, J. A., Emerging plume management technologies: In situ treatment zones, *Eighth Ann. Air Waste Manage. Mtg.*, Kansas City, MO, June 1992.
3. Evans, J. C., Slurry trench cutoff walls for waste containment, *Int. Symp. Environ. Geotechnology*, Vol. 1, Envo Publishing, 1986.
4. Boscardin, M. D. and Ostendorf, D. W., Cutoff walls to contain petroleum contaminated soils, in *Petroleum Contaminated Soils*, Calabrese, E. J. and Kostecki, P. T., Eds., Lewis Publishers, Boca Raton, FL, 1989.
5. C³ Environmental, Waterloo Barrier, (marketing brochure), 1995.
6. Gillham, R. W., O'Hannesin, S. F., and Orth, W. S., Metal enhanced abiotic degradation of halogenated aliphatics: Laboratory tests and field trials, *Haz. Mat. Central Conf.*, Chicago, IL, March 1993.
7. Wilson, E. K., Zero-valent metals provide possible solution to groundwater problems, *Chem. Eng. News*, July 3, 1995.
8. Sweeny, K. H. and Fischer, J. R., Reductive Degradation of Halgenated Pesticides, U.S. Patent 3,640,821, 1972.
9. Matheson, L. J. and Tratnyek, P. G., Reductive dehalogenation of chlorinated methanes by iron metal, *Environ. Sci. Tech.*, 28, 2045, 1994.
10. Jones, D. A., *Principles and Prevention of Corrosion*, Macmillan, New York, 1992.
11. Johns, F., (internal communication) Geraghty & Miller, Inc., 1995.
12. Keyes, G., (internal communication) Geraghty & Miller, Inc., 1995.
13. O'Brien, K., (internal communication) Geraghty & Miller, Inc., 1995.
14. U.S. Environmental Protection Agency, In Situ Remediation Technology Status Report: Treatment Walls, EPA 542-K-94-004, April 1995.

8

IN SITU REACTIVE ZONES

8.1 INTRODUCTION

The concept of *in situ* reactive zones is based on the creation of a subsurface zone where migrating contaminants are intercepted and permanently immobilized or degraded into harmless end products. The successful design of these reactive zones requires the ability to engineer two sets of reactions between (1) the injected reagents and the migrating contaminants and (2) the injected reagents and the subsurface environment to manipulate the biogeochemistry to optimize the required reactions, in order to effect remediation. These interactions will be different at each contaminated site and, in fact, may vary within a given site. Thus, the major challenge is to design an engineered system for the systematic control of these reactions under the naturally variable or heterogeneous conditions found in the field.

The effectiveness of the reactive zone is determined largely by the relationship between the kinetics of the target reactions and the rate at which the mass flux of contaminants passes through it with the moving groundwater. Creation of a spatially fixed reactive zone in an aquifer requires not only the proper selection of the reagents, but also the proper mixing of the injected reagents uniformly within the reactive zone. Furthermore, such reagents must cause few side reactions and be relatively nontoxic in both its original and treated forms.

When dealing with dissolved inorganic contaminants, such as heavy metals, the required process sequence in a pump and treat system to remove the dissolved heavy metals present in the groundwater becomes very complex and costly (Figure 8.1). In addition, the disposal of the metallic sludge, in most cases as a hazardous waste, is also very cost prohibitive. Therefore, *in situ* treatment methods capable of achieving the same mass removal reactions for dissolved contaminants in an *in situ* environment are evolving and gradually gaining prominence in the remediation industry.

The advantages of an *in situ* reactive zone to address the remediation of groundwater contamination are as follows:

- An *in situ* technology: eliminates the expensive infrastructure required for a pump and treat system; no disposal of water or wastes.
- Inexpensive installation: primary capital expenditure for this technology is the installation of injection wells.
- Inexpensive operation: reagents are injected at fairly low concentrations, and hence the cost should be insignificant; only sampling required is groundwater quality monitoring; management of large volumes of contaminated water without any disposal needs.
- Can be used to remediate deep sites: cluster injection wells can be installed to address deeper sites.
- Unobtrusive: once the system is installed, site operations can continue with minimal obstructions.

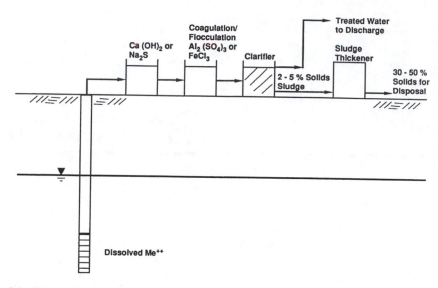

Figure 8.1 Schematic of process train for dissolved heavy metals present in groundwater.

- *In situ* degradation of contaminants: organic contaminants can be degraded by implementing the appropriate reactions.
- Immobilization of contaminants: utilizes the capacity of the soils and sediments to adsorb, filter out, and retain inorganic contaminants.

8.2 TYPES OF *IN SITU* REACTIONS

Manipulation of the oxidation–reduction (redox) potential of an aquifer is a possible approach for *in situ* remediation of redox-sensitive groundwater contaminants. In addition, various microbially induced or chemically induced reactions also can be achieved in an *in situ* environment. As noted earlier, creation of spatially fixed reactive zones to achieve these reactions is very cost-effective in comparison to treating the entire plume as a reaction zone.

8.2.1 Heavy Metals Precipitation

The mechanisms that can be used to reduce the toxicity of heavy metals dissolved in groundwater are *transformation* and *immobilization*. These mechanisms can be induced by both abiotic and biotic pathways. Abiotic pathways include oxidation, reduction, sorption, and precipitation. Examples of biotically mediated processes include reduction, oxidation, precipitation, biosorption, bioaccumulation, organo-metal complexation, and phytoremediation. In this chapter, immobilization mechanisms induced by precipitation only will be discussed.

Dissolved heavy metals can be precipitated out of solution through various precipitation reactions shown below. A divalent metallic cation is used as an example in these reactions.

$$\text{Hydroxide precipitation:} \quad Me^{++} + 2OH^- \rightarrow Me(OH)_2 \downarrow \tag{8.1}$$

$$\text{Sulfide precipitation:} \quad Me^{++} + S^{2-} \rightarrow MeS \downarrow \tag{8.2}$$

$$\text{Carbonate precipitation:} \quad Me^{++} + CO_3^{--} \rightarrow MeCO_3 \downarrow . \tag{8.3}$$

Theoretical behavior of solubility of these precipitation mechanisms is shown in Figure 8.2.

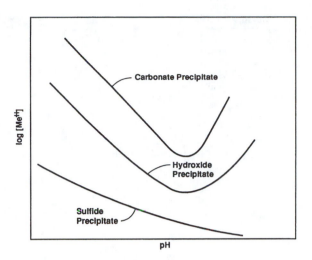

Figure 8.2 Theoretical pathways for solubility of heavy metal cations.

Hydroxide and sulfide precipitation of heavy metals have been used successfully in conventional industrial wastewater systems. Lime $(Ca(OH)_2)$ or other alkaline solutions such as potash (KOH) are used as reagents for hydroxide precipitation. Sodium sulfide (Na_2S) is normally used as the reagent to form extremely insoluble metallic sulfide precipitates. Injection of these chemical reagents into the contaminated aquifers, to create a reactive zone, will precipitate the heavy metals out of solution. However, injection of a reactive, pH-altering chemical reagent into the groundwater may be objectionable from a regulatory point of view. Obtaining the required permits to implement chemical precipitation may be difficult. Furthermore, the metallic cations precipitated out as hydroxide could be resolubilized slightly as a result of any significant shift in groundwater pH.

Under reducing conditions, heavy metal cations can be removed from solution as sulfide precipitates if sufficient sulfur is available. The solubility pattern of heavy metals as a function of pH and redox potential is summarized for Cd, Ni, Zn, As, and Cr in Figure 8.3. In systems containing a sufficient supply of sulfur, neutral to mildly alkaline pH and low redox conditions are most favorable for the precipitation of many heavy metals. Chromium is insoluble under reducing conditions, as Cr(III) hydroxide, but only at neutral to mildly acidic and alkaline pH values.

Precipitation as sulfides is considered the dominant mechanism limiting the solubility of many heavy metals. Sulfide precipitation is particularly strong for "chalcophilic" metals exhibiting "B-character," such as Cu(I), Ag, Hg, Cd, Pb, and Zn; it also is an important mechanism for transition elements such as Cu(II), Ni(I), Co(II), Fe(II), and Mn(II).[1] Two situations can be distinguished in natural systems during sulfide precipitation conditions: the existence of a certain sulfide precipitation capacity (SPC), or (when exceeding the SPC) the accumulation of free sulfide (as H_2S or HS^-) in the aqueous phase.[1] At excess sulfide concentrations, solubility of some metals can be increased by the formation of thio complexes. However, the stability of these complexes is still questionable.

The sulfide ions necessary to mediate sulfide precipitation can be directly injected into a reactive zone in the form of sodium sulfide (Na_2S). However, the sulfide ion (S^{2-}) is one of the most reduced ions and its stability within the reactive zone is short-lived. It will be converted to sulfate(SO_4^{--}) very quickly in the presence of oxidizing conditions within the contaminated plume. Addition of a very easily biodegradable organic substrate, such as carbohydrates, will enhance the formation of reduced, anaerobic conditions by depleting the available oxidation potential. The presence of carbohydrates serves two purposes: microorganisms use it as their growth substrate by depleting the available oxygen, and use it as an energy source for the reduction of sulfate to sulfide.

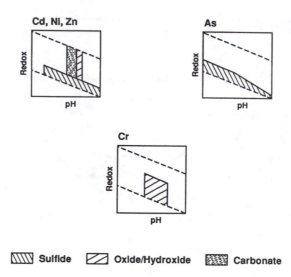

Figure 8.3 Typical patterns for redox–pH diagrams for Cd, Ni, Zn, As, and Cr. (From Schulin, R., Geiger, G., and Furrer, G., Heavy metal retention by soil organic matter under changing environmental conditions, in *Biogeodynamics of Pollutants in Soils and Sediments*, Salomons, W. and Strigliani, W. M., Eds., Springer, Berlin, 1995.)

Indirect microbial transformation of metals can occur as a result of sulfate reduction when anaerobic bacteria oxidize simple carbon substrates with sulfate serving as the electron acceptor. The net result of the process is the production of hydrogen sulfide (H_2S) and alkalinity (HCO_3^-). Sulfate reduction is strictly an anaerobic process and proceeds only in the absence of oxygen. The process requires a source of carbon to support microbial growth, a source of sulfate, and a population of sulfate-reducing bacteria. Dilute black strap molasses solution is an ideal feed substrate for this purpose, since typical black strap molasses contains approximately 20% sucrose, 20% reducing sugars, 10% sulfated ash, 20% organic nonsugars, and 30% water.[2]

The H_2S formed in the reaction can react with many types of contaminant metals to precipitate metals as insoluble metallic sulfides. The process follows the reaction

$$H_2S + Me^{++} \rightarrow MeS \downarrow + 2H^+ . \tag{8.4}$$

As noted earlier, Cu, Ag, Cd, Pb, Zn, Ni, Co, in addition to Fe and Mn can be precipitated as metallic sulfides. Precipitated metallic sulfides will remain in an insoluble, stable form,[3] unless the subsurface redox conditions change dramatically. The solubility products of various metallic precipitates are shown in Table 8.1. It can be seen from Table 8.1 that metallic sulfides are much more stable precipitates than hydroxides.

Table 8.1 Comparison of Solubility Products for Various Metallic Precipitates

	$-\log K_{sp}$ at 25°C	
Metal	Hydroxide	Sulfide
Cu	$Cu(OH)_2 = 17$	$CuS = 36.1$
Zn	$Zn(OH)_2 = 16.9$	$ZnS = 24.7$
Cd	$Cd(OH)_2 = 20.4$	$CdS = 27.0$
Hg	$Hg(OH)_2 = 21.1$	$HgS = 52.7$
Cr	$Cr(OH)_3 = 31$	—

From Stum, W. and Morgan, J. J., *Aquatic Chemistry*, John Wiley & Sons, New York, 1981.

The production of alkalinity from sulfate reduction reactions causes an increase in pH, which can result in metal precipitation through the formation of insoluble metal hydroxides or oxides. This process follows the reaction

$$Me^{2+} + 2H_2O \rightarrow Me(OH)_2 \downarrow + 2H^+ . \tag{8.5}$$

8.2.1.1 Chromium Precipitation

In situ microbial reduction of dissolved hexavent chromium Cr(VI) to trivalent chromium Cr(III) yields significant remedial benefits, because Cr(III) is less toxic, less mobile, and precipitates out of solution much more readily. In fact, it has been stated that the attenuation of Cr(VI) in the reduced Cr(III) from within an aquifer is a viable groundwater remediation technique.[4]

In situ microbial reduction of Cr(VI) to Cr(III) can be promoted by injecting a carbohydrate solution, such as dilute molasses solution. The carbohydrates, which consist mostly of sucrose, are readily degraded by the heterotrophic microorganisms present in the aquifer, thus depleting all the available dissolved oxygen present in the groundwater. Depletion of the available oxygen present causes reducing conditions to develop. The mechanisms of Cr(VI) reduction to Cr(III), under the induced reducing conditions can be (1) likely a microbial reduction process involving Cr(VI) as a terminal electron acceptor for the metabolism of carbohydrates, by species such as *Bacillus subtilis*,[5] (2) an extracellular reaction with by-products of sulfate reduction such as H_2S,[4] and (3) abiotic oxidation of the organic compounds including the soil organic matter such as humic and fulvic acids.

The primary end product of Cr(VI) to Cr(III) reduction process is chromic hydroxide [Cr(OH)$_3$], which readily precipitates out of solution under alkaline to moderately acidic and alkaline conditions.[4] To ensure that this process will provide both short-term and long-term effectiveness in meeting groundwater cleanup objectives, the chromium precipitates must remain immobilized within the soil matrix of the aquifer, and shall not be subject to Cr(OH)$_3$ precipitate dissolution or oxidation of Cr(III) back to Cr(VI) once groundwater conditions revert back to natural conditions. Based on the results of significant research being conducted on the *in situ* chromium reduction process,[4] it is readily apparent that the Cr(OH)$_3$ precipitate is essentially an insoluble, stable precipitate immobilized in the soil matrix of the aquifer.

Contrary to the numerous natural mechanisms that cause the reduction of Cr(VI) to Cr(III), there appear to be only a few natural mechanisms for the oxidation of Cr(III). Indeed, only two constituents in the subsurface environment (dissolved oxygen and manganese dioxide) are known to oxidize Cr(III) to Cr(VI).[7] The results of studies conducted on the potential reaction between dissolved oxygen and Cr(III) indicate that dissolved oxygen will not cause the oxidation of Cr(III) under normal groundwater conditions.[7,8] Studies have shown that Cr(III) can be oxidized by manganese dioxides, which may be present in the soil matrix.[7] However, only one phase of manganese dioxides is known to oxidize appreciable amounts of Cr(III), and this process is inversely related to groundwater pH.[7] Hence, the oxidation of Cr(III) back to Cr(VI) in a natural aquifer system is highly unlikely.

The Cr(OH)$_3$ precipitate has an extremely low solubility (solubility product K_{sp} = 6.7×10^{-31}),[9] and thus, very little of the chromium hydroxide is expected to remain in solution. It has been reported that aqueous concentration of Cr(III), in equilibrium with Cr(OH)$_3$ precipitates, is around 0.05 mg/l within the pH range of 5 to 12 (Figure 8.4).[4] The pH range of natural aquifer systems will be within 5 to 12, and hence the potential for the chromic hydroxide to resolubilize is unlikely. Furthermore, the potential for coprecipitation with ferric ions will further decrease the solubility of Cr(OH)$_3$.

Dissolved Cr(VI) can be also precipitated as Cr(OH)$_3$ in a reactive zone, also by the injection of ferrous sulfate solution into a reactive zone at appropriate concentrations. Cr(VI)

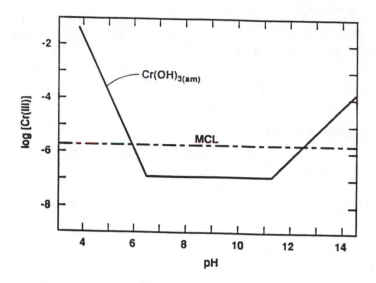

Figure 8.4 Cr(III) concentration in equilibrium with Cr(OH)$_3$.

exists as chromate, CrO_4^{2-}, under neutral or alkaline conditions and dichromate, $Cr_2O_7^{2-}$, under acidic conditions. Both species react with ferrous ion:

$$\text{Acidic conditions:} \qquad Cr_2O_7^{2-} + 6Fe^{2+} + 14H^+ \rightarrow 2Cr^{3+} + 6Fe^{3+} + 7H_2O \qquad (8.6)$$

$$\text{Neutral or alkaline conditions:} \quad CrO_4^{2-} + 3Fe^{2+} + 4H_2O \rightarrow Cr^{3+} + 3Fe^{3+} + 8OH^-. \qquad (8.7)$$

Both Cr(III) and Fe(III) ions are highly insoluble under natural conditions of groundwater (neutral pH or slightly acidic or alkaline conditions).

$$Fe^{3+} + 3OH^- \rightarrow Fe(OH)_3 \downarrow \qquad (8.8)$$

$$Cr^{3+} + 3OH^- \rightarrow Cr(OH)_3 \downarrow. \qquad (8.9)$$

The addition of ferrous sulfate into the reactive zone may create acidic conditions, and hence the zone downgradient of the ferrous sulfate injection zone may have to be injected with soda ash or caustic soda to bring the pH back to neutral conditions.

8.2.1.2   Arsenic Precipitation

Soluble arsenic occurs in natural waters only in the pentavalent, As(V), and trivalent, As(III), oxidation states. Although both organic and inorganic forms of arsenic have been detected, organic species (such as methylated arsenic) are rarely present at concentrations greater than 1 ppb and are generally considered of little environmental significance compared with inorganic arsenic species. Thus, this discussion focuses exclusively on the behavior of inorganic arsenic.

Thermodynamics provides useful insight into the equilibrium chemistry of inorganic arsenic species. In oxygenated waters, As(V) is dominant, existing in anionic forms of either $H_2AsO_4^-$, $HAsO_4^{2-}$, or AsO_4^{3-} over the pH range of 5 to 12, which covers the range encountered in natural groundwater.[10] Under anoxic conditions, As(III) is stable, with nonionic (H_3AsO_3)

and anionic ($H_2AsO_3^-$) species dominant below and above pH 9.22, respectively. In the presence of sulfides, precipitation of AsS (realgar) or As_2S_3 (orpiment) may remove soluble As(III) and exert considerable control over trace arsenic concentrations.[11,12]

The thermodynamic reduction of As(V) to As(III) in the absence of oxygen could be chemically slow and may require bacterial mediation.[13] As noted in the previous section, injection of dilute solution of blackstrap molasses will create the reducing conditions for As(V) to be reduced to As(III) and also provide the sulfide ions for As(III) to precipitate as As_2S_3. These reactions are described by the following equations.[10]

Reduction of As(V) to As(III) under anaerobic conditions:

$$HAsO_4^{2-} \rightarrow HAsO_2 .$$ (8.10)

In the presence of S^{--} under anaerobic conditions:

$$HAsO_2 + S^{--} \rightarrow As_2S_3 \downarrow .$$ (8.11)

Within oxygenated zones in the aquifer, oxidation of ferrous ion (Fe(II)) and Mn(II) leads to formation of hydroxides that will remove soluble As(V) by coprecipitation or adsorption reactions. The production of oxidized Fe–Mn species and subsequent precipitation of hydroxides are analogous to an *in situ* coagulation process for removing As(V).[10]

8.2.2 *In Situ* Denitrification

In situ denitrification can be accomplished by organisms belonging to the genera *Micrococcus*, *Pseudomonas*, *Denitrobacillus*, *Spirillum*, *Bacillus*, and *Achromobacter*, which are present in the groundwater environment. Denitrifying organisms will utilize nitrate or nitrite in the absence of oxygen as the terminal electron acceptor for their metabolic activity. If any oxygen is present in the environment, it will probably be used preferentially. The energy for the denitrifying reactions is released by organic carbon sources that act as electron donors. The microbial pathways of denitrification include the reduction of nitrate to nitrite and the subsequent reduction of nitrite to nitrogen gas.

$$NO_3^- \rightarrow NO_2^- \rightarrow N_2 \uparrow .$$ (8.12)

In biological wastewater treatment processes employing denitrification, a cheap, external carbon source such as methanol is added as the electron donor. It has long been known that NO_3^- can be converted to N_2 gas in anaerobic groundwater zones in the presence of a labile carbon source.[14]

In situ microbial denitrification is based on the same principle as conventional biological wastewater treatment systems, except that it is carried out in the subsurface by injecting the appropriate organic carbon source. Since methanol could be an objectionable substrate from a regulatory point of view, sucrose or sugar solution is an optimum substrate to be injected.

It should be noted that in the hierarchy of redox reactions, NO_3^- is the most favored electron acceptor after dissolved oxygen. Hence, considerable attention should be focused in maintaining the redox potential in the optimum range, so that Mn(IV), Fe(III), sulfate reduction conditions, or methanogenic conditions are not formed in the subsurface. Furthermore, since denitrification is a reduction reaction, alkalinity and pH tend to increase in the aquifer. Since the end product N_2 gas will escape into the vadose zone, and hence the aquifer system is not a closed system, increased alkalinity will be observed in the groundwater. If the NO_3^- concentration is not very high, this concern will be short-lived.

8.2.3 Abiotic Reduction by Dithionite

The dithionite or hydrosulfite ion, $S_2O_4^{2-}$, may be conceptualized as two sulfoxyl (SO_2^-) radicals joined by a sulfur–sulfur bond. The S–S bond in $S_2O_4^{2-}$ is considerably longer (and hence weaker) than typical S–S bonds, and this weak link in the $S_2O_4^{2-}$ ion is the key to its chemistry, because the ion reversibly dissociates to form two SO_2^- radicals that are strong and highly reactive reductants.[15] Since the SO_2^- radicals are extremely reactive, the dissociation of the $S_2O_4^{2-}$ is the slow and rate-limiting step in the reactions.

Injection of sodium thionite was found to reduce structural ferric ion in the soils, thus producing the reduced ferrous iron which remains available to react with soluble oxidized compounds present in the groundwater.[15,16] Further experiments indicated that dithionite could abiotically reduce carbon tetrachloride (CCl_4) to achieve more than 90% removal.[15]

It was reported that the half life of dithionite ion is about 2 or 3 days and this half life is adequate for reducing the contaminants within the induced reactive zone, while ensuring that dithionite does not remain as a contaminant in the groundwater for an extended time.[16]

8.2.4 *In Situ* Chemical Oxidation

During chemical oxidation of an organic compound, the compound is converted by an oxidizing agent into harmless end products, typically having either a higher oxygen or lower hydrogen content than the original compound. Typical oxidizing agents that have been used in the past under varying circumstances include chlorine dioxide (ClO_2), hypochlorite (either sodium (NaOCl) or calcium ($Ca(OCl)_2$), hydrogen peroxide, and ozone. Recently potassium permanganate (KMnO4) has been considered as a strong chemical oxidant to address chlorinated organic compounds. Injection of chlorine-based oxidants into a reactive zone may be objectionable from a regulatory point of view, due to the potential of forming various chlorinated by-products.

Hydrogen peroxide is a powerful oxidizing agent whose decomposition products, water and oxygen, are nontoxic.

$$H_2O_2 \rightarrow H_2O + \ddot{O} \qquad\qquad (8.13)$$

$$2H_2O_2 \rightarrow 2H_2O + O_2 . \qquad\qquad (8.14)$$

Solutions of hydrogen peroxide are relatively safe, effective, and easy to use. Hydrogen peroxide can be a source of hydroxyl radicals, one of the most potent oxidizers known. The hydroxyl radical is second only to fluorine in oxidation potential among the common oxidants.

There are two main methods of producing hydroxyl radicals (OH.) with hydrogen peroxide in an *in situ* reactive zone. In the first method, Fenton's reagent can be used to enhance the production of hydroxyl radicals. Fenton's reagent is a mixture of hydrogen peroxide and ferrous salts.

$$Fe^{2+} + H_2O_2 \rightarrow Fe^{3+} + OH^- + OH. \qquad\qquad (8.15)$$

Hydroxyl free radical (OH.) generated as an intermediate reaction product is a very strong, nonselective oxidant for a wide range of organic compounds, in addition to being a very strong oxidant. The magnitude of the rate constant for the reactions between OH. and organic compounds generally lies within a range of 10^9 to 10^{11} M^{-1} s^{-1}, which for practical purposes is an "instantaneous" rate for oxidation.[17]

The second main method is by an advanced oxidation process involving the mixed injection of hydrogen peroxide and ozone into the reactive zone.

$$H_2O_2 + O_3 \rightarrow 3O_2 + 2OH. \tag{8.16}$$

The emergence of *in situ* ozonation techniques for groundwater remediation is a direct result of advances made in air sparging technology in the last few years. For *in situ* ozonation, a properly designed air sparging system is an effective way to deliver ozone to the reactive zone. Furthermore, ozone can oxidize organic contaminants by direct oxidation in addition to the oxidation by free hydroxyl radical. Ozone as a direct oxidant is the third strongest oxidant after fluorine and hydroxyl radical.

Contaminants most amenable to oxidation by H_2O_2/O_3 injection include polynuclear aromatic hydrocarbons (PAHs), chlorinated ethenes such as trichloroethylene (TCE) or dichloroethylene (DCE), and petroleum-related compounds such as benzene, toluene, ethyl benzene, and xylenes.

Injection of H_2O_2/O_3 into a reactive zone may be integrated with other remediation technologies to achieve the highest possible treatment efficiencies. The amount of oxygen generated as a result of H_2O_2/O_3 injection will increase the dissolved oxygen content, and thus enhance the rate of aerobic biodegradation. Used as a pretreatment step before bioremediation, H_2O_2/O_3 injection can break down some complex organics, presenting simpler organic molecules that are more amenable to biodegradation. However, it should be noted that both H_2O_2 and O_3 at high enough concentrations or long enough residence times can act as sterilizing agents, killing the microbial population. Thus, when designing a system that integrates bioremediation and oxidation within the reactive zone, special consideration should be given to injection flow rate, concentration, and residence time.

Potassium permanganate ($KMnO_4$) is used as an oxidizing agent to control odor in sewage treatment plants and industrial wastewater treatment. Potassium permanganate is available either as granular or needle-shaped crystals, both of dark purple color. The solubility in water is strongly influenced by temperature, and at high concentrations the solution has a characteristic purple color.

Oxidations with permanganate can occur via several different reaction paths:[18] electron abstraction, hydrogen atom abstraction, hydride-ion abstraction, and direct donation of oxygen to the organic substrate. The pH of the system will determine whether the oxidation will involve one, three, or five electron exchange and whether the reaction will be fast or slow. It should be noted that oxidation potential increases with decreasing pH due to the capacity to accept a higher number of electrons.

Under extremely alkaline conditions (pH > 12), one electron is transferred:

$$MnO_4^- + e^- \rightarrow MnO_4^{2-}. \tag{8.17}$$

In the pH range of 3.5 to 12, three electrons are transferred:

Acidic conditions: $\qquad MnO_4^- + 4H^+ + 3e^- \rightarrow MnO_2 + 2H_2O \tag{8.18}$

Alkaline conditions: $\qquad MnO_4^- + 2H_2O + 3e^- \rightarrow MnO_2 + 4OH^-. \tag{8.19}$

Under more strongly acidic conditions (pH < 3.5), five electrons are transferred:

$$MnO_4^- + 8H^+ + 5e^- \rightarrow Mn^{2+} + 4H_2O. \tag{8.20}$$

Besides pH, permanganate oxidation reaction rate and degree depends on temperature, time, and concentrations. Organic compounds that contain carbon–carbon double bonds, primary alcohols, polynuclear aromatic hydrocarbons (PAHs), phenols, amines, and organic sulfur compounds can be potentially oxidized by $KMnO_4$ under optimum conditions in an *in situ* reactive zone.[18]

In situ chemical oxidation within a reactive zone may also oxidize compounds other than the target contaminants. Humic substances, carbonates, bicarbonates, organic content of the soil, and inorganic ions may also be oxidized in addition to the contaminants. Hence, efficiency of utilization of the oxidizing agent in targeting the contaminant mass may be significantly lower than stoichiometric predictions.

8.2.5 *In Situ* Microbial Mats

Inoculation of specialized microbial cultures for the *in situ* degradation of specific target compounds has been tried in the past.[19] Usually, naturally occurring microbes are grown in surface bioreactors, separated from their growth medium, resuspended in groundwater from the site, and then reinjected into the reactive zone. The habitat for the inoculated microbial population can be the native soils or specially designed porous filters. Concentration of target contaminants is gradually increased in the surface bioreactor to enhance the acclimatization and species selection of the microbial population.

The microbial mat concept is based upon intercepting the migrating plume within the reactive zone. The survivability of the injected microbial culture with the highly specific species of microorganisms may not be long enough within the reactive zone. Thus, replenishment of the specialized, selective microbial culture should be accomplished by reinoculation.

The microbial mat concept can be implemented for various types of biodegradable contaminants. Both aerobic and anaerobic reactions can be accomplished within these reactive zones. However, it should be noted that this concept is still considered experimental and has to go through extensive field testing.

8.3   AQUIFER PARAMETERS AND TRANSPORT MECHANISMS

Redox processes can induce strong acidification or alkalinization of soils and aquifer systems. Oxidized components are more acidic (SO_4^{2-}, NO_3^-) or less basic (Fe_2O_3) than their reduced counterparts (H_2S, NH_3). As a result, alkalinity and pH tend to increase with reduction and decrease with oxidation. Carbonates are efficient buffers in natural aquifer systems in the neutral pH range.

Many events can cause changes in redox conditions in an aquifer. Infiltration of water with high dissolved oxygen concentration, fluctuating water table, excess organic matter, introduction of contaminants that are easily degradable, increased microbial activity, and deterioration of soil structure can impact the redox conditions in the subsurface. However, there is an inherent capacity to resist redox changes in natural aquifer systems. This inherent capacity depends on the availability of oxidized or reduced species. Redox buffering is provided by the presence of various electron donors and electron acceptors present in the aquifer.

An engineered *in situ* reactive zone has to take into consideration how the target reactions will impact the redox conditions within and downgradient of the reactive zone, in addition to degrading the contaminants with the available residence time. Furthermore, careful evaluation should be performed regarding the selectivity of the injected reagents toward the target contaminants and the potential to react with other compounds or aquifer materials. Careful monitoring, short-term and long-term, should be performed to determine whether the natural equilibrium conditions can be restored at the end of the remediation process. In some cases modified biogeochemical equilibrium conditions may have to be maintained over a long period of time to prevent the reoccurrence of contaminants.

8.3.1 Contaminant Removal Mechanisms

As noted earlier, the mechanisms used to reduce the toxicity of dissolved contaminants can be grouped into two major categories: transformation and immobilization. Examples of some of these mechanisms have been discussed in Section 8.2. Conversion of chlorinated organic compounds to innocuous end products such as CO_2, H_2O, and Cl^- either by biotic or abiotic reaction pathways is an example of transformation mechanisms. Precipitation of Cr(VI) as $Cr(OH)_3$ by either abiotic or biotic reaction pathways and subsequent filtration by the soil matrix is an example of immobilization mechanisms.

It can be assumed, in most cases, that the end products of transformation mechanisms will result in dissolved and gaseous species. It can also be assumed that the impact of these end products on the natural redox equilibrium will be short-term. If the impact is expected to be significant, it can be controlled by limiting the reaction kinetics and the transport of the end products away from the reaction zone. Dilution and escape of dissolved gases will also help in restoring the natural equilibrium conditions in the reaction zone.

Immobilization mechanisms, which include heavy metals precipitation reactions, in reality transform the contaminant into a form (precipitate) that is much less soluble. In addition, transport of dissolved heavy metals in groundwater should also be considered as a two-phase system in which the dissolved metals partition between the soil matrix and the mobile aqueous phase.

Metal precipitates resulting from an *in situ* reactive zone may move in association with colloidal particles or as particles themselves of colloidal dimensions.[20] The term colloid is generally applied to particles with a size range of 0.001 to 1 μm. The transport of contaminants as colloids may result in unexpected mobility of low solubility precipitates. It is important to remember that the transport behavior of colloids is determined by the physical/chemical properties of the colloids as well as the soil matrix.

Metal precipitates may be pure solids (e.g., PbS, ZnS, $Cr(OH)_3$) or mixed solids (e.g., $(Fe_x, Cr_{1-x}) (OH)_3$, $Ba(CrO_4, SO_4)$). Mixed solids are formed when various elements coprecipitate or due to interaction with aquifer materials. The potential for many interactions of heavy metal cations in the aquifer matrix is shown in Figure 8.5.

Colloidal precipitates larger than 2 μm in the low flow conditions common in aquifer systems will be removed by sedimentation.[22] Colloidal precipitates are more often removed mechanically in the soil matrix. Mechanical removal of particles occurs most often by straining, a process in which particles can enter the matrix, but are caught by the smaller pore spaces as they traverse the matrix.[22]

Colloidal particles below 0.1 μm will be subjected more to adsorptive mechanisms than mechanical processes. Adsorptive interactions of colloids may be affected by the ionic strength of the groundwater; ionic composition; quantity, nature, and size of the suspended colloids; geologic composition of the soil matrix; and flow velocity of the groundwater. Higher levels of total dissolved solids (TDS) in the groundwater encourages colloid deposition.[22]

In aquifer systems with high Fe concentrations, the amorphous hydrous ferric oxide can be described as an amphoteric ion exchange media. As pH conditions change, it has the capacity to offer hydrogen ions (H^+) or hydroxyl ions (OH^-) for cation or an ion exchange, respectively.[23] Adsorption behavior is primarily related to pH (within the typical range of 5.0 to 8.5), and at typical average concentrations in soil, the iron in a cubic yard of soil is capable of adsorbing from 0.5 to 2 lb of metals as cations or metallic complexes.[23] This phenomenon is extremely useful for the removal of As and Cr.

8.4 DESIGN OF *IN SITU* REACTIVE ZONES

The optimization of subsurface environmental conditions to implement target reactions for remediating groundwater plumes holds a lot of promise. Application and treatment

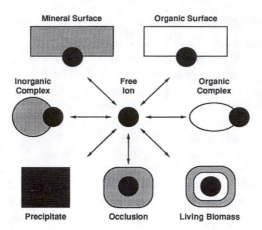

Figure 8.5 Heavy metal interactions in an aquifer matrix. From Schulin, R., Geiger, G., and Furrer, G., Heavy metal retention by soil organic matter under changing environmental conditions, in *Biogeodynamics of Pollutants in Soils and Sediments*, Salomons, W. and Strigliani, W. M., Eds., Springer, Berlin, 1995.

efficiencies of these same processes in above-ground pump and treat systems were impacted by the required residence times and the infrastructure required to implement these processes. However, when dealing with slow-moving groundwater plumes, long available residence times can be utilized as an advantage to implement cost-effective remediation strategies.

In situ reactive zones can be designed as a curtain or multiple curtains to intercept the moving contaminant plume at various locations. An obvious choice for the location of an intercepting curtain is the downgradient edge of the plume. This curtain will act as a containment curtain to prevent further migration of the contaminants (Figure 8.6A). A curtain can be installed slightly downgradient of, or within, the source area to prevent the mass flux of contaminants migrating from the source (Figure 8.6B). This will shrink the size of the contaminant plume faster. If the duration of remediation is a critical factor, another curtain can be installed between the above two curtains for further interception at the middle of the plume (Figure 8.6C).

Another approach to designing an *in situ* reactive zone is to create the reactive zone across the entire plume. The injection points can be designed on a grid pattern to achieve the reactions across the entire plume. However, it should be noted that the cost of installation of injection wells constitutes the biggest fraction of the system cost, looking at both capital and operational costs. Hence, it becomes very clear that the reduction of the total number of injection wells will significantly reduce the system costs, and this leads to the conclusion that the curtain concept will be the preferred and most cost-effective approach to implement *in situ* reactive zones.

The three major design requirements for implementing an *in situ* reactive zone are (1) creation and maintenance of optimum redox environment and other biogeochemical parameters such as pH, presence or absence of dissolved oxygen, and temperature, etc., (2) selection of the target process reactions and the appropriate reagents to be injected to achieve these reactions, and (3) delivery and distribution of the required reagents in a homogeneous manner across the entire reactive zone, both in the lateral and vertical directions.

8.4.1 Optimum Pore Water Chemistry

The composition of interstitial water is the most sensitive indicator of the types and the extent of reactions that will take place between contaminants and the injected reagents in the aqueous phase. Determination of the baseline conditions of the appropriate biogeochemical parameters is a key element for the design of an *in situ* reactive zone. This evaluation will

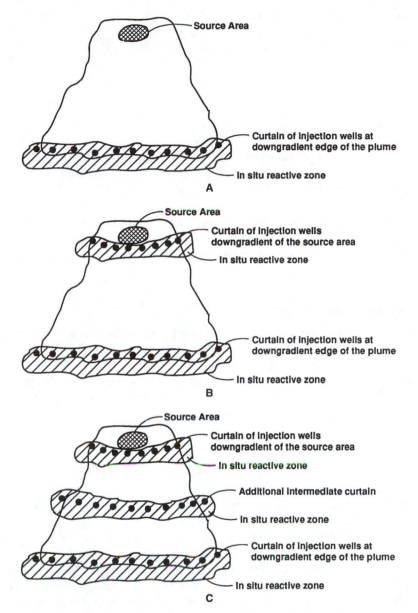

Figure 8.6 *In situ* reactive zones based on the curtain concept. A. One curtain at downgradient edge. B. Two curtains at downgradient edge and at source area. C. Three curtains to remediate the plume faster.

give a clear indication of the existing conditions and the necessary steps to be taken to optimize the environment to achieve the target reactions. A potential list of the biogeochemical parameters is presented below:

- dissolved oxygen
- pH
- temperature
- redox potential
- total organic carbon (dissolved and total)
- total dissolved solids
- total suspended solids

- NO_3^-
- NO_2^-
- SO_4^{2-}
- S^{--}
- Fe (total and dissolved)
- Mn (total and dissolved)
- carbonate content
- alkalinity
- concentration of dissolved gases (CO_2, N_2, CH_4, etc.)
- microbial population enumeration (total plate count and specific degraders count)
- any other organic or inorganic parameters that have the potential to interfere with the target reactions.

It should be noted that the number of parameters that need to be included in the list of baseline measurements will be site-specific and will be heavily influenced by the target reactions to be implemented within the reactive zone. The above list is a universal list and should be used as a reference.

8.4.2 Reactions and Reagents

Based on the preliminary evaluation of the existing subsurface environment, appropriate reagents have to be selected to optimize the environment as well as to achieve the target reactions. The reactions and reagents and their interaction has been discussed in detail in Section 8.2. Injection of dilute blackstrap molasses solution is an example, where precipitation of metals can be achieved in an anaerobic environment due to the reaction between the heavy metal cations and the sulfide ions.

8.4.3 Injection of Reagents

Design of the reagents injection system requires an extensive evaluation and understanding of the hydrogeologic conditions at the site and specifically within the plume and the location of the reactive zones. This understanding has to include both a macroscopic, site-wide pattern and at microscopic levels between layers of varying geologic sediments. Specific geologic/hydrogeologic parameters required for the design of an *in situ* reactive zone are presented in Table 8.2.

As noted earlier, delivery, distribution, and proper mixing of the injected reagents is a key element to the success of remediation within an *in situ* reactive zone. Location and spacing of the injection wells and the placement of screens within each well (cluster) is critical to achieve the above objective.

Injection of reagents can be implemented in two ways: (1) gravity feed, and (2) pressure injection deeper into the well. Gravity feed is feasible only under conditions when the depth of contamination is very shallow (Figure 8.7). Under gravity feed conditions, injected reagents will tend to spread over the water table as a sheet flow, and the mixing within the reactive zone will be dominated by diffusion, rather than advective flow.

When the depth of contamination is deeper, multiple injection points may be required within a well cluster at each injection point (Figure 8.8). The reagent solution will have to be injected under pressure into the injection well. It is preferable to release the reagents at the bottom of each screen. If needed, mixing within each well can be provided by recirculating pumps placed in each well. Under this configuration, mixing within the reactive zone will be influenced by both advective and diffusional transport of the reagents. Concentration of the injected feed solution should be dilute enough to avoid any downward migration due to density differences between the reagent and groundwater.

Table 8.2 Impacts of Various Geologic/Hydrogeologic Parameters on the Design of an *In Situ* Reactive Zone

Geologic/hydrogeologic parameter	Design impact
Depth to water table	Injection well depth and screen locations
Width of contaminant plume	Number of injection wells
Depth of contaminant plume	Number of injection points within a well cluster
	Pressure injection vs. gravity feed
Groundwater velocity	Injection flow rate, residence time for the target reactions
	Dilution of end products
Hydraulic conductivity (horizontal and vertical)	Mixing zones of reagents, extent of reactive zone
	Number of injection points within a well cluster
Geologic variations, layering of various soil sediments	Location of well screens within injection points
Soil porosity and grain size distribution	Removal of end products resulting from immobilization reactions (such as heavy metals precipitation)

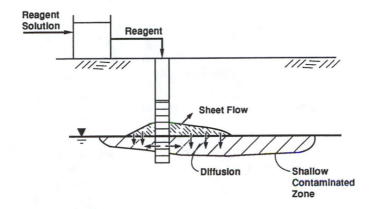

Figure 8.7 Gravity feed of reagents when the contamination is shallow.

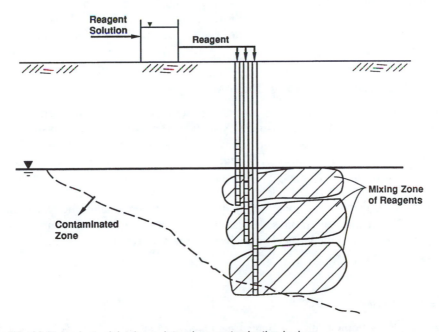

Figure 8.8 Multiple cluster injection points when contamination is deep.

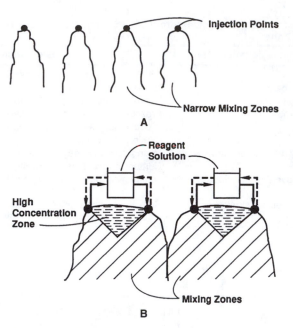

Figure 8.9 Injection point configurations. A. Narrow mixing zones downgradient of injection points. B. Cyclic extraction and injection of adjoining wells.

During gravity feed of the reagents, the lateral spread of the injected solution will be significant due to the sheet flow effect. However, under pressure injection conditions, down-gradient migration of the injected reagents and thus the mixing zone could be very narrow, depending on the hydrogeologic conditions within the reactive zone (Figure 8.9A.) One way to overcome this problem is to install closely spaced injection points. This option, even though easier to implement, will significantly increase the cost of the system. Cyclic extraction and injection of adjoining wells, treated as a pair, will create a wider mixing zone downgradient of the injection wells, and thus will eliminate the need to install closely spaced injection points (Figure 8.9B). Extracted groundwater can be used as the dilution water to maintain the feed injection solution concentrations.

8.4.4 Laboratory Bench-Scale Studies

It is always preferable to perform a laboratory study to determine whether the proposed target reactions are achievable. The laboratory study can be used to obtain data on (1) reagent chemistry in the subsurface, (2) intra-aqueous redox kinetics and manipulation, (3) required residence times for the target reactions, (4) required acclimatization period for any microbially induced reactions, (5) need for any system enhancements during scale-up to the field scale, and (6) fate of end products and side effects of the reaction on the aquifer.

The best results from a laboratory study will be obtained when the test is run with samples collected from the proposed location of the reactive zone. Column studies per-formed with core, soil samples, and groundwater obtained from the site will yield the most reliable results.

8.5 REGULATORY ISSUES

In most cases, implementation of an *in situ* reactive zone requires injection of appropriate reagents and manipulation of the redox and biogeochemical environment within the reactive

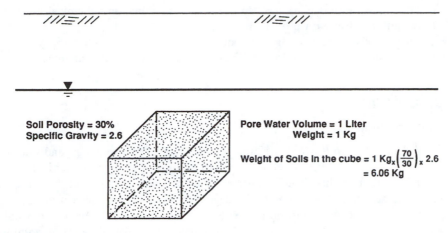

Figure 8.10 Unit cube with pore water volume of 1 l.

zone. Injection of reagents, albeit innocuous, nonhazardous, and nonobjectionable, may raise some alarms regarding the short-term and long-term effects within the aquifer.

During immobilization reactions—for example, heavy metals precipitation—the contaminant is immobilized within the soil matrix below the water table. As noted earlier, under natural conditions, this immobilization will be irreversible in most cases. Hence, the cleanup objective for the dissolved contaminant will be based upon the groundwater standard (for example, Cr(VI) = 10 μg/l, and when the contaminant is immobilized in the soil matrix the cleanup standard will be based upon the soil standards (Cr(III) = 100 mg/kg). The huge difference in the two standards for Cr (10 ppb vs. 100 ppm) in the two phases is a significant benefit and provides a major advantage for achieving remediation objectives through an *in situ* reactive zone.

In addition, consider a unit volume of the soil matrix below the water table, which has 1 l of water in its pore spaces (Figure 8.10). Assuming a porosity of 30% and soil specific gravity of 2.6, the same cube will have about 6.0 kg of soil. If the dissolved Cr(VI) concentration within the cube is 5 mg/l (ppm) the pore water within the cube contains 5 mg of Cr(VI) mass. When all this chromium is immobilized within the soil matrix of the cube, the concentration of the Cr(III) in the soil is equal to 0.83 mg/kg (ppm) (i.e., 5 mg divided by 6.5 kg). It becomes very clear that in addition to the much less stringent standards, the concentration itself is reduced significantly during immobilization within a reactive zone.

8.6 FUTURE WORK

The *in situ* reactive zone is an innovative and emerging technology in the remediation industry. Furthermore, implementation and wide acceptance of this technology is still in its infancy, and thus the experience and knowledge of this technology is very much empirically based. Substantial amount of developmental work needs to be done on this technology before it becomes widely accepted.

Future work should be focused on

- tools to design the appropriate specification of injection rates, durations, and concentrations to achieve optimal control at the field scale
- tools to predict/estimate and measure the target reaction kinetics in an *in situ* environment
- tools to quantify reagent and pore water chemistry at the field scale

- reactive transport modeling tools to couple the microbial and chemical reactions to the physical transport processes
- better methods to measure the intra-aqueous redox and biogeochemical kinetics
- better understanding of the long-term fate of the immobilized contaminants.

8.7 CASE STUDY

8.7.1 Introduction

A field demonstration test was performed to evaluate an innovative groundwater remediation technique involving the *in situ* reduction of chromium at an industrial facility in the midwestern U.S. To date, this evaluation has involved conducting a 6-month *in situ* test near the source area at the site to determine the degree to which hexavalent chromium can be reduced and precipitated out within the aquifer due to the development of biologically induced reducing conditions. The test was developed to evaluate innovative, low-cost *in situ* remediation techniques that could be used to potentially augment or replace the conventional groundwater pump and treat system which they currently operate at the facility. Although the existing pump and treat system provides containment of the chromium plume and a certain degree of chromium mass removal, the system is expensive to operate (hundreds of thousands of dollars annually) and does not provide any means for source reduction and active groundwater remediation. Thus, a low-cost *in situ* remediation technique that would achieve source reduction and active groundwater remediation may provide a high degree of remedial benefits, either in conjunction with the continued operation of the existing pump and treat system or as a stand-alone remediation approach.

The field test required the installation of three injection wells and five monitoring wells. These wells augment the existing monitoring well network at the facility. The three injection wells have been installed within the vacant facility building and the five monitoring wells have been installed along the eastern edge of the facility building (Figure 8.11). The newly installed injection and monitoring wells are shallow monitoring wells screened over the approximate interval of 10 to 15 ft below grade.

To promote the *in situ* biological reduction of hexavalent chromium Cr(VI) to trivalent chromium Cr(III), a dilute black strap molasses solution (which contains readily degradable carbohydrates and sulfur) has been periodically injected (at a batch feed rate of approximately 40 gal every 2 weeks per injection well) into the shallow portion of the impacted aquifer via the three injection wells. The carbohydrates, which consist mostly of sucrose, are readily degraded by the indigenous heterotrophic microorganisms present in the aquifer. This metabolic degradation process utilizes all of the available dissolved oxygen contained in the groundwater. Depletion of the available oxygen present in the groundwater causes reducing conditions to develop. Under the induced reducing conditions, the Cr(VI) is reduced to Cr(III). The actual mechanism of chromium reduction is likely a biotic oxidation–reduction process involving the Cr(VI) serving as a terminal electron acceptor for the catabolized carbohydrates. The primary end product of the Cr(VI) to Cr(III) reduction process is chromic hydroxide [$Cr(OH)_{(3)}$], which readily precipitates out of solution under alkaline to moderately acidic conditions. These precipitates are then retained (i.e., filtered out) by the soil particles within the aquifer.

8.7.2 Injection/Monitoring Well System

As stated previously, a total of three injection wells and five monitoring wells were installed to facilitate implementation and assessment of the biological *in situ* chromium reduction process. The three injection wells were installed within the former production area

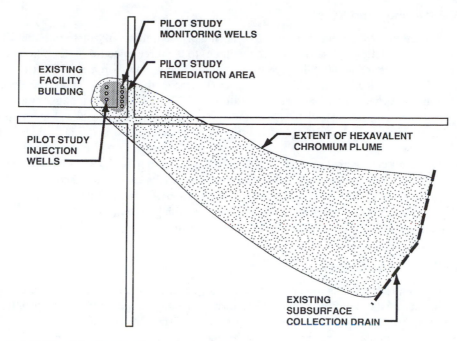

Figure 8.11 Site of *in situ* chromium reduction pilot study.

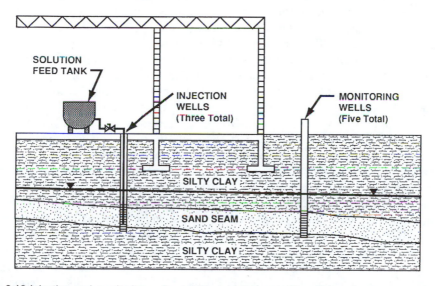

Figure 8.12 Injection and monitoring wells, *in situ* chromium reduction pilot study.

of the facility, while the five monitoring wells were installed just outside of the eastern edge of the existing facility building (Figure 8.11). The monitoring wells are positioned approximately 35 to 40 ft to the east of the injection wells. Each of the injection and monitoring wells were installed using the hollow stem auger method and consist of 4 in. diameter PVC casing and 5 ft long, 4 in. diameter, ten-slot PVC well screens. Each of the injection wells include a 1 in. PVC drop pipe for directing the injection solution to the middle of the screened interval. The installed injection and monitoring wells are shallow wells screened across a 1 to 3 ft thick sand seam at an approximate interval of 10 to 15 ft below grade. Although it is not known with certainty, it is assumed that this sand seam is continuous between the injection wells and the monitoring wells (Figure 8.12).

8.7.3   Solution Feed System

The solution feed system is designed to accurately and efficiently distribute the dilute water/molasses solution (200:1 dilution, by volume) to the three injection wells. The system components consist of a 300-gal polyethylene storage tank, approximately 100 ft of $\frac{1}{2}$ in. reinforced polyvinyl chloride (PVC) tubing with associated fittings to make the desired connections, and several PVC ball valves and stopcocks. The distribution tubing is connected to a hose barb fitting at each of the three injection wellheads. Each wellhead is equipped with a combination well seal/drop tube assembly through which solution is fed to the saturated zone.

Each month approximately 200 to 250 gal of water and 1 to $1\frac{1}{4}$ gal of molasses are added to the solution storage tank and mixed thoroughly. Biweekly batch feeding of 40 gal of solution to each injection well is performed (i.e., 80 gal per well per month). Each injection well is fed individually through the PVC distribution tubing system, and flow is controlled by manually adjusting the various ball valves and stopcocks.

8.7.4   Monitoring Events

Seven monitoring events were performed since system installation and start-up in December 1994. An initial monitoring event was performed in conjunction with system installation to determine baseline conditions. Subsequent to system start-up one monitoring event was performed each month for the first 6 months of the pilot study.

Monthly monitoring events included a series of measurements in each of the three injection wells and five monitoring wells to determine groundwater elevation, dissolved oxygen (DO) concentration, oxidation–reduction potential (ORP), pH, temperature, and hexavalent chromium [Cr(VI)] concentration. Depth to water measurements were performed to determine groundwater elevation and well water volume in each respective well. A YSI 6000 down-hole probe was used to measure DO, ORP, pH, and temperature in each injection well prior to bailing. Teflon bailers were used to bail three well volumes from each of the injection wells and monitoring wells in turn. Subsequent to the extraction of three well volumes from a respective well, the DO, ORP, pH, temperature, and Cr(VI) concentration were measured and recorded. A Hach Model CH-12 colorimetric test kit was used to determine hexavalent chromium concentrations in each well by mixing a 5 ml groundwater sample with a chromium reagent and comparing the sample to a concentration/color chart. Colorimetric field analysis results have agreed closely with laboratory analytical results. Groundwater samples were collected for laboratory analysis during the baseline monitoring event and then following the third month of system operation. These samples were analyzed for hexavalent chromium, total chromium, sulfate, sulfide, and total organic carbon concentrations.

Prior to initiating the *in situ* biological reduction pilot study, an initial microbiological enumeration task was performed. This initial assessment task was conducted to confirm that there is an adequate population of indigenous heterotrophic microorganisms present in the groundwater at the site. Groundwater samples were collected from existing monitoring wells and also from the influent to the existing groundwater treatment system. The collected samples were submitted for heterotrophic plate count (HPC) analyses. The results of the HPC analyses confirmed that there is an adequate population of heterotrophic bacteria indigenous to the aquifer and that it is possible to stimulate the microbial activity necessary to induce the required reducing conditions within the aquifer.

As mentioned earlier, monthly monitoring was performed to determine groundwater elevation, DO concentration, ORP, pH, temperature, and Cr(VI) concentration in each injection well. In addition, groundwater samples collected during the baseline monitoring event and then 3 months following process initiation were analyzed for hexavalent chromium, total

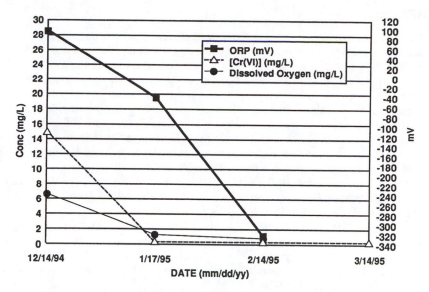

Figure 8.13 Graph of Cr(VI), DO, and ORP levels in injection well 1 (IW-1).

chromium, sulfate, sulfide, and total organic carbon concentration. Based on laboratory analysis, the initial Cr(VI) concentrations (baseline) in injection wells IW-1, IW-2, and IW-3 were 16, 1.2, and 0.02 mg/l, respectively. Field determined baseline Cr(VI) concentrations (15, 1, and <0.2 mg/l for IW-1, IW-2, and IW-3, respectively) agreed closely with the laboratory results.

The down-hole ORP data indicated that strong reducing conditions developed in and around the three injection wells within 1 month of process initiation and reached near steady-state conditions with 2 months of process initiation. Baseline ORP levels within the injection wells were within the range of +96 to +147 mV prior to initiating the carbohydrate solution injection process (indicating relatively moderate oxidizing conditions) and have been reduced down to a steady-state level of approximately −150 to −200 mV (indicating strong reducing conditions).

Because of the rapid inducement of reducing conditions, the concentration of hexavalent chromium in the injection wells decreased from a high of 15 mg/l to below 0.2 mg/l during the first month of process operation (Figure 8.13). The levels of hexavalent chromium measured in the injection wells (based both on field and laboratory analyses) remained below 0.2 mg/l through the first 6 months of monitoring. The laboratory analytical results for hexavalent chromium measured in the injection well samples collected 3 months following process initiation were all below the detection level of 0.05 mg/l. In these same samples, the levels of total chromium were slightly higher than the levels of hexavalent chromium but were still below the 0.2 mg/l groundwater cleanup objective for this site. This indicates that what chromium remains in the groundwater is in the trivalent form, rather than in the more toxic and mobile hexavalent form. In addition, it is important to note that the analytical data is based on unfiltered groundwater samples. Because only trace amounts of trivalent chromium have been detected in the unfiltered groundwater samples it is readily apparent that the chromium precipitates are being retained by the aquifer materials and are not subject to colloidal transport through the aquifer.

As expected based on the introduction of relatively high concentrations of carbohydrates and the development of strong reducing conditions, the levels of TOC have increased in the groundwater samples collected from the injection wells and the pH levels within the injection wells have decreased slightly since the start of the pilot study.

REFERENCES

1. Hong, J., Forstner, U., and Calmano, W., Effects of redox processes on acid producing potential and metal mobility in sediments, in *Bioavailability: Physical, Chemical and Biological Interactions*, Hamelink, J. L., et al., Eds., Lewis Publishers, Boca Raton, FL, 1994.

2. Sax, N. I. and Lewis, R. J., *Hamley's Condensed Chemical Dictionary*, 11th ed., Van Nostrand Reinhold, New York, 1987, 791.

3. Ganze, C. W., Wahlstrom, J. S., and Turner, D. R., Fate of heavy metals in sludge disposal landspread operation—A case history, *Water Sci. Technol.*, 19(8), 19, 1987.

4. Palmer, C. and Puls, R., Natural Attenuation of Hexavalent Chromium in Groundwater and Soils, U.S. Environmental Protection Agency, Office of Research and Development, Office of Solid Waste and Emergency Response, EPA/540/S-94/505, 1994.

5. Melhorn, R. J., Buchanan, B. B., and Leighton, T., Bacterial chromate reduction and product characterization, in *Emerging Technology for Bioremediation of Metals*, Means, J. L. and Hinchee, R. E., Eds., Lewis Publishers, Boca Raton, FL, 1994.

6. Stum, W. and Morgan, J. J., *Aquatic Chemistry*, John Wiley & Sons, New York, 1981.

7. Eary, L. and Rai, D., Kinetics of chromium (III) oxidation to chromium (VI) by reaction with manganese dioxides, *Environ. Sci. Technol.*, 21 (12), 1187, 1987.

8. Schroeder, D. and Lee, G., Potential transformations of chromium in natural waters, *J. Water Air Soil Pollution*, 4, 355, 1975.

9. DeFilippi, L., Bioremediation of hexavalent chromium in water, soil and slag using sulfate reducing bacteria, in *Remediation of Hazardous Waste Contaminated Soils*, Wise, D. and Trantolo, D., Eds., Marcel Dekker, New York, 1994.

10. Edwards, M., Chemistry of arsenic: Removal during coagulation and Fe-Mn oxidation, *J. Am. Water Works Assoc.*, 86 (9), 64, 1994.

11. Ferguson, J. F. and Garvis, J., A review of the arsenic cycle in natural waters, *Water Res.*, 6, 1259, 1972.

12. Lipton, D. S., et al., Combined removal of arsenic, VOCs, and SVOCs from groundwater using an anaerobic/aerobic bioreactor, in *Emerging Technology for Bioremediation of Metals*, Means, J. L. and Hinchee, R. E., Eds., Lewis Publishers, Boca Raton, FL, 1994.

13. McBride, B. C. and Wolfe, R. S., *Biochemistry*, 10, 4312, 1971.

14. Delwiche, C. C., The nitrogen cycle and nitrous oxide, in *Denitrification, Nitrification and Atmospheric Nitrous Oxide*, Delwiche, C. C., Ed., Wiley-Interscience, New York, 1981.

15. Amonette, J. E., et al., Abiotic reduction of aquifer materials by dithionite: A promising in situ remediation technology, in *In Situ Remediation: Scientific Basis for Current and Future Technologies*, Gee, G. W. and Wins, N. R., Eds., Battelle Press, Columbus, OH, 1994.

16. U.S. Environmental Protection Agency, In Situ Remediation Technology Status Report: Treatment Walls, EPA 542-K-94-004, 1995.

17. Rarikuma, J. X. and Gurol, M. D., Fenton's reagent as a chemical oxidant for soil contaminants, in *Chemical Oxidation*, Vol. 2, Eckenfelder, N. W., Bowers, A. R., and Roth, J. A., Eds., Technomic Publishing, Lancaster, PA, 1994.

18. Walton, J., Labine, P., and Reidies, A., The chemistry of permanganate in degradative oxidations, in *Chemical Oxidation*, Eckenfelder, W. W., Bowers, A. R., and Roth, J. A., Eds., Technomic Publishing, Lancaster, PA, 1994.

19. Taylor, R. T., et al., In situ bioremediation of trichloroethylene contaminated water by a resting cell methanotrophic microbial filter, *Hydrolog. Sci. J.*, 38 (4), 323, 1993.

20. Puls, R. W., Groundwater Sampling for Metals, Ch. 14, U.S. Environmental Protection Agency EPA/600/A-94/172, 1994.

21. Schulin, R., Geiger, G., and Furrer, G., Heavy metal retention by soil organic matter under changing environmental conditions, in *Biogeodynamics of Pollutants in Soils and Sediments*, Salomons, W. and Strigliani, W. M., Eds., Springer, Berlin, 1995.

22. Vance, D. B., Particulate transport in groundwater—I: Colloids, *Natl. Environ. J.*, Nov/Dec, 1994.

23. Vance, D. B., Iron: The environmental impact of a universal element, *Natl. Environ. J.*, May/June, 1994.

9   HYDRAULIC AND PNEUMATIC FRACTURING

9.1   INTRODUCTION

Low-permeability, fine-grained soils such as clay, and silt and rock represent a significant challenge to *in situ* remediation due to the meager rates of fluid flow extractable under these conditions. Despite the low permeability of clays, silts, and competent rock, these formations can still become impacted. The density of these formations often renders conventional cleanup techniques, such as soil vapor extraction and bioremediation, ineffective.

In the recent past, hydraulic and pneumatic fracturing methods have been developed for creating fractures in dense soils and making existing fractures larger to enhance the mass transfer of contaminants. The fractures created increase the effective permeability and change paths of fluid flow, thus making *in situ* remediation more effective and economical. Fracturing also reduces the number of extraction wells required, trimming labor and material costs.

Pneumatic fracturing injects highly pressurized air or other gas into consolidated, contaminated sediments to extend existing fractures and to create a secondary network of fissures and channels. This process accelerates the removal of contaminants by soil vapor extraction, bioventing, enhanced *in situ* biodegradation, and *in situ* electrokinetics. Typically, pneumatic fracturing is used in formations where the fractures will remain open for a long time without support.

Hydraulic fracturing involves injecting a fluid, usually water, at modest rates and high pressures into the soil matrix to be fractured. High pressure water is used to cut a disc-shaped notch at the bottom of a borehole: the notch becomes the starting point for the fracture. A slurry mixture of sand and biodegradable gel is then pumped at high pressure to create a distinct fracture. As the gel degrades, it leaves a highly permeable sand-lined fracture with the sand acting as a propping agent preventing the fracture from collapsing. The fractures, thus formed, can be utilized to augment various other *in situ* technologies discussed previously.

The utility of hydraulic fractures is by no means limited to well stimulation. Relatively large volumes of solid compounds can be delivered to the subsurface as granular materials, filling the fractures. The capability to deliver solid compounds, which previously required excavation techniques, presents a variety of possible new applications. These new *in situ* applications include injection of solid compounds that slowly release oxygen and nutrients to enhance *in situ* aerobic degradation; filling the fractures with electrically conductive material such as graphite to enhance electroosmosis and perhaps electrical heating for *in situ* vitrification; and filling the fractures with metal catalysts, such as elemental iron, to degrade a wide range of chlorinated organic compounds.

9.2   APPLICABILITY

Almost any rock or soil formation can be fractured, given enough time, energy, and effort. The key aspects that have to be considered for remediation purposes are: will the benefit derived from fracturing offset the cost of the process, and what are the risks and benefits of the process? Armed with the answers to these questions, the decision to proceed with testing and, ultimately, full-scale application of the technique can be made on an informed basis.

Fracturing is most appropriately applied to soils where the natural permeability is insufficient to allow adequate movement of fluids to achieve the remediation objectives in the desired time frame. The following soil types and rocks are generally suitable for applying fracture techniques:[1]

- silty clay/clayey silt
- sandy silt/silty sand
- clayey sand
- sandstone
- siltstone
- limestone
- shale

Fracturing a sand or gravel formation, while possible, is probably not justified because the benefits derived from the increase in soil permeability will not match the cost of the process.

Fracturing, by itself, is not a remediation technique. Fracturing has to be combined with other technologies to facilitate the reduction of contaminant mass and concentration. Fracturing techniques are equally applicable to both vadose zone (unsaturated) soils and saturated zone soils to improve the flow of air and water, respectively. Fracturing should be considered primarily to overcome the poor accessibility to the contaminants for extraction and also to overcome the difficulty in uniform delivery of treatment reagents.

By fracturing, not only are higher permeability zones created for enhancement of advective flow through the contaminated zone, but the pathways for diffusion-controlled migration of the contaminants are also created. The creation of advective flow channels and shortened diffusive pathways result in enhanced mass removal rates during soil vapor extraction. Diffusion-limited extraction will still influence the rate of contaminant recovery even after fracturing, and the properties of the contaminant and the media will still influence the residual concentrations.

As noted earlier, fracturing can also expand the applicability of other *in situ* remediation technologies beyond enhanced vapor and liquid extraction in low permeability soils. These technologies include

- *in situ* biodegradation (by enhancing the delivery of oxygen and nutrients into inaccessible locations)
- *in situ* electrokinetics (enhancing the fluid flow in the fractured zones)
- *in situ* vitrification (creating heating zones by injecting graphite into the fractures)
- *in situ* air sparging (by creating fractured pathways to collect the injected air laden with contaminants)

9.2.1 Geologic Conditions

As with all subsurface remediation techniques, fracturing is applicable only for a range of site conditions. In addition to the consideration of soil/rock types, described in the previous section, the mode of deposition of the sediments, and the changes that took place after deposition affect the effectiveness of fracturing. Most notably, the state of *in situ* stresses has long been characterized as the primary variable in the orientation of fracture formation.[2]

Fractures can be generated in geologic formations if the pressurized fluid is injected at a pressure which exceeds the natural strength, as well as the *in situ* stresses present in the

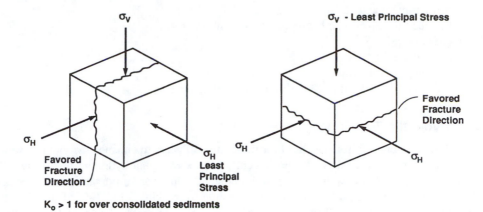

Figure 9.1 Directions of fracture formation as influenced by the least principal stress.

formation. It must also be injected at a flow rate that exceeds the natural permeability of the formation so that sufficient "back" pressure can be developed. Fractures will tend to propagate in the direction normal to the least principal stress in the formation, with propagation following the path of least resistance.

In situ stress fields are subdivided into horizontal and vertical components (σ_x, σ_y, and σ_z). When initially deposited, the three principal stresses are in equilibrium and are equal to the overburden pressure. External forces (tectonics, burial/excavation, glaciation, and cycles of desiccation/wetting) after deposition modify these stress fields.

Overconsolidation is defined as compaction of sedimentary materials exceeding that which was achieved by the original overburden. The formation and later melting of glaciers is one condition that results in overconsolidation. The weight of the ice on the soil initially compacts the sedimentary grains. When the ice melts, the vertical stress is relaxed but the horizontal stress still maintains a residual component of the loaded conditions. Erosion or overburden removal by excavation also present conditions which relax the vertical stress field. Additionally, the cyclic swelling and desiccation of clay rich formations can also create conditions of over consolidation.

In overconsolidated formations where the least principal stress is vertical, fractures will tend to propagate horizontally (Figure 9.1). Conversely, in normally consolidated or under consolidated formations, fractures will tend to propagate vertically (Figure 9.1). Since most contaminated sites have overconsolidated formations, it is expected that pneumatic and hydraulic fracture propagation will be predominantly horizontal. In stratified formations, which have natural weakness along the bedding planes, the tendency toward horizontal fracture patterns is even more accentuated. For implementing soil vapor extraction technology, horizontal fractures will be favored, since vertical fractures will create a significant amount of short circuiting of the extracted air.

Geotechnical engineers express the ratio of the horizontal to vertical stress, K_o. In general, values of K_o greater than 1.0 will favor flat-lying fractures, and the larger the value of K_o the more the flat-lying orientation will be favored.

9.3 DESCRIPTION OF THE PROCESS

Currently there are two types of fracturing methodologies employed for environmental applications. Hydraulic (water-based) and pneumatic (air-based) fracturing are the two variants of this technology. The selection between these two types of fracturing are based on the following considerations:

- soil structure and stress fields
- the need to deliver solid compounds into the fractures
- target depth
- desired areal influence
- contractor availability
- acceptability of fluid injection by regulatory agencies

9.3.1   Hydraulic Fracturing

Hydraulic fracturing has been used for more than 50 years to enhance the yield of wells recovering oils at great depth, and it has recently been shown that hydraulic fracturing will also enhance the yield of wells recovering liquids and vapors from contaminated zones in the subsurface.[3] The process is reportedly responsible for making 25 to 30% of the U.S. oil reserves economically viable. The parallels between economic recovery of petroleum hydro-carbons and viability of *in situ* treatment alternatives are very evident.

Hydraulic fracturing may be defined as the process of creating a fracture or fracture system in a porous medium by injecting a fluid under pressure through a well bore in order to overcome native stresses. To fracture a formation, energy must be generated by injecting a fluid down a well and into the formation. Effectiveness of hydraulically created fractures is measured both by the orientation and areal extent of the fracture system and by the postfracture enhancement of vapor or liquid recovery.

Hydraulic fracturing begins by injecting a fluid into a borehole at a constant rate until the pressure exceeds a critical value and a fracture is nucleated. The properties that a fracturing fluid should possess are low leak-off (fluid loss) rate, the ability to carry a propping agent, and low pumping friction loss. The fluid also should break down easily after the fracture formation.

Low leak-off (fluid loss) rate is the property that permits the fluid to physically open the fracture and one that controls its areal extent. The rate of leak-off to the formation is dependent upon the viscosity and the wall-building properties of the fluid. Postfracture breakdown is necessary such that the injected fluids do not "clog" the formation.

Cross-linked guar gum is an example of a common fracture fluid used for environmental application. The most widely used form is the continuous mix grade of gum, referred to as such because it hydrates rapidly and reaches a usable level of viscosity fast enough that it can be used continuously. Since guar gum is a food-grade compound, it minimizes the potential for regulatory objections for the process.

Because of the characteristic high viscosity, guar gum is capable of transporting coarse-grained silica sand or other granular material, as a slurry, into the fracture. The coarse-grained silica sand is called a propping agent to keep the fracture open upon relaxation of the injection pressure, when the guar gum gel is decomposed by an enzyme added during injection. Pumps specifically designed for high-viscosity, high-solids fluid handling should be selected to inject the slurry at the required pressures.

Hydraulic fractures are generally created beneath a casing into which a lance is advanced and withdrawn to the required depth with a hammer. Lateral pressure of the soil seals the casing during the controlled injection of the fracturing fluid and the proppants. The casing can be driven deeper to create another fracture (Figure 9.2). Stacks of gently dipping hydraulic fractures can be created with vertical spacing of 0.5 to 1 ft using the driven casing method; vertical spacings of less than 0.5 ft tend to result in fractures that merge at short distances from the borehole.[3] A high-pressure water nozzle is used to cut a disc-shaped notch with the preferred horizontal orientation at the bottom of the casing, and the notch becomes the starting point for the fracture.

The injection pressure required to create hydraulic fractures is remarkably modest (less than 100 psi). For example, at the beginning of injection during a test at 5 ft depth, the pressure increased abruptly to 64 psi, but then decreased sharply when the fracture began to

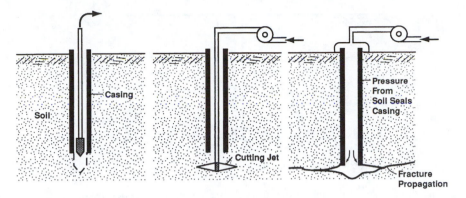

Figure 9.2 Method for creating hydraulic fractures in soil.

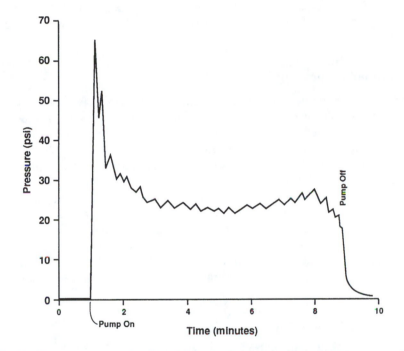

Figure 9.3 Injection pressure as a function of time during hydraulic fracturing.

propagate (Figure 9.3).[3] Injection pressure was between 15 and 20 psi during propagation.[3] Slightly greater pressures are required to create fractures at greater depth.

In some cases, the fracture is nearly flat-lying in the vicinity of the borehole and the dip increases to approximately 20° at some distance away, whereas in other cases the fractures appear to maintain a roughly uniform dip from the borehole to the point of termination (Figure 9.4). In nearly every case, the fracture has a preferred direction of propagation so that the borehole is off the center of the fracture. The preferred direction of propagation is commonly related to distribution of vertical load at the ground surface, with the fractures propagating toward regions of diminished vertical load. Beneath sloping ground, therefore, it is possible to anticipate the preferred direction; it is typically downslope.

It has been reported that radial dimensions of 20 to 35 ft have been achieved during hydraulic fracturing at depths of up to 30 ft.[3] The average thickness of the fracture ranged from 0.2 to 0.4 in. The largest fracture that has been characterized was 55 ft in the radial direction and the thickest was 1 in.[3] The maximum dimension of a hydraulic fracture depends on the volume of fluid injected into it. But this dimension is not without bounds, because

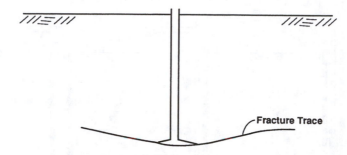

Figure 9.4 Trace of idealized hydraulic fracture.

the fracture climbs and will vent at the ground surface with continued injection. The volume of injected slurry then becomes critical, and currently, empirical methods that make use of observations and field measurements serve to develop an initial design. The empirical design is developed by creating a fracture in an uncontaminated area and then adjusting the design based on collected data.

9.3.2 Pneumatic Fracturing

Pneumatic fractures can be generated in geologic formations if air or any other gas is injected at a pressure that exceeds the natural strength as well as the *in situ* stresses present in the formation. As noted earlier, pneumatic fracture propagation will be predominantly horizontal at overconsolidated formations. However, in shallow recent fills, some upward inclination of the fractures has been observed, the reason for which is attributed to the lack of stratification and consolidation in these formations.[4]

The amount of pressure required to initiate pneumatic fractures is dependent on the cohesive or tensile strength of the formation, as well as the overburden pressure (dependent upon the depth and density of the formation). An expression for predicting pneumatic fracture initiation pressure has been developed by assuming the geological material to be brittle, elastic, and overconsolidated.[5] Assuming the formation has an effective unit weight, δ', and an apparent tensile strength, t_a, the fracture initiation pressure, P_i, may be estimated by

$$P_i = C\delta'Z + t_a + P_o \tag{9.1}$$

where C = coefficient (ranging from 2.0 to 2.5)
         Z = overburden depth
         P_o = hydrostatic pressure.

Substituting typical values for clay soil and shale bedrock at a depth of 20 ft, the above expression yields initiation pressures of 100 psi and 200 psi, respectively. Fracture initiation pressures are, therefore, relatively modest at shallow depths (where most of the contamination occurs).

The most important system parameter for efficient pneumatic fracturing is injection flow rate, as it largely determines the dimensions of a pneumatic fracture. Once a fracture has been initiated, it is the high volume airflow which propagates the fracture and supports the formation. The design goal of a pneumatic fracturing system therefore becomes one of providing the highest possible flow rate. Field observations indicate that pneumatic fractures reach their maximum dimension in less than 20 seconds, after which continued injection simply maintains the fracture network in a dilated state (in essence, the formation is "floating" on a cushion of injected air).[4] Pneumatically induced fractures continue to propagate until

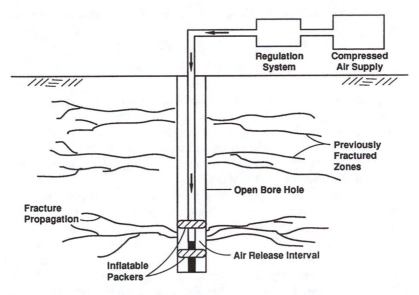

Figure 9.5 Schematic of pneumatic fracturing process.

they intersect a sufficient number of pores and existing discontinuities, so that leak-off (fluid loss) rate into the formation exactly equals the injection flow rate.

In general, injection rates of up to 1000 scfm are sufficient to create satisfactory fracture networks in low permeability formations. To date, the radii of pneumatic fractures have ranged from 10 to 25 ft from the injection point.[6]

Figure 9.5 shows the major components including a compressed air supply, a flow-pressure system, a flow-pressure regulation system, and an injector. The compressed air supply may consist of either a bank of compressed gas cylinders, or an air compressor and receiver tank. The regulation system allows control of the two parameters critical to successful fracturing—injection pressure and injection flow rate—which are adjusted according to the geology of the site and the depth of fracture. Air is injected into the formation through an injector which is placed in the borehole.

An individual pneumatic fracture is accomplished by (1) advancing a borehole to the desired depth of exploration and withdrawing the auger, (2) positioning the injector at the desired fracture elevation, (3) sealing off a discrete 1 or 2 ft interval by inflating the flexible packers on the injector with nitrogen gas, (4) applying pressurized air for approximately 30 s, and (5) repositioning the injector to the next elevation and repeating the procedure. A typical fracture cycle takes approximately 15 min, and a production rate of 15 to 20 fractures per day is attainable with one rig.

The pneumatic fracturing procedure typically does not include the intentional deposition of foreign propping agents to maintain fracture stability. The created fractures are thought to be "self propping," which is attributed to both the asperities present along the fracture plane as well as the block shifting which takes place during injection. The aperture or thickness of a typical pneumatically induced fracture is approximately 0.5 to 1 mm.[4] Testing to date has confirmed fracture viability in excess of 2 years, although the longevity is expected to be highly site-specific.[1]

Without the carrier fluids used in hydraulic fracturing, there are no concerns with fluid breakdown characteristics for pneumatic fracturing. There is also the potential for higher permeabilities within the fractures formed pneumatically, in comparison to hydraulic fractures, as these are essentially air space and are devoid of propping agents. The open, self-propped fractures resulting from pneumatic fracturing are capable of transmitting significant amounts of fluid flow. The high flow potential for even small fractures may be explained by the "cubic law," which states that flow rate in planar fractures is proportional to the cube of the aperture.

9.4 FEASIBILITY EVALUATION

Fracturing success is dependent on the application of both sound engineering and sound judgment. The knowledge base and literature information on successful cleanup of contaminated sites where fracturing has been applied for testing and more so for full-scale remediation is very limited. With continued testing and reporting of both successes and failures, understanding of the technology will develop to the point where geologic conditions favoring the technology will become better understood.

Screening a site for possible application of fracturing first requires the understanding of the mechanics and applicability of fracturing to enhance permeability, but also the integration with the selected cleanup technology. Based on the site investigations and the extent of contamination, a preliminary estimate of the number of fractures necessary to provide adequate coverage can be determined. As a general rule of thumb, fracture formation in the range of from 20 to 35 ft or more is possible for near-surface soils and, with all other factors remaining the same, increased radius with depth may be possible. The relationship between depth (loading) of fracture location and fracture dimensions needs to be considered for a full-scale application. Specifically, more closely spaced shallow fractures may need to be created to achieve the desired end result. Fracture propagation in rock formations has been found to be greater than in soil formations, primarily due to the competence and cohesion of these units.

9.4.1 Geologic Characterization

A primary step in the evaluation of feasibility of fracturing is an examination of detailed and accurate geologic cross sections illustrating sediment layering and grain sizes in the target zone, and the contaminant characteristics present in the target zone. Because contaminants often reside within low-permeability, fine-grained soils, it is important to understand this relationship.

At least one continuous core boring should be installed to characterize major and minor changes in lithology. Cores collected during continuous and depth-specific sampling should be examined for factors contributing to secondary permeability such as coarse-grained sediment inclusions and naturally occurring fractures. These secondary permeability characteristics of the soil or rock formation may influence the creation of engineered fractures. Pneumatic fractures, in particular, may propagate along existing fracture patterns. Hydraulic fractures have been found to be less influenced by existing fractures.[3] The site and geologic parameters to be evaluated are summarized below.

- Type of soil/rock
- Type of deposition
- Groundwater depth
- Perched water level (if any)
- Type of contamination (e.g., VOCs, hydrocarbons, etc.)
- Depth of contamination

9.4.2 Geotechnical Characterization

In addition to the qualitative evaluations described above, target zone soil samples should be submitted for geotechnical evaluations of grain size analysis, liquid and plastic limits of soil, moisture content, and unconfined compressive strength. Details and implications of these tests for a candidate site for fracturing are as follows:

- Grain size analysis: Although fractures can be created in sediments and rock of nearly any grain size, the highest degree of permeability improvement can be expected from the finer grained soils. Grain size analysis can be performed by using the sieve analysis method (ASTM methods D421 and D422) and/or the hydrometer analysis method (ASTM methods D421 and D422).
- Liquid and plastic limits of soil: This parameter is also known as the Atterberg limits and characterizes the plasticity of a soil. In general, fractures created in highly plastic clays will not propagate as well as in more brittle materials. Formations having $W_n < W_l$ are most suitable for artificial fracturing, where W_n is the natural moisture content and W_l is the liquid limit. Soils having $W_n > W_l$ (or liquidity index > 0) may liquefy under a sudden shock imparted during the fracturing process. The estimation of W_n and W_p (plastic limit) would also give an indication of the degree of consolidation of soil. If W_n is closer to W_p than to W_l, the soil may be over consolidated. If W_n is closer to W_l (or larger), the soil may be normally consolidated. Liquid and plastic limits of soil can be measured by ASTM method 4318.
- Soil moisture content: Overall soil permeability improvements are achievable with fracturing; however, vapor flow in particular is also controlled by soil moisture. Improvements in vapor flow through highly saturated soils (at or near field capacity) will not be achieved by the production of fracturing alone. Additional means of moisture removal may be required to obtain the desired effect through fracturing under these circumstances. ASTM method D2216 may be used to estimate the soil moisture content.
- Unconfined compressive strength: The unconfined compressive strength can be used for predicting the orientation and direction of propagation of fractures. As noted earlier, the state of *in situ* stresses plays a key role in the orientation and ultimate effect on permeability enhancement. The artificially induced fractures are assumed to be vertical in normally consolidated soil and horizontal in over consolidated deposits. ASTM method 2166 is used to measure the unconfined compressive strength of soils.
- Permeability: As discussed previously, fracturing is generally applied at sites with characteristically low permeability. A baseline estimate of permeability (vapor and/or liquid) is often available from testing concluded at the site during site investigations. This baseline estimate of permeability provides a basis for evaluating the necessity, benefit, and effectiveness of the fracturing process. In general, greater improvement of vapor or fluid flow and radial influence is observed in formations with lower initial permeability.
- Cohesion: The more cohesive the soil is, more amenable it will be to fracturing. Longevity of the fractures, upon relaxation of fracture stress, is high in cohesive soils. Fracturing in cohesive soils such as silty clays has been particularly successful.

9.5 PILOT TESTING

Upon completion of the preliminary screening and geotechnical testing, pilot testing is typically conducted for further performance evaluation and to provide a design basis for a full-scale system. Pilot testing is by far the most powerful and useful means of screening a site for a full-scale remediation incorporating fracturing, since experience has shown that preliminary screening of a site cannot always accurately predict the performance of either hydraulic or pneumatic fracturing.

The pilot test plan should incorporate the following steps:

- area selection
- baseline permeability/mass recovery estimation
- fracture point installation
- test method and monitoring.

9.5.1 Area Selection

Selection of the area for the pilot test within the contaminated site is the first step in designing the pilot test. The decision must be made whether to test the technology within the impacted area(s) of the site or to conduct testing outside the contaminated zone. It is generally preferred to test within the contaminated zone to reduce the impact of lateral heterogeneities and to collect data on contaminant recovery rates prior to and after fracturing.

For pilot testing of a single fracture well, an area of approximately 4000 ft^2 should be sufficient. This area should encompass the anticipated maximum limits of fracture propagation.

9.5.2 Baseline Permeability/Mass Recovery Estimation

To aid in the evaluation of fracturing benefits vs. the costs and risks of the technology, a baseline estimate of soil permeability and contaminant mass recovery rates is typically conducted prior to implementing the fracture formation. Because fracturing is generally considered for low-permeability formations ($K_{air} < 1$ Da, $K_H < 10^{-5}$ cm/s, where K_{air} is pneumatic permeability, and K_H is hydraulic conductivity in the horizontal direction), careful evaluation of the location-specific permeabilities will enhance the success of fracture formation.

After a geologic formation has been fractured, the ability to treat and/or remove the contaminants will depend on the flow and transport characteristics of the artificially fractured medium. The two general approaches for analyzing flow in fractured media include the equivalent porous medium and the dual porosity approaches.[7]

As the name implies, the *equivalent porous medium approach* assumes that the fractures are distributed sufficiently throughout the formation so that it can be analyzed with standard porous media methods. The applicability of this approach largely depends on the scale of the domain under study. For example, if the fractures are very closely spaced and/or the area under study is very large, the porous media method will yield satisfactory results.[7]

Many situations require the use of the *dual porosity approach* to analyze flow and transport in the fractured media. In the dual porosity approach, the fractured media is assumed to be a superposition of two flow systems over the same volume, consisting of a porous matrix and the open fracture network. As a special case of the dual porosity method, it is often useful to analyze the discrete fractures only and ignore the flow and storage characteristics of the porous matrix blocks. It can be concluded that the vast majority of the flow in an engineered fracture formation occurs as discrete fracture flow.[7]

9.5.3 Fracture Point Installation

Specifically designed fracture point installation is required for pilot testing of the fracturing technologies. The fracture intervals are selected to coincide with the target zone of contamination. Fracture locations are also targeted for the low-permeability sediments or rock within a layered setting.

The multistage processes of implementing hydraulic and pneumatic fractures have been discussed in previous sections. Upon reaching the desired maximum depth of fracture formation through the processes described earlier, the borehole is often completed using conventional well installation techniques. The placement of a well central to the point of radiating fractures allows for the withdrawal of vapors and liquids through the relatively permeable zones containing secondary permeability.

9.5.4 Test Method and Monitoring

Pilot testing of the fracturing technologies is generally a two-step process. The first step is conducted during the actual formation of the fractures. During this step, the approximate dimensions and the orientation of the fracture pattern are determined. The second step in the testing process is to determine the increase in vapor or fluid movement within and beyond the area of fracture propagation and the corresponding increase in contaminant mass removal rates.

9.5.4.1   Fracture Aperture

Fracture aperture is the perpendicular distance between the adjacent walls of a fracture which is air- or water-filled. Fracture aperture is the major controlling factor for fluid flow through a fractured media. It is very difficult to define apertures in terms of true width, since the asperities that create fracture surface roughness also affect the fracture aperture.

Field measurement of fracture aperture is most commonly performed indirectly, using borehole hydraulic tests. Assuming only one fracture intersects the test interval, a packer test will yield an aperture thickness as a function of the hydraulic conductivity by using the cubic law. The cubic law states that the functional relationship between flow, Q, and fracture aperture thickness, b, can be represented by

$$Q \propto b^3 . \tag{9.2}$$

If more than one fracture intersects the test interval, then this method will overestimate the aperture of either fracture. A borehole camera and ground surface heave measurements also can be used to estimate the fracture aperture. A high-resolution borehole video camera can be lowered into the borehole to obtain insight into the effects of fracturing by comparing the films from before and after conditions.

9.5.4.2   Fracture Spacing

Fracture spacing is the perpendicular distance between adjacent fractures. Fracture spacing is influenced by the soil or rock composition, texture, structural position, and bed thickness. As a general trend, fracture density decreases with depth, as does fracture porosity.

9.5.4.3   Fracture Orientation

The orientations of fractures, though not regular, is not purely random. For soils, the loading history, and thus the degree of consolidation, is assumed to govern the orientation of the artificially engineered fractures. Orientation of a fracture can be expressed by its strike and dip. The change in ground surface elevation during fracturing has been found to provide a reasonable approximation of the fracture locations in the subsurface.

Ground surface displacement (heave) is generally recorded during fracturing by an array of survey points that are monitored in real time. Heave detection can be used to estimate the dimensions of both aperture and length of the fractures. Simplistically, the heave can be measured with an engineering level and graduated rods driven into the ground. A limitation of this method for surface heave measurement is apparent when the observed surface effects

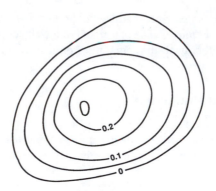

Figure 9.6 Typical tiltmeter ground surface heave contour.

become smaller as the depth of fracture formation becomes deeper. Tiltmeters can be used to collect the dynamic time history data of fracture propagation. Tiltmeters sense the tilting of the surface, and the array of tiltmeters can generate the data to develop the contours of the surface deformation caused by fracturing (Figure 9.6).

Pressure and/or flow measurement devices also can be used at existing or installed monitoring wells for estimating the horizontal extent of the fractures as they establish fluid communication with outlying monitoring wells. During fracture injection, evidence of direct communication is often observed in the form of air rushing out of the monitoring wells. Additional evidence can be collected in the form of negative pressure in the outlying wells during vacuum extraction. Air communication measurements are valuable since they not only provide absolute confirmation of whether fractures have intersected a particular well, but also provide useful data for evaluating the postfracture enhancement of permeability of the formation.

For pneumatic fracturing, the surface heave during pressure application is substantially higher than the residual heave after pressure relaxation. The residual heave is generally 10 to 20% of the maximum displacement (typically less than a few inches). For hydraulic fracturing, the ground displacement is directly related to the volume of the injected slurry, and the thickness of the fractures decreases with distance from the point of injection, following the path of least resistance. Heterogeneity within the soil matrix, naturally occurring fracture patterns and, to a lesser degree, bedding planes appear to influence the orientation of the created fractures.

Surface loading also influences the pathway of the fracture font. High surface loading created by manmade structures or changes in topography can also influence the fracture patterns. If needed, temporary surface loading can be used to "steer" the fractures toward a desired location. Vehicles have been used successfully for this application.

Because of the displacement caused by the fracture formation, care must be exercised when working adjacent to buildings or other structures. While some structures can withstand these moderate displacements, the integrity of others may be compromised. A careful evaluation of the structure's strength and stability must be performed prior to implementing a fracture test near a building.

9.5.4.4   Enhancement of Vapor or Fluid Movement

During the second step of pilot testing, the permeability enhancement resulting from the fractures formed should be evaluated. The enhanced flow characteristics should be compared with the baseline measurement or estimate of vapor or fluid movement prior to fracturing. The applicability of the process and, ultimately, the number, locations, and depth intervals of a full-scale fracture system will depend heavily on this evaluation.

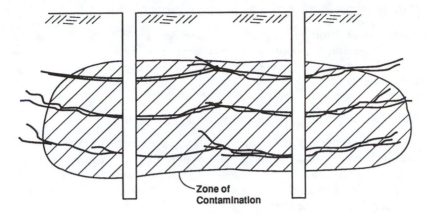

Figure 9.7 Successive fracturing of target zone with overlapping fractures.

The second step of the pilot testing program may entail a soil vapor extraction test or groundwater pumping test to determine the enhancement in permeability. In addition, it is also important to monitor the changes in chemical composition of the soil gas as a result of the access to new pockets of contamination caused by fracturing.

9.6 SYSTEM DESIGN

Upon completion of preliminary screening and pilot testing, design of a full-scale fracturing system can be initiated. The implementation of a full-scale program should be based on economic and feasibility evaluations. In essence, fracturing should be selected as a component of the final remediation system, only if the cost of integrating fracturing is less than alternative methods such as multiple wells with closer well spacings or excavation and above-ground treatment and disposal.

Based on the results of field testing, a fracture wells location plan should be selected to encompass the area of known contaminant impact (Figure 9.7). The fracture wells locations plan must take into consideration the asymmetric orientation of fracture propagation. For instance, the fracture points may be more closely spaced north to south than east to west due to asymmetry. To control the role of diffusion and the possible creation of low flow zones, an engineering safety factor should be applied such that the fracture zones overlap in the plan view.

The depth intervals for the fractures should correspond with the known distribution of contaminants. This requirement again emphasizes the importance of site characterization.

Because of geologic heterogeneity present at almost every site, the full-scale fracturing plan should be designed with some flexibility in mind. In most instances, it would be wise to specify a range of possible fracture point locations with field adjustments made during installation to optimize the overall system performance. Depending on the size of the site and number of fracture points, it may also be advisable to implement the fracturing program in a phased approach. For example, fracture wells could be installed on a 1-week cycle. During the first week, fracture points could be installed, followed by testing of these points for performance (e.g., enhancement of vapor or liquid extraction rates). Adjustments can then be made for the next cycle of fracture installations.

Even with fracturing, contaminant removal rates will be rate-limited by diffusional flow between the areas of high, advection-controlled flow. When compared to contaminant removal rates before fracturing, postfracture rates will be higher, if the process is successful and applied under the right conditions. Eventually, however, diffusion-controlled mass transfer will influence the time required to reach the cleanup standards. The diffusive distances will be shortened significantly due to the fracture network formed.

9.7 INTEGRATION WITH OTHER TECHNOLOGIES

As noted earlier, hydraulic or pneumatic fracturing are not "stand-alone" remediation techniques. Once a fracture network is established in a low-permeability formation, the gaseous, adsorbed, and liquid contaminants are more easily accessed by complimentary remediation technologies.

9.7.1 Soil Vapor Extraction Combined with Fracturing

A major obstacle for the application of soil vapor extraction (SVE) as a remediation technique is permeability of the formation. Low-permeability formations, such as fractured shales, silts, and clays, usually do not allow sufficient subsurface airflow for conventional SVE to be effective. Fracturing of such formations will help in overcoming the difficulties in implementing SVE at these sites.

The increase in extraction airflow rate provided by both pneumatic and hydraulic fracturing means that contaminants can be removed faster by volatilization. The formation permeability increase created by fracturing also allows for a much greater vacuum radius of influence to be induced from an extraction well. Since the spacing between extraction wells is significantly increased, the total number of wells needed to remediate a site is reduced. This leads to a substantial costs savings.

It is noted that often the highest contaminant concentrations occur within and adjacent to existing structural discontinuities in low-permeability formations (e.g., joints, cracks, bedding planes). Since fracturing dilates and interconnects existing discontinuities, direct access is provided to a majority of the contaminant mass. Even the small airflows through the smaller fracture network are capable of volatilizing and removing contaminants, thereby causing an outward diffusive gradient of the contaminant from the matrix block to the larger fractures.

The following two case studies illustrate the efficacy of fracturing in enhancing the mass removal rates during SVE.

1. The impacted zone at this site (in the northeastern U.S.) was characterized as siltstone and shale with naturally occurring fractures. Pneumatic fractures were installed between 9 and 16 ft below grade. Before fracturing, the vapor extraction rates from each of the tested wells was below the sensitivity of the measuring instrument (less than 0.6 scfm) at an applied vacuum of 136 in. of water. A single fracture well was installed central to the monitoring points, as shown in Figure 9.8.[8] The distances between the fracture well and the monitoring points were 7.5 to 20 ft.

    Based on elevation measurements recorded by an electronic tiltmeter during fracturing, surface heave was observed up to 35 ft from the fracturing well. The flow rates from each of the test wells surrounding the fracture well increased substantially after fracturing. Specifically, the flow rate increased by more than 15-fold after fracturing. Vacuum measurements within the monitoring points also increased after fracturing by 4 to 100 times in comparison to prefracture conditions. Pneumatic fracturing improved access to the contamination substantially by increasing the mass removal rate by approximately 25 times (Figure 9.9).[4] It is also interesting to note the change in chemical composition of the soil gas summarized in Table 9.1. Before fracturing, TCE was the predominant component of the soil gas, representing approximately 84%. However, after fracturing, other compounds became more dominant, even though the removal rate of TCE had increased substantially. This shift in soil gas composition indicates that new pockets of contamination were accessed by pneumatic fracturing.[4]

2. Hydraulic fracturing tests were conducted on vadose zone soils at a site in the midwestern U.S. The site was contaminated with TCE, 1,1,1-TCA, 1,1-DCA and PCE.[9] The soils were characterized as a silty/clayey till to a depth of approximately 20 ft below grade. The permeability of the soil was estimated to be 10^{-7} to 10^{-8} cm/s. The pilot-scale demonstration created six fractures in two wells at depths of 6, 10, and 15 ft below grade over a 1 day period. At an applied vacuum of 240 in. of water, the vacuum influence in unfractured soil

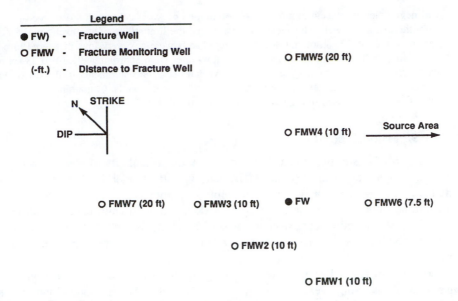

Figure 9.8 Well location plan for pneumatic fracturing.

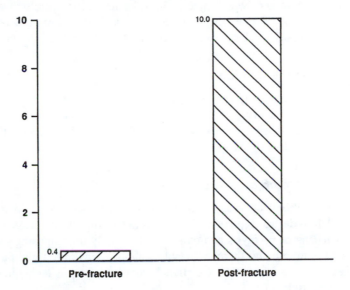

Figure 9.9 Comparison of TCE mass removal under pre- and postpneumatic fracturing by SVE.

Table 9.1 Volatile Organic Compounds Present in Extracted Soil Gas at Pneumatic Fracturing Site

Compound	Prefracture % of total	Postfracture % of total
TCE	84	15
Benzene	6	37
PCE	5	33
Chloroform	4	13
Methylenechloride	1	2
Total mass removal rate of all compounds	0.78×10^{-5} lbm/min	113.6×10^{-5} lbm/min

Note: lbm/min = pound mass per minute.

Adapted from Schuring, J. R., Pneumatic fracturing to remove soil contaminants, *NJIT Res.*, 2, Spring 1994.

was negligible, decreasing to a few tenths of an inch of water column at a distance of 5 ft from the extraction well. This clearly demonstrated the limitations of a conventional SVE system at this site. The flow rates in unfractured soils were also very low, measured to be approximately 1 scfm. In the fractured soils, flow rates from approximately 14 to 23 scfm were achieved under similar vacuum levels. Vacuum level measurements in the fractured soil also increased dramatically up to 25 ft from the fracture wells. Contaminant recovery rates similarly increased in the fractured soil by 7 to 14 times.

9.7.2   *In Situ* Bioremediation

The success of *in situ* bioremediation depends on the availability of electron acceptors, such as O_2, and nutrients such as N and P. The delivery and transport of these nutrients may become the rate-limiting factor in low-permeability formations.

Fracture networks formed during hydraulic and pneumatic fracturing can be utilized as delivery pathways for introducing reagents as sources of O_2, N, and P. These reagents can be introduced both in the saturated and unsaturated zones in the form of gaseous or liquid reagents.

In addition, slowly reactive materials containing O_2 and nutrients can be used as proppants during hydraulic fracturing to enhance *in situ* biodegradation both in the vadose and saturated zones. A "time-release" oxygen source such as sodium percarbonate[10] or magnesium peroxide can be used as proppants and, when injected into the impacted soils, will slowly release oxygen over a long period. The advantage of this process is that aerobic conditions can be locally maintained in the subsurface, which might not have been possible otherwise.

9.7.3   Reductive Dechlorination

Testing is currently underway for the injection of elemental iron filings for the creation of subsurface conditions favoring reductive dechlorination of chlorinated aliphatic compounds. A horizontal "flat-lying" reactive wall can be thus created to promote the accelerated attenuation of chlorinated compounds.

9.7.4   *In Situ* Vitrification or *In Situ* Heating

The *in situ* vitrification technology uses electric power to heat contaminated soil past its melting point and, thus, destroy organic contaminants in the soil. The process destroys organic contaminants by means of pyrolysis and oxidation, thermally decomposing some inorganic contaminants, and immobilizing thermally stable compounds within a glass and crystalline vitrified material. The most important operational parameter for this technology is the electrical input to the melting zone.

In situ vitrification technology uses graphite electrodes to implement the input of electrical energy to heat the soil. A graphite-based proppant can be used during hydraulic fracturing to install the graphite as electrodes to conduct electricity. In addition, flaked graphite and glass frit can be used as a mixture of proppants to act as a starter path since dry soil is usually not electrically conductive.

In the recent past, electrical soil heating has been considered as a means of enhancement to soil vapor extraction. If graphite-based proppants are used during hydraulic fracturing, the graphite can be used as the electrodes to implement electrical soil heating to increase the contaminants' vapor pressure.

9.7.5   *In Situ* Electrokinetics

Recently, the use of electrokinetics as an *in situ* method for soil remediation has received increasing attention due to its unique applicability to low-permeability soils. Electrokinetics

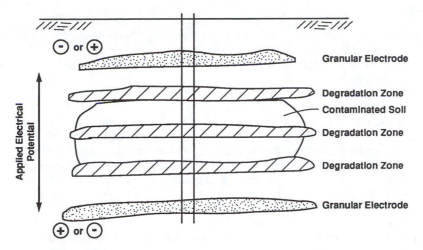

Figure 9.10 Schematic diagram of the "lasagna process."

includes the transport of water (electroosmosis) as well as ions (electromigration) as a result of an applied electric field.[11] Electroosmosis, in particular, has been used since the 1930s for dewatering clays, silts, and fine sands.[11] For remedial applications, water is typically introduced into the soil at the anode to replenish the water flowing toward the cathode due to electroosmosis. The water flow is utilized to flush and/or degrade the contaminants in the subsurface soil. The contaminants flushed from the subsurface soil to the ground surface at the cathode region can be collected for further treatment and disposal, if needed (Figure 9.10).

Advantages with electroosmosis include uniform water flow through heterogeneous soil, high degree of control of the flow direction, and very low power consumption. There are, however, several major drawbacks associated with electroosmosis for remedial applications. These include low liquid velocities induced by electroosmosis, (typically about 1 in./day for clay soils), additional above-ground treatment, steep pH gradient in the soil bed, and precipitation of metals near the cathode.

An integrated approach coupling electrokinetics and hydraulic fracturing with complimentary *in situ* technologies to eliminate or minimize the drawbacks associated with the use of electrokinetics has been developed recently. This process, called the "lasagna process" by its developers, is so named for its layers of electrodes and treatment zones.

The general concept is to use electrokinetics to move contaminants from the soils into "treatment zones," where the contaminants are removed from the groundwater by adsorption, immobilization, or degradation. Hydraulic fracturing can provide an effective and low-cost means for creating such zones horizontally in the subsurface soil within the contaminated zone. A graphite-based proppant can be used to install the horizontal electrodes above and below the contaminated zone (Figure 9.10). Hydraulically fractured zones also will create much more permeable zones than the native soils to enhance the liquid velocities induced by electroosmosis.

The treatment zones can also be vertical, which can be constructed using sheet piling, trenching, slurry walls, or deep soil mixing techniques. The treatment zones can also be continuous instead of being discrete.

Liquid flow can be periodically reversed, if needed, simply by the cyclic application of low-voltage DC current to the electrodes. This mode will enable multiple passes of the contaminants through the treatment zones for complete sorption or degradation. The polarity reversal also serves to minimize complications associated with long-term operation of unidirectional electrokinetic processes. For example, the cathode effluent (high pH) can be recycled directly back to the anode side (low pH), which provides a convenient means for pH neutralization as well as more simple water management.

REFERENCES

1. Schuring, J. R. and Chan, P. C., Pneumatic Fracturing of Low Permeability Formations — Technology Status Paper, unpublished paper, 1993.
2. Hubbert, M. K. and Willis, D. G., Mechanics of hydraulic fracturing, *Petrol. Trans. AIME*, 210, 153, 1957.
3. Murdoch, L., Hydraulic and Impulse Fracturing Techniques to Enhance the Remediation of Low Permeability Soils, unpublished paper, 1993.
4. Schuring, J. R., Pneumatic fracturing to remove soil contaminants, *NJIT Res.*, 2, Spring 1994.
5. King, T. C., Mechanism of Pneumatic Fracturing, M. S. thesis, Department of Civil and Environmental Engineering, New Jersey Institute of Technology, Newark, 1993.
6. Nautiyal, D., Fluid Flow Modeling for Pneumatically Fractured Formations, M. S. thesis, Department of Civil and Environmental Engineering, New Jersey Institute of Technology, Newark, NJ, 1994.
7. Gale, J. E., Assessing the Permeability Characteristics of Fractured Rock, Geological Society of America, Special Paper 189, 1982.
8. U.S. Environmental Protection Agency, Accutech Pneumatic Fracturing Extraction and Hot Gas Injection, Phase I, Applications Analysis Report, EPA/540/AR-93/509, 1993.
9. U.S. Environmental Protection Agency, Hydraulic Fracturing Technology, Applications Analysis and Technology Evaluation Report, Risk Reduction Environmental Laboratory, EPA/540/AR-93/505, 1993.
10. Vesper, S. J., Murdoch, L. C., Hayes, S., and Davis-Hooper, W. J., Solid oxygen source for bioremediation in subsurface soils, *J. Hazardous Materials*, 1993.
11. Ho, S. V., Sheridan, P. W., Athmev, C. J., Heitkamp, M. A., Brackin, J. M., Weber, D., and Brodsky, P. H., Integrated in situ soil remediation technology: The lasagna process, *Environ. Sci. Technol.*, 29, 2528, 1995.

10

PHYTOREMEDIATION

10.1 INTRODUCTION

Phytoremediation is the use of plants to remediate contaminated soil or groundwater. This technique can be used for the remediation of inorganic contaminants as well as organic contaminants. Most of the activity in phytoremediation takes place in the rhizosphere—in other words, the root zone. Phytoremediation of inorganic contaminants can be further categorized into phytostabilization and phytoextraction.[1]

Phytostabilization is the use of plants to stabilize contaminated soil by decreasing wind and water erosion and also decreasing water infiltration and the subsequent leaching of contaminants. *Phytoextraction* is the removal of inorganic contaminants by above-ground portions of the plant. When the shoots and leaves are harvested, the inorganic contaminants are reclaimed or concentrated from the plant biomass or can be disposed.

Plants have been used for remediation in the past. A number of free-floating aquatic and aquatic emergent plant species and their associated microorganisms have been used for more than a decade in constructed wetlands for municipal and industrial wastewater treatment.[2] Several fast-growing tree plantations have been established and are under active study for their potential use in wastewater cleanup in land discharge systems.[3,4]

Plant species can be selected to extract and assimilate or extract and chemically decompose target organic contaminants. Heavy metals can be taken up and bioaccumulated in plant tissues. Many inorganic compounds considered to be contaminants are, in fact, vital plant nutrients that can be absorbed through the root system for use in growth and development. Heavy metals can be taken up and bioaccumulated in plant tissues. Organic chemicals such as PAHs and pesticides can be absorbed and metabolized by plants and trees.

The advantages of phytoremediation are the low capital costs, aesthetic benefits, minimization of leaching of contaminants, and soil stabilization. The operational cost of phytoremediation is also substantially less and involves mainly fertilization and watering for maintaining plant growth. In the case of heavy metals remediation, additional operational costs will also include harvesting, disposal of contaminated plant mass, and repeating the plant growth cycle.

The limitations of phytoremediation are that the contaminants present below rooting depth will not be extracted and that the plant or tree may not be able to grow in the soil at every contaminated site due to toxicity. In addition, the remediation process can take years for contaminant concentrations to reach regulatory levels and thus requires a long-term commitment to maintain the system.

Phytoremediation is most suited for sites with moderately hydrophobic contaminants such as benzene, toluene, ethylbenzene, xylenes, chlorinated solvents, PAHs, nitrotoluene ammunition wastes, excess nutrients such as nitrate, ammonium, and phosphate, and heavy metals.

10.2   PHYTOREMEDIATION MECHANISMS OF ORGANIC CONTAMINANTS

Plants and trees remove organic contaminants utilizing two major mechanisms: (1) direct uptake of contaminants and subsequent accumulation of nonphytotoxic metabolites into the plant tissue, and (2) release of exudates and enzymes that stimulate microbial activity and the resulting enhancement of microbial transformations in the rhizosphere (the root zone).[5]

10.2.1   Direct Uptake

Not all organic compounds are equally accessible to plant roots in the soil environment. The inherent ability of the roots to take up organic compounds can be described by the hydrophobicity (or lipophilicity) of the target compounds. This parameter is often expressed as the log of the octanol–water partitioning coefficient, K_{ow}. Direct uptake of organics by plants is a surprisingly efficient removal mechanism for moderately hydrophobic organic compounds. There are some differences between the roots of different plants and under different soil conditions, but generally the higher a compound's log K_{ow}, the greater the root uptake.

Hydrophobicity also implies an equal propensity to partition into soil organic matter and onto soil surfaces. Root absorption may become difficult with heavily textured soils and soils with high native organic matter. There are several reported values available in the literature regarding the optimum log K_{ow} value for a compound to be a good candidate for phytoremediation (as an example, log K_{ow} = 0.5 to 3.0,[5] log K_{ow} = 1.5 to 4.0.)[1] It was also reported that compounds that are quite water soluble (log K_{ow} <0.5) are not sufficiently sorbed to the roots or actively transported through plant membranes.

From an engineering point of view, a tree could be thought of as a shell of living tissue encasing an elaborate and massive chromatography column of twigs, branches, trunk, and roots. The analogous resin in this system is wood, the vascular tissue of the tree, and this "resin" is replenished each year by normal growth.[3] Wood is composed of thousands of hollow tubes, like the bed of a hollow fiber chromatography column, with transpirational water serving as the moving phase. The hollow tubes are actually dead cells, whose death is carefully programmed by the tree to produce a water-conducting tissue which also functions in mechanical support. A complex, cross-linked, polymeric matrix of cellulose, pectins, and proteins embedded in lignin forms the walls of the tubes.[3] The cell wall matrix is chemically inert, insoluble in the majority of solvents, and stable across a wide range of pH.

Once an organic chemical is taken up, a plant can store (sequestration) the chemical and its fragments in new plant structures via lignification, or it can volatilize, metabolize, or mineralize the chemical all the way to carbon dioxide, water, and chlorides (Figure 10.1). Detoxification mechanisms may transform the parent chemical to nonphytotoxic metabolites, including lignin, that are stored in various places in plant cells. Many of these metabolic capacities tend to be enzymatically and chemically similar to those processes that occur in mammalian livers, and one report equated plants to "green livers" due to the similarities of the detoxification process.[7]

Different plants exhibit different metabolic capacities. This is evident during the application of herbicides to weeds and crops alike. The vast majority of herbicidal compounds have been selected so that the crop species are capable of metabolizing the pesticide to nontoxic compounds, whereas the weed species either lack this capacity or metabolize it at too slow a rate. The result is the death of the weed species without the metabolic capacity to rid itself of the toxin.[1]

The shear volume and porous structure of a tree's wood provides an enormous surface area for exchange or biochemical reactions. Some researchers are attempting to augment the inherent metabolic capacity of plants by incorporating bacterial, fungal, insect, and even mammalian genes into the plant genome.[1,3]

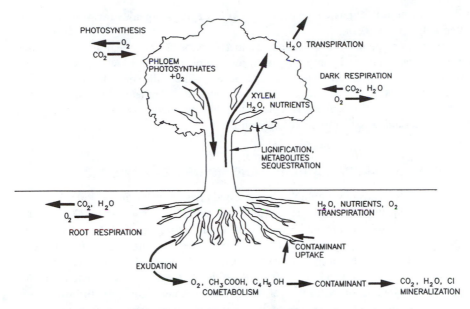

Figure 10.1 Oxygen, CO_2, water, and contaminate cycling through a tree.

10.2.2 Degradation in Rhizosphere

In natural plant ecosystems, indigenous soil microorganisms present in the rhizosphere (root zone of the plants) are found in mutual relationships with plants. The microflora that responds to the presence of living roots is distinctly different from the characteristic soil population, due to the plant creating a unique subterranean habitat for microorganisms. The plant, in turn, is markedly affected by the population it has stimulated, since the root zone is the area from which mineral nutrients are obtained. The rhizosphere is often divided into two general areas, the inner rhizosphere at the very root surface and the outer rhizosphere embracing the immediately adjacent soil. The microbial population is larger in the inner zone where the biochemical interactions are most pronounced and root exudates are concentrated.

Different plant species often establish somewhat different subterranean floras. The differences are attributed to variations in rooting habits, tissue composition, and excretion products of the plant. The primary root population is determined by the habitat created by the plant; the secondary flora, however, depends upon the activities of the initial population. The age of the plant also alters the microbial population in the rhizosphere.

The roots of the plants exude a wide spectrum of compounds including sugars, amino acids, carbohydrates, and essential vitamins that may act as growth and energy-yielding substrates for the microbial consortia in the root zone[8] (Figure 10.1). Exudates may also include compounds such as acetates, esters, benzene derivatives, and enzymes.[8] This process enriches the *in situ* microbial populations present in the rhizosphere for enhanced degradation of organics by the provision of appropriate beneficial primary substrates for cometabolic transformations of the target contaminants. It was also reported that wherever significant transformation of contaminants was evident, the following enzyme systems were present: dehalogenase, nitroreductase, peroxidase, laccase, nitylase, and oxygenase.[5]

In addition to the plant exudates, the rapid decay of fine-root biomass can become an important addition of organic carbon to soils. The additional organic carbon, in turn, may increase microbial mineralization rates. The increase in organic carbon levels also serves to retard the contaminant migration into groundwater.

Roots also harbor mycorrhizae fungi, which metabolize organic contaminants. These fungi, growing in symbiotic association with the plant, have unique enzymatic pathways, similar to white rot fungus enzymes, that help to degrade organics that could not be transformed solely by bacteria.

In summary, plants provide exudates that provide an excellent habitat for increased microbial populations and pump oxygen to roots, a process that ensures aerobic transformations near the root that otherwise may not occur in the bulk soil. Due to the presence of certain primary substrates in the exudate system, anaerobic cometabolic transformations may also take place in the rhizosphere. A typical microbial population in the rhizosphere comprises 5×10^6 bacteria, 9×10^5 actinomycetes, and 2×10^3 fungi per gram of air-dried soil.[5]

10.3 PHYTOREMEDIAITON MECHANISMS OF HEAVY METALS

Most heavy metals have multiple chemical and physical forms in the soil environment. All forms are not equally hazardous, nor are all forms equally amenable to uptake by plants. The chemistry of the metal and its mobility will inherently impact the toxicity in the environment.

Metal fractionation or sequential extraction schemes (such as TCLP) sometimes are used to describe metal behavior in soils. Most metals interact with the inorganic and organic matter that is present in the root-soil environment. Potential forms of metals include those dissolved in the soil solution, adsorbed to the vegetation's root system, adsorbed to insoluble organic matter, bonded to ion exchange sites on inorganic soil constituents, precipitated or coprecipitated as solids, and attached to or inside the soil biomass.

Phytoremediation of heavy metal contaminated soils can be divided into phytostabilization and phytoextraction approaches.[1]

10.3.1 Phytostabilizaiton of Heavy Metals

Implementation of phytostabilization involves the reduction in the mobility of heavy metals by minimizing soil erodibility, decreasing the potential for wind-blown dust, and reduction in contaminant solubility by the addition of soil amendments.

Erosion leads to the concentration of heavy metals because of the selective sorting and deposition of different size fractions of the soil. Eroded material is often transported over long distances, thus selectively extending the effects of contamination and increasing the risk to the environment. Erosion can, therefore, cause the buildup of concentrations of normally nontoxic contaminants to toxic levels at locations where transported material is deposited.

Planting of vegetation at contaminated sites will significantly reduce the erodibility of the soils both by water and wind. Density of vegetation will effectively hold the soil and provide a stable cover against erosion.

Another element of phytostabilization is to supplement the system with a variety of alkalizing agents, phosphates, organic matter, and biosolids to render the metals insoluble and unavailable to leaching. Materials with a calcareous character or a high pH can be added to influence the acidity: compounds such as lime and gypsum. Specific binding conditions can be influenced by adding concentrated Fe, Mn, or Al compounds. To maintain or raise the organic matter content in the soils, various materials such as humus or peat materials, manure or mulch can be added.

This chemical alteration should be quickly followed by establishing a plant cover and maximizing plant growth. The amendments sequester the metals into the soil matrix and plants keep the stabilized matrix in place, minimizing wind and water erosion.

10.3.2 Phytoextraction of Heavy Metals

The use of unusual plants that have the ability to accumulate very high (2 to 5%) concentrations of metals from contaminated soils in their biomnass provides the basis for this phytoremediation technique. The metals are translocated to the shoot and tissue via the roots. These plants are called hyperaccumulator plants and they exhibit the ability to tolerate high concentrations of toxic metals in above-ground plant tissues; these species contain toxic element levels in the leaf and stalk biomass (LSB) of about 100 times those of nonaccumulator plants growing in the same soil, with some species and metal combinations exceeding conventional plant levels by a factor of more than a thousand.[9]

Many hyperaccumulator plants that are nonwoody (not a tree) have been identified to have the capacity to accumulate metals. *Thlaspi caerulescens* was found to accumulate Zn up to 2000 to 4000 mg/kg.[10] The Indian mustard plant *Brassica juncea*, grown throughout the world for its oil seed, was found to accumulate a significant amount of lead.[11] One planting of mustard in a hectare of contaminated land was found to soak up 2 t of lead. If three plantings could be squeezed in per year, 6 t of lead per hectare can be extracted. Both hemp dogbane (*Apocynum* sp.) and common ragweed also have been observed to accumulate significant levels of lead. *Aeollanthus subcaulis* var. *lineris* and *Paspalum notatum* are other hyperaccumulator plants known to accumulate Cu and Cs, respectively. Hyperaccumulator plants can address contamination in the shallow soils only up to 24 in. in depth. If contamination is deeper, 6 to 10 ft, deep-rooted poplar trees can be used for phytoextraction of heavy metals. These trees can accumulate the heavy metals by sequestration. However, there are concerns specifically for trees that leaf litter and associated toxic residues will be blown off site. This concern may be tested in the laboratory to see whether uptake and translocation of the metals into the leaves exceed the standards.

Hyperaccumulators have the metal-accumulating characteristics that are desirable, but lack the biomass production, adaptation to current agronomic techniques, and physiological adaptations to the climatic conditions at many contaminated sites. It has been reported that harvesting done at different seasons in a year had pronounced differences in accumulation levels.[10] In the future, genetic manipulation techniques may provide better hyperaccumulator species. The success of phytoextraction depends on the use of an integrated approach to soil and plant management. The disciplines of soil chemistry, soil fertility, agronomy, plant physiology, and plant genetic engineering are currently being used to increase both the rate and efficiency of heavy metal phytoextraction.

The schematic of the process involved in heavy metal phytoextraction is shown in Figure 10.2. Translocation from the root to the shoot must occur efficiently for the ease of harvesting. After harvesting, a biomass processing step or disposal method that meets regulatory requirements should be implemented.

10.3.3 Phytosorption and Phytofiltration of Heavy Metals

Aquatic plants and algae are known to accumulate metals and other toxic elements from solution.[12] There are large differences in bioremoval rates due to species and strain differences, cultivation methodology, and process control techniques. In the past, commercial systems have used immobilized algal biomass for removing radionuclides and other heavy metals in the aqueous phase.[13]

Plants that are naturally immobilized, such as attached algae and rooted plants, and those that could be easily separated from suspension, such as filamentous microalgae, macroalgae, and floating plants, have been found to have high adsorption capacities. In a recent study, one blue-green filamentous algae of the genus *Phormidium* and one aquatic rooted plant, water milfoil (*Myriophyllum spicatum*) exhibited high specific adsorption for Cd, Zn, Ph, Ni, and Cu.[12]

It has been reported that porous beads containing immobilized biological materials such as sphagnum peat moss can be used for extracting metals dissolved in the aqueous phase.[14] The beads designated as BIO-FIX beads readily adsorbed Cd, Pb, and other toxic metals from dilute waters.

In a recent study, it was reported that *Saccharomyces cerevisiae* yeast biomass, when treated with a hot alkali, exhibited an increase in its biosorption capacity for heavy metals.[15]

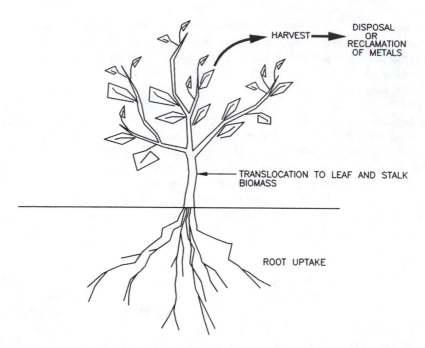

HARVEST ➡ DISPOSAL
OR
RECLAMATION
OF METALS

TRANSLOCATION TO LEAF AND STALK
BIOMASS

ROOT UPTAKE

Figure 10.2 Process schematic describing the various processes during phytoextraction of heavy metals.

It was also reported that caustic treated yeast immobilized in alginate gel could be reactivated and reused to remove Cu, Cd, and Zn in a manner similar to the ion exchange resin.

10.4 PHYTOREMEDIATION OF NITROGEN COMPOUNDS

Nitrogen is an important nutrient for plant growth. When nitrates or other inorganic nitrogen species are present above the allowable limits in groundwater, phytoremediation is a viable remediation alternative under shallow water table conditions. Corn and deep-rooted poplar trees have been found to be very effective in reducing nitrate concentrations in groundwater.[16,17] Poplar trees were most effective at reducing the nitrate concentrations, and the perennial nature of the trees ensures prolonged protection throughout the year. Deeper planting of the trees, closer to the water table, resulted in better growth and greater nitrogen accumulation in leaves.[17]

Munitions wastes such as 2,4,6-trinitrotoluene are of considerable interest as candidates for phytoremediation.[5] An excreted nitroreductase activity in the root zone has been identified as the catalyst for degradation of this compound.

10.5 FIELD APPLICATIONS OF PHYTOREMEDIATION

Many successful field-scale applications of phytoremediation have been reported in the literature. Each successful application reported has site-specific conditions unique to that particular site as well as different remediation objectives. Although phytoremediation is not a panacea for hazardous waste problems, it has proven effective in several applications for remediation of shallow contaminated sites. Before the technology can mature, we need a better understanding of the role of metabolites, enzymes, and the selection of plant and tree systems for various wastes and hydrogeologic conditions.

Trees potentially are the lowest-cost plant type to be used for phytoremediation. A number of trees can grow on land of marginal quality, and this will allow establishment of trees on sites with low fertility and poor soil structure, thus keeping costs low for plant establishment. Since trees are perennial plants with long life spans, site remediation can continue for decades with little or no maintenance costs. Trees of the Salicaceae family (willow and poplar) have been planted at several locations because of their flood tolerance and fast growth.[5] Various grass species such as prairie grass, crested wheat grass, and cattails have been used in phytoremediation. Alfalfa has been used widely for its deep rooting and root zone metabolic activity. Parrot feather and Eurasian water milfoil have been applied to break down ammunition wastes.[5]

A proprietary technique call TreeMediation™ has been developed with a specific deep root system.[17] Such root systems are known to facilitate the uptake of contaminated groundwater by hybrid poplar trees up to 30 ft below land surface. In addition, the hybrid poplar trees can transpire a quantity of water sufficient to impact the flow of groundwater when planted in sufficient density. Such hydraulic control in shallow groundwater conditions is valuable in that the migration of the contaminant plume can be reduced or eliminated.

One recent study demonstrated that poplar trees, which possess cytochrome P-450s analogous to the oxygenases responsible for transformation of compounds such as TCE in the mammalian liver, exposed to 100 mg/l of TCE did uptake and chemically alter this contaminant. TCE and its metabolites were found in the roots and tissue of the study trees, but not in control trees or in the soil used for potting the trees.[18] In a subsequent study, popular seedlings exposed to ^{14}C-labeled TCE were found to generate ^{14}C-labeled carbon dioxide. Intermediate compounds generated during oxidation are thought to be 2,2,2-trichloroethanol, and di- and trichloroacetic acid.[18] Similar studies have shown positive results for toluene and benzene.

Poplars are phreatophytic, capable of extending their roots into aerobic water tables. For example, the roots of poplars growing alongside streams can easily be observed intertwined in the stream bottom. The degree to which poplar roots would penetrate the saturated zone cannot be easily estimated. However, if their access to soil moisture from precipitation is limited, poplars will draw large amounts of water from the top of the saturated aquifer. Evapotranspiration will draw down the water table below the trees similar to a pump and treat system. Under optimal conditions, a hybrid poplar occupying 4 m^2 of ground can cycle 100 l of groundwater per day; this translates to approximately 30,000 gal of water per acre of trees per day.[3] Poplar trees have been reported to grow 6 to 8 ft per year. Figure 10.3 shows the development of the root zone with time. Predicted impact on groundwater flow at a shallow water table site is shown in Figure 10.4. Simulations of a proposed design can be carried out based on extent of contamination, hydrogeological data, and past precipitation and infiltration records.

A big advantage of phytoremediation over conventional pump and treat systems is the ability of the roots to penetrate the microscopic scale pores in the soil matrix. Contaminants adsorbed or trapped in these micropores are minimally or not impacted by the pump and treat system. In the case of phytoremediation, the roots can penetrate these micropores for contaminant removal.

Another reported application of phytoremediation is the planting of 10,000 poplar trees per hectare as the final cap on a landfill in Oregon.[5] Treatment of organic wastes is not the main goal at this site, but rather to keep the site natural and free from infiltration. This proprietary technique is known as Ecolotree™ cap, and installation involves the placement of a thick cover with soil amendments for storage of water within the root zone.[19] Densely planted poplar trees uptake water through their extensive root systems as a pump and transpire the water back to the atmosphere. Advanced moisture control systems using time domain reflectometry (TDR) technique have been installed at this site. Although hybrid poplars seem to tolerate organic chemicals quite well, high concentrations of metals, salts, and ammonia are toxic.

For long-term closure of landfills in temperate climates, it might be desirable to use higher value hardwood trees such as walnut or pecan, in addition to fast-growing poplars,

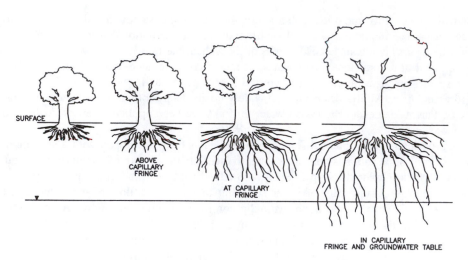

Figure 10.3 Placement of root ball with time due to maturation of the tree.

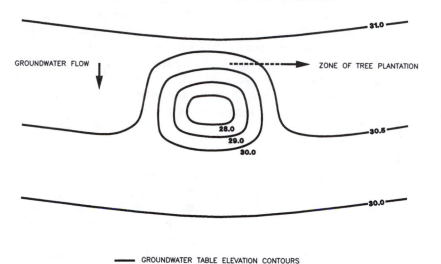

Figure 10.4 Predicted groundwater flow conditions at maturation stage of TreeMediation™ system.

unless there is a ready market for poplar products. Walnuts or pecans under good fertility conditions use large amounts of water and produce both nuts and wood. In subtropical regions, eucalyptus would be more appropriate than poplars. One would use alfalfa only in a situation where a short-lived perennial is satisfactory. Replanting would be necessary after a few years, if the remediation was not completed in that time. It is essential to choose climatically adapted species from among those with desired growth characteristics.

Degradation of petroleum contaminants in the rhizosphere has been reported in the literature. Microbial numbers were substantially greater in soil with plants when compared to soil containing no plants, indicating that plant roots enhance microbial populations in contaminated soil for enhanced biodegradation of contaminant.[20]

Another reported study investigated the uptake of two pesticides, alachlor and atrazine, present in the soils.[21] It was found that plant uptake is an influential process on the fate of alachlor and atrazine.

A recent report considers some strategies for engineering plants to improve bioremediation in the root zone.[22] One of the simpler approaches is to make use of the organism *Agrobacterium rhizogenes* to induce a state called hairy root disease. Depending on the virulence of the strain used, the extent of root production is variable, but generally infection leads to a significant enhancement of rooting without obvious detrimental effects on the host plant. Increased root mass has the apparent advantage of increasing the surface area available for microbial colonization. Root exudation may be increased in proportion to increase in root area. Such rhizosphere enhancements could improve bioremediation potential of the plant–microbial system. It is suggested that when water is not freely available in unlimited quantities, increased root mass could lead to greater water uptake, and hence greater contaminant mobilization and potential degradation.

Genetic engineering of plants by insertion of genes for chlorinated phenolics catabolism is in progress.[22] These enzymes may also allow metabolism of TCE. A number of companies have introduced genes for degradation of herbicides into crop plants.[22] Some of these approaches could prove useful for remediation of other contaminants.

10.6 LIMITATIONS AND KNOWLEDGE GAP

- Phytoremediation is most effective only at sites with shallow contamination in the soils and/or sites with shallow water table.
- Can be applied only under warmer climates for 12 months per year remediation.
- This technique may not be applicable for highly hydrophobic contaminants due to the tendency of the contaminants to remain adsorbed to the soil particles.
- The question as to whether the contaminants can accumulate in leaves and be released as litter or accumulate in the wood and mulch has not been resolved from a regulatory point of view.
- The possibility of binding or complexation of some of the contaminants with the exudates and subsequent transport by the groundwater.
- Evaluation and development of proper handling and disposal methods for the harvested hyperaccumulator plants. The feasibility of cost-effective metal recovery techniques should be evaluated further.
- Enhancement of tree root mass, via genetic engineering techniques, for increased rhizosphere detoxification and contaminant translocation.
- Enhancement of detoxification in the plant by cloning the plant with bacterial genes.

REFERENCES

1. Cunningham, S. D., Berti, W. R., and Huang, J. W., Remediation of contaminated soils and sludges by green plants, in *Bioremediation of Inorganics*, Hinchee, R. E., Means, J. L., and Burns, D. R., Eds., Battele Press, Columbus, OH, 1995.
2. Hammer, D. A., *Constructed Wetlands for Wastewater Treatment*, Lewis Publishers, Boca Raton, FL, 1989.
3. Stomp, A. M., Han, K. H., Wilbert, S., and Gordon, M. P., Genetic improvement of tree species for remediation of hazardous wastes, *In Vitro Cell. Dev. Biol.*, 29, 227, 1993.
4. Menge, D. and Frederick, D., Potential for Renovation of Municipal Wastewater Using Biomass Energy Hardwood Plantations, in *Proc. Natl Bioenergy Conf.*, Coeur D'Alene, ID, 1991.
5. Schnoor, J. L., Licht, L. A., McCutcheon, S. C., Wolfe, N. L., and Carreira, L. H., Phytoremediation of organic and nutrient contaminants, *Environ. Sci. Technol.*, 29(7), 1995.
6. Bell, R. M., Higher Plant Accumulation of Organic Pollutants from Soils, U.S. Environmental Protection Agency, EPA600/R-92/138, 1992.
7. Sandermann, J., Plant metabolism of xenobiotics, *Trends Biotechnol.*, 17, 82, 1992.
8. Narayanan, M., Davis, L. C., and Erickson, L. E., Fate of volatile chlorinated organic compounds in a laboratory chamber with alfalfa plants, *Environ. Sci. Technol.*, 29, 2437, 1995.

9. Cornish, J. E., Goldberg, W. C., Levine, R. S., and Benemann, J. R., Phytoremediation of soils contaminated with toxic elements and radionuclides, in *Bioremediation of Inorganics*, Hinchee, R. E., Means, J. L., and Burns, D. R., Eds., Battelle Press, Columbus, OH, 1995.

10. Brown, S. L., Chaney, R. L., Angle, J. S., and Baker, A. J. M., Zinc and cadmium uptake by hyperaccumulator *Thlaspi caerulescens* and metal tolerant *silene vulgaris* grown on sludge amended soils, *Environ. Sci. Technol.*, 29, 1581, 1995.

11. Bishop, J. E., Pollution fighters hope a humble weed will help reclaim contaminated soil, *Wall Street J.*, Aug. 7, 1995.

12. Wang, T. C., Weissman, J. S., Ramesh, G., Varadarajan, R., and Benemann, J. R., Bioremoval of toxic elements with aquatic plants and algae, in *Bioremediation of Inorganics*, Hinchee, R. E., Means, J. L., and Burns, D. R., Eds., Battelle Press, Columbus, OH, 1995.

13. Feiler, H. D. and Darnall, D. W., Remediation of Groundwater Containing Radionuclides and Heavy Metals using Ion Exchange and the Alga SORD Biosorbent System, Final Report under Contract No. 02112413, DOE/CH-9212, U.S. Department of Energy, 1991.

14. Jeffers, T. H., Bennett, P. G., and Corwin, R. R., Biosorption of Metal Contaminants Using Immobilized Biomass: Field Studies, Report of Investigations 9461, Bureau of Mines, U.S. Department of the Interior, 1993.

15. Lu, Y. and Wikins, E., Heavy metal removal by caustic-treated yeast immobilized in alginate, in *Bioremediation of Inorganics*, Hinchee, R. E., Means, J. L., and Burns, D. R., Eds., Battelle Press, Columbus, OH, 1995.

16. Paterson, K. G. and Schnoor, J. L., Vegetative alteration of nitrate fate in unsaturated zone, *J. Environ. Eng.*, 119(5), 986, 1993.

17. Gatliff, E. G., Vegetative remediation process offers advantages over traditional pump and treat technologies, *J. Remediation*, Summer 1994.

18. Strand, S. E., Wilmoth, J., Newman, L. A., Ruszaj, M., Duffy, J., Heilman, P. E., Ekuan, G., Choe, N., Shurtleff, B., Brandt, M., Wilbert, S. M., and Gordon, M., Transpiration and Transformation of Trichloroethylene in Poplar, University of Washington, Seattle, 1993.

19. Licht, L. A. and Madison, M., Using poplar trees as a landfill cover: Experiences with the Ecolotree™ cap, presented at *SWANA 11th Ann. Northwest Regional Solid Waste Symp.*, Portland, OR, April 1995.

20. Lee, E. and Banks, M. K., Bioremediation of petroleum contaminated soil using vegetation: A microbial study, *J. Environ. Sci. Health*, A28 (10), 2187, 1993.

21. Paterson, K. G. and Schnoor, J. L., Fate of alachlor and atrazine in a riparian zone field site, *Water Environ. Res.*, 64 (3), 274, 1992.

22. Stomp, A. M., Han, K. H., Wilbert, S., Gordon, M. P., and Cunningham, S. D., Genetic strategies for enhancing phytoremediation, *Ann. N.Y. Acad. Sci.*, 721, 481, 1994.

11

PUMP AND TREAT SYSTEMS

11.1 INTRODUCTION

Until the very recent past, almost all groundwater cleanup systems installed involved variations of the technology called "pump and treat." Between 1982 and 1992, 73% of the cleanup agreements at Superfund sites where groundwater is contaminated specified the use of pump and treat technology.[1,2] At most of these sites, the cleanup goal is to restore the aquifer so that the water extracted from it will be suitable for drinking without further treatment. Yet, within the past few years, the experience in the industry indicates that drinking water standards may be essentially impossible to achieve in a reasonable time frame at certain sites.

Pump and treat systems operate by pumping groundwater to the surface, removing the contaminants, and then either recharging the treated water back into the ground or discharging it to a surface water body or municipal sewage plant. Once groundwater has been pumped to the surface, contaminants can be removed to very low levels with established technologies used to treat drinking water and wastewater. However, pumping the contaminated water from the aquifer does not guarantee that all of the contaminants have been removed from beneath the site. Contaminant removal is limited by the behavior of contaminants in the subsurface (a function of contaminant characteristics), site geology and hydrogeology, and extraction system design.

The experience accumulated in the industry indicates that the performance of pump and treat systems can be depicted as in Figure 11.1. The theoretical removal curve shows the number of pore volumes of groundwater that must be pumped to remove the contamination, assuming all of it dissolves readily. The removal with tailing curve shows the number of aquifer volumes of groundwater that must be pumped to remove the contamination when significant undissolved sources of contamination are present. Examples of such sources are adsorbed phase contamination in the saturated zone and capillary fringe, nonaqueous phase liquids (NAPLs) floating on the water table or present below the water table.

Pump and treat systems can be designed to meet two very different objectives: (1) containment, to prevent the contamination from spreading, and (2) restoration, to remove the contaminant mass. In pump and treat systems designed for containment, the extraction rate is generally established as the minimum rate sufficient to prevent enlargement of the contaminated zone. In systems designed for restoration, the pumping rate is generally established to be much larger than that required for containment so that clean water will flush through the contaminated zone at an expedited rate. Because of their reduced pumping requirements, pump and treat systems designed for containment are much less costly to operate than systems designed for restoration. In all other fundamental ways, the two types of systems are identical.

This chapter presents a general survey of design and application issues associated with groundwater pump and treat systems. It is recognized that a brief chapter such as this cannot provide comprehensive coverage of all the design issues associated with groundwater pump and treat systems. The intent is to provide the reader with a structured thought process for

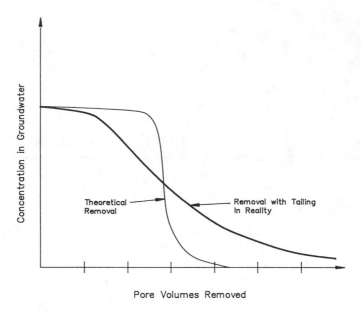

Figure 11.1 The effect of tailing on cleanup time.

addressing a generic groundwater treatment problem and identifying the various steps of the design process and the associated issues that need to be addressed.

For the purposes of this chapter, the design of a groundwater pump and treat system is divided into the following steps:

- Definition of the problem
- Screening of options
- Engineering
- Permitting

11.2 DEFINITION OF THE PROBLEM

In general, a proper and thorough site investigation has to be completed at any contaminated site prior to initiating the design effort. A properly conducted site investigation will be able to define the problem at hand. The definition of the problem that requires a pump and treat solution can be divided into two areas: (1) determination of relevant hydrogeologic and hydraulic parameters and (2) identification of contaminants of concern (COCs).

11.2.1 Hydrogeologic and Hydraulic Parameters

There is an abundance of discussion and information available in the literature on this subject. Table 11.1 presents only a summary of these parameters and their importance for groundwater cleanup.

11.2.2 Contaminants of Concern

Contaminants of concern (COC) are those chemicals in the groundwater that present a risk and are creating the need for the pump and treat system. Selecting the remedial action alternative depends on site conditions, contaminant properties, and discharge criteria. Contaminant properties such as solubility, strippability, adsorbability, and biodegradability, and air and water discharge criteria ultimately dictate the most appropriate technology.

Table 11.1 Hydrogeologic and Hydraulic Parameters Important for Groundwater Cleanup

Parameter	Importance for groundwater cleanup
• Hydraulic conductivity	Ease with which water can move through a formation and influences the rate at which groundwater can be pumped for treatment. Influences the total flow rate of the system.
• Hydraulic gradient	Influences the direction of contaminant movement based upon the elevation and pressure differences.
• Transmissivity	Influences the rate at which groundwater can be pumped and, thus, influences the total flow rate of the system.
• Groundwater velocity	Influences the direction and velocity of dissolved contaminant movement. Plays an important role when designing a containment system.
• Porosity	Pores store water and contaminants. Influences the hydraulic conductivity and impacts the fate of the contaminants due to various physical, chemical, and biological processes that take place in the saturated zone.
• Effective porosity	Has an impact on the groundwater velocity.
• Storage coefficient	Influences the quantity of groundwater that can be obtained by pumping.
• Specific yield	Fraction of total pore volume released as water by gravity drainage during pumping of an unconfined aquifer and influences the quantity of groundwater that can be obtained by pumping.

Table 11.2 Treatability Aspects of Contaminants

Question	Underlying issue
• Are the COCs organics?	The suite of treatment technologies applicable to organics includes various separation and destruction techniques, both biological and physical/chemical.
• If organic, are they chlorinated?	The presence of chlorinated organics suggests that aerobic biological treatment process may not be applicable.
• If organic, are they highly water soluble?	Water-soluble organics (such as alcohols and ketones) are difficult to remove using common separation technologies (i.e., carbon adsorption or air stripping); however, they are quite amenable to biological or chemical techniques.
• If organic, are they extremely insoluble?	The presence of extremely insoluble organics (such as PCBs or multi-ring PAHs) in groundwater, at concentrations above their solubility limit, suggests that the organics are adsorbed to colloidal solids, and that thorough solids removal will be required.
• Are the COCs inorganic?	The suite of technologies applicable to inorganics are generally limited to physical/chemical separation techniques. Some nonmetals (i.e., nitrates, ammonia) can be treated biologically.
• Are both organics and inorganics present?	Separate treatment technologies will have to be incorporated in the treatment train. Consideration will need to be given to the preferred order of treatment.
• Is the contaminant present as nonaqueous phase liquid (NAPL)?	Priority should be given to remove the separate phase contaminants (either as LNAPL or DNAPL) to minimize the continuous solubilization of contaminants into the aqueous phase.

Table 11.2 presents typical questions that should be asked as part of understanding the nature of the contaminants of concern, along with the underlying issues.

11.2.3 Water Chemistry

Water chemistry refers to the properties of the water that impact the way it "behaves" in the treatment system. The parameters that define water chemistry, other than contaminant concentrations, include the following:

- pH
- total dissolved solids

- conductivity
- total suspended solids
- total and dissolved iron
- total and dissolved manganese
- calcium hardness
- total hardness
- total alkalinity
- dissolved oxygen
- temperature

Evaluation and analysis of the above parameters will allow the designer to address the issues of fouling during the operational phase. Common operational problems fall into two categories: (1) clogging due to precipitation of metals and scaling, and (2) electrical/mechanical problems with pumps, level switches, and other components of process control system. In some instances, the water quality may be so extreme that certain technologies will be ruled out during the technology screening phase.

11.2.4 Flow Rate

The significance of the flow rate as it relates to equipment sizing during the design phase is obvious. However, there are some other flow-related issues that the designer should consider. These include the following:

- The duration of the pump and treat program should be estimated. Long-term programs warrant more durable equipment, more sophisticated instrumentation and control systems, and more permanent structures.
- The consistency of the flow rate over the short term should be considered. For example, if the pumping wells are low yield, the flow rates from these wells may vary and it may not be practical to design a system that operates on a continuous basis. Equalization of the flow will be necessary.
- The consistency of the flow over the long term should also be considered. It is often the case that both the flow rate and the contaminant concentrations decrease significantly over the life of a groundwater remediation system. Designing a treatment system for maximum flow rates and contaminant concentrations often results in a system that, after a relatively short period of operation, is significantly oversized. Serious consideration should be given to incorporate the life cycle concept in which flow rates and concentrations expected over the life of the project should be emphasized. Life cycle design can be accomplished by designing a flexible modular system in which treatment processes or components can be installed and replaced as modules. Accepting an undersized system during the initial phases of operation is another option.
- To obtain the required number of pore volume exchanges in a reasonable period of time, extraction of groundwater at a reasonable rate becomes necessary. To achieve the required flow rate, extraction points must be of sufficient number, adequately located, properly maintained, and pumped as continuously as possible.

11.2.5 Physical and Regulatory Constraints

The most obvious regulatory constraint pertains to the discharge of the treated water. Unless the water is to be used for some useful purpose such as boiler feedwater or noncontact cooling water, some form of a discharge permit will normally be required. Some treatment systems also generate air emissions. If so, air permitting must be addressed. While it is normally impractical to get final discharge limits from the appropriate regulatory agencies at this early stage of a design, a review of the appropriate regulations and discussions with the appropriate regulatory agency personnel will provide the designer with a good intitial approximation of the regulatory constraints pertaining to air and water discharges.

Physical constraints are those factors associated with the physical setting and location of the treatment system that will impact the design. The following are examples of questions that should be considered in evaluating and understanding the physical constraints of a treatment system design:

- What are the treated water discharge options? Is there a suitable surface water discharge point nearby? Is discharging to the POTW an option, and what is the associated cost? Is the treatment system located at an active industrial facility? If so, can the facility use the treated water as process water, cooling water, boiler feed water, etc.? Is there sufficient open area and are the soil conditions suitable for an infiltration gallery? Is the depth to groundwater and the aquifer characteristics suitable for reinjection wells?
- What type of enclosure will be necessary for the treatment facility? Is freeze protection an issue? Will explosion-proof equipment be necessary?
- Is there adequate space to locate the treatment facility?
- Are utilities readily available? If so, are there any "free" energy sources such as waste steam?
- Are there existing design and operating standards at the facility, and is consistency and/or compatibility with existing facilities an issue?

11.2.6 Design Objectives

Upon completion of the above thought processes, the designer should have a thorough understanding of the design objectives. The quality of the water to be treated will have been analyzed and understood, both from a chemistry standpoint and from a flow standpoint. The regulatory environment will have been considered, along with site-specific factors, and the air and water discharge options and limits will have been identified and quantified. It is strongly recommended that the design objectives be documented in a design basis memorandum, and that this document be updated as the design proceeds through the subsequent phases.

11.3 SCREENING OF OPTIONS

The next step, following the definition of the problem, is the identification and screening of the various treatment options that will enable the designer to achieve the design objectives. The screening process is comprised of evaluating the identified technologies from a technical viability and cost standpoint. When evaluating the cost, capital costs and operation and maintenance costs over the predicted life of the system should be taken into consideration.

The end result of the screening process is the selection of the lowest cost, most technically viable approach, taking into consideration all of the factors discussed in the previous sections. Table 11.3 provides general observations regarding the technical viability and cost of some of the more common treatment technologies for groundwater contaminated with *organic* compounds. A brief description of these technologies is provided in later sections. Table 11.4 provides general observations regarding the technical viability and cost of some of the more common treatment technologies for groundwater contaminated with *inorganic* compounds, especially heavy metals. A brief description of these technologies is provided in later sections.

Table 11.3   Technical and Cost Viability of Various Technologies for Treating Organic Compounds Present in Groundwater

Technology	Technical issues	Cost drivers
Air stripping	• Vendor quotes and performance guarantees readily available. • 99.9% removal efficiencies are achievable. • Phase transfer process (contaminants are not destroyed). • Option of packed tower vs. low profile. • Fe and Ca fouling can be a significant issue. Bacterial fouling can also be a significant issue. • Air emissions control is generally required.	• Air emission controls. • Chemical pretreatment to inhibit metal precipitation, scaling, and bacterial fouling. • Operation and maintenance associated with fouling.
Carbon adsorption	• Vendor quotes and performance guarantees readily available. • 99.9% removal efficiencies are achievable. • Simple technology • Not applicable for highly soluble and miscible compounds. • Phase transfer process (contaminants are not destroyed).	• Need to use pressure rated vessels. • Cost of carbon (includes new carbon and cost of disposal of spent carbon). • Pretreatment and operation and maintenance associated with fouling.
Steam stripping	• Overhead can be condensed and product recovered. • Calcium and iron fouling more severe at elevated temperatures. • Vacuum steam stripping operates at lower temperature. • Considerable engineering required for the system design and operation. • Phase transfer process (contaminants are not destroyed).	• Value of recovered product. • Cost of steam (energy and capital cost). • Additional process equipment (condensers, heat exchangers, vapor recompression equipment, etc.). • Operation and maintenance associated with fouling. • More intense operator attention required.
Chemical oxidation	• Oxidants include $UV/H_2O_2/O_3$, UV/H_2O_2, UV/O_3, potassium permanganate, Fenton's reagent, hypochlorite, etc. • Destruction technique. • Oxygen effective under high temperature and pressure (wet air oxidation). • Various reactor configurations possible (batch, plug flow, CSTR). • Potential side benefit associated with oxidation (inorganics destruction cyanide, sulfide, iron, and manganese).	• Capital cost process equipment. • O & M cost associated with energy consumption. • Chemical requirement exceeds the stoichiometric requirement based on contaminant levels (all impurities also will be oxidized). • O & M cost associated with fouling.
Biodegradation	• Applicable only for biodegradable compounds. • Operational modes include attached growth, suspended growth and fluidized bed reactors. • Influent consistency is critical to maintaining a healthy biomass. • Operational parameters have to be maintained at consistent levels. • Aerobic processes have faster degradation rates than anaerobic processes.	• Cost process equipment to support the bioreactor operation. • O & M problems associated with changes in pH, temperature, and toxic loadings, etc. • Loss of biomass and drop in efficiency.
Membrane filtration	• Contaminants are separated and can be reused. • Pretreatment for removal of suspended solids is required. • Based on the waste stream, microfiltration or ultrafiltration or nanofiltration or hyperfiltration (reverse osmosis) can be employed. Selection of the type of filtration depends on the type of contaminants and the molecular size of contaminants.	• Higher capital and O & M cost. • O & M cost associated with membrane fouling. • Cost of disposal of brine as a waste.

Table 11.4   Technical and Cost Viability of Various Technologies for Treating Inorganic Compounds Present in Groundwater

Technology	Technical issues	Cost drivers
Precipitation	• Most heavy metals are removed by pH adjustment and precipitation as hydroxides. Some metals can be removed as metallic sulfides too. • Coprecipitation with iron effective for some heavy metals (arsenic). • Optimum pH different for different metals. • Solids separation system is required. • Dewatering of the sludge needs optimum operating conditions. 30 to 50% solids content is the maximum amount of solids achievable by dewatering and sludge thickening processes.	• Capital cost of process equipment (chemical addition, coagulation, flocculation, clarification, sludge thickening equipment). • Disposal of potentially hazardous sludge. • Sludge volume is increased due to impurities present, such as iron and manganese and also by the chemicals added as coagulants and flocculants. • Constant operator attention required.
Ion exchange	• Capable of high removal efficiency. • Contaminants are concentrated in regenerant stream, which is approximately 10% of total influent stream. • Fouling of ion exchange resin by naturally occurring ions (iron, calcium, manganese, and magnesium) and competition by these ions for ion exchange sites. • Optimum pH conditions required. • Competition by impurities present such as sulfate, nitrate, etc. • Presence of chelation and sequestration diminishes ion exchange efficiency.	• Capital costs of process equipment, including ion exchange vessels, and regenerant handling equipment, etc. • Disposal cost of residual stream from regeneration process.
Adsorption	• Carbon (conventional activated carbon and various impregnated carbons) are effective at removing some metals. • Several proprietary adsorption processes applicable to select metals. • Presence of chelates diminishes adsorption efficiency. • Best applied as polishing step to streams with low influent concentrations.	• Low capital cost. • Potentially high O & M costs due to high adsorbent use and associated disposal and replacement costs.
Reverse osmosis	• Capable of high removal efficiency. • A reject stream is generated. • Optimum pH required. • Membrane fouling could be a constant problem. • Membranes are damaged by oxidants such as chlorine.	• Capital cost of process equipment. • Disposal cost of reject stream. • Development of lower pressure membranes applicable to multivalent ions is making reverse osmosis more economical for heavy metals. • O & M costs associated with fouling.
Steam stripping	• Same issues as discussed in the previous table. • Removal of ammonia is possible.	• Same issues as discussed in the previous table.
Chemical oxidation	• Contaminants such as cyanide, sulfide, etc. can be oxidized. • No need for UV light.	• Moderate capital and O & M costs.

11.3.1   Oil/Water Separation

When lighter-than-water, nonaqueous phase liquids (LNAPLs) such as petroleum products are floating on top of the groundwater table, recovery of such contamination can be accomplished in two ways: (1) recovery of the LNAPL separately from the contaminated groundwater, or (2) recovery of the LNAPL and contaminated water as total fluids. When the LNAPL thickness is small and the site hydrogeologic conditions are less permeable, total fluids recovery will be the preferred technique for recovering the floating separate phase contamination. Under such circumstances separation of the recovered oil and water becomes necessary prior to further treatment of the contaminated groundwater. The oil separated from the total fluids stream can be stored and recycled or disposed.

When total fluids are recovered, oil can exist in water in several forms:

- *Free oil*, which rises quickly to the water surface when given a short, quiescent settling period
- *Mechanical dispersions*, which are distributions of fine oil droplets ranging in size from microns to fractions of a millimeter and having stability due to electrical charges and other forces, but not due to the presence of surface active materials
- *Chemically stabilized*, which are distributions of oil droplets similar to mechanical dispersions, but which have additional stability due to chemical interactions typically caused by surface active agents present at the oil/water interface
- *Dissolved oil*, which is either dissolved in a chemical sense, or dispersed in such fine droplets (typically less than 5 μm) that removal by normal physical means is impossible
- *Oil that adheres to the surface of particulate materials*, referred to as oil-wet solids.

The degree of an oil/water separation problem depends on the oil particulate size distribution. To limit the dispersion and emulsification of entrained oil, centrifugal pumps and other equipment that have strong shearing forces should not be used. To limit chemical emulsification, wastes containing surface active agents should not be mixed with other oil-laden contamination.

In most groundwater pump and treat applications, separation of oil is primarily based on gravity separation techniques. For highly emulsified oil, dissolved air flotation and centrifugation techniques also can be used. Secondary treatment such as chemical treatment and filter coalescence and tertiary treatment such as ultrafiltration, biological treatment, and carbon adsorption may be required for extreme cases of emulsification and dissolved oil.

Gravity separation, the primary and most common treatment, is based on the specific gravity difference between water and immiscible oil globules, and is used to move free oil to the surface of a water body for subsequent skimming of oil. A commonly used gravity separator is also known as an oil/water separator and is shown in Figure 11.2.

The effectiveness of an oil/water separator depends on proper hydraulic design and the retention time provided for a given rise velocity. Longer retention times in a quiescent zone generally increase separation efficiency. The liquid retention must be sufficient to permit oil droplets rising at a given velocity to reach the fluid boundary, where they can be removed by skimming. Effective removal of oil droplets with a given rise velocity is also a function of the system geometry.

Design of an oil/water separator is based on the Stokes' law for terminal velocity of spheres in a liquid medium that is applicable to the rise of oil globules larger than 150 μm in diameter.[3] In addition to the concept of the rising oil globule, the design of an oil/water separator is also based upon the following three parameters:

- a minimum horizontal area,
- a minimum vertical cross-sectional area,
- a minimum depth to width ratio of 0.3.

The above parameters take into account the ratio between the rise velocity and the horizontal velocity and the effects of turbulence and short-circuiting.

Coalescing media can be incorporated in oil/water separators for enhanced oil separation. The function of coalescing media is to act as a receptive medium and increase the agglomeration of oil globules (down to 20 μm in size) and allow them to coalesce, therefore increasing oil globule size and thus increasing the rise velocity of oil.

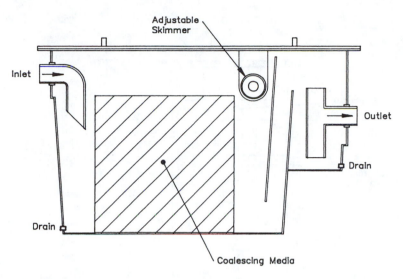

Figure 11.2 Schematic description of an oil/water separator.

Coalescing media vary in the materials used and the effective pore size. In some coalescing media, a fibrous material such as nylon or propylene is wound about a rigid spool to form a cartridge. The tightness of the wrap and the fiber diameter largely control the effective porosity for these devices. Other coalescing media incorporate the use of tightly woven or tightly wrapped sheets of fiberglass. Since coalescing media does tend to plug with particulates, less costly pleated-paper type elements are often used as coalescing media or as prefiltration media.

In selecting the appropriate oil/water separator equipment, the specifics of the oily waste problem should be carefully evaluated. Typical characteristics of the oily water mixture that should be determined include the oil and bulk fluid densities, viscosity of the oil, the oil rise velocity, the oil droplet size distribution, the presence of emulsifying agents, and the suspended solids content and distribution. Low shear positive displacement pumps should be used to prevent shear of the fluid. If the oil is present in quantities greater than roughly 1%, gravity separation or similar methods should be used to achieve bulk separation. If chemical emulsions are present, they should be treated before contacting coalescing media. De-emulsifying agents should be evaluated by jar testing.

11.3.2 Air Stripping

Air stripping is a physical mass transfer process and is generally considered as the best available technology for many volatile organic compounds (VOCs) present in contaminated groundwater. Air stripping uses relatively clean air to remove contaminant VOCs dissolved in water and transfers the contaminants into the gaseous phase. In drinking water treatment, the air stripping concept has actually been employed for a long time and was often referred to also as aeration and degasification. There are various aeration options that have been used in water treatment for quite some time. They are as follows:

- Open air storage
- Free fall
- Spray nozzle
- Venturi
- Cascade aeration
- Inclined cascade
- Hydraulic jet
- Membrane stripping
- Rotary stripper
- Cross current
- Countercurrent cascade
- Countercurrent packed

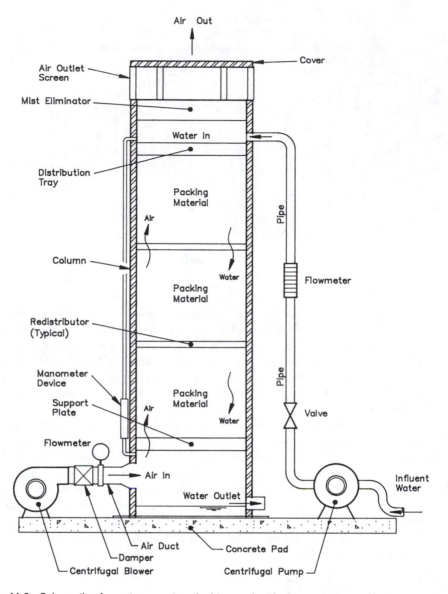

Figure 11.3 Schematic of countercurrent packed tower air stripping system.

- Atomerator
- Stat tray
- Catenary grid

- Multi-stage fine bubble
- Multi-stage cross flow
- Sieve tray

The above aeration devices can be divided into two major classes: packed columns and packing less systems.

11.3.2.1   Countercurrent Packed Columns

The conventional air stripper configuration used in groundwater treatment is a countercurrent packed column. In this configuration, contaminated groundwater is pumped to the top of a packed column, and simultaneously, clean air is blown from the base of the

Table 11.5 Air Stripper Design Factors

Parameter	Impact
Influent water flow rate	Influences tower cross-sectional area based on allowable hydraulic loading to avoid flooding.
Treatment efficiency	Influences packed bed depth.
Compounds being treated	Influences air to water ratio.
Water temperature	Influences Henry's law constant and stripping efficiency; also influences tower height to minimize residence time to avoid freezing.
Aesthetics	Influences tower height and any limitations on height.
Inorganic water chemistry	Influences fouling potential and subsequent operation and maintenance problems.

column (Figure 11.3). Air and water flow countercurrently over loosely dumped or structured media. Water is evenly distributed at the top and flows by gravity over the inert media bed. Airflows from the bottom and up through the cascading water. The mixing of the air and water over the media provides the opportunity for the mass transfer process to take place.

The mass transfer process is governed by physical and chemical properties of each contaminant, the water temperature, the air to water ratio, the height of the air/water contact column, and the physical properties of packing media used in the column. The mass transfer theory has been elucidated in detail in the literature.[4-6]

The basic equation for a stripping system is based on the assumptions that the air stream is free of any organics, that the volume of organics removed from the water is too small to affect the volume of air or water, and that Henry's law applies. Air stripping system design can be accomplished by algebraic manipulations of equations that mathematically describe the physical process of air stripping. Theoretical correlations available in the literature as well as readily available air stripper design software are the preferred choices for designing a countercurrent packed air stripping column. The basic factors that must be considered in the design of an air stripper are summarized in Table 11.5.

The packing media may take on many different forms, as many different types and sizes are available commercially. Each piece of packing media is typically from 1 to 4 in. in diameter and is cylindrical or spherical in shape. Each piece contains many internal ribs and braces to maximize the surface area available for the air–water interface. Maximization of air and water contact is always desirable for maximum mass transfer efficiency but interferes, in general, with requirements for low airflow pressure losses and the potential for flooding.

The key to the feasibility of air stripping a given chemical compound depends on the Henry's law constant of that compound. Henry's law constant is strongly influenced by the temperature. However, Henry's law constant alone does not necessarily relate to the efficiency with which the compound can be stripped. As noted earlier, the stripping efficiency depends on many factors such as depth of packed bed, air to water ratio, and the type of packing media. The efficiency reaches a point of diminishing return since each added foot of packing is less effective than the previous foot. Packing depths can range from a few feet for a very low removal efficiencies (50%) up to 35 ft for very high removal efficiencies (99.99%).

While countercurrent packed columns are by far the most common air stripping configuration applied for groundwater treatment systems, packing less low profile air strippers should be considered when contaminated sites have any of the following characteristics:

- height limitations—due to issue of winterization or aesthetics
- limited site accessibility—indoor vs. outdoor requirements
- low concentrations of contaminants—packing media may not be required
- high levels of inorganic compounds—need to constantly clean the stripping system

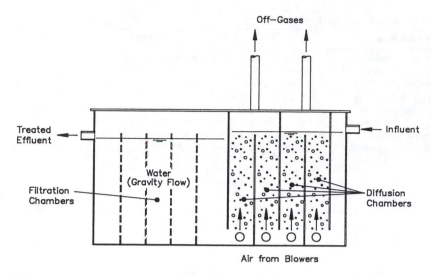

Figure 11.4 Schematic of multichamber fine bubble aeration system.

- high levels of suspended solids—need to constantly clean the stripping system
- need for minimal visibility of equipment—aesthetics and public relations

11.3.2.2   Multiple Chamber Fine Bubble Aeration System

In a diffused bubble aeration system, fine or coarse bubble diffusers are used to produce thousands of air bubbles in the water being treated. As the bubbles move through the water, the mass transfer process takes place at the air–water interface of each bubble. This process occurs until the bubbles either leave the water or become saturated with contaminants. Thus, the finer the bubbles are, the greater the opportunity for transfer of contaminants due to the increased total bubble surface area. The location, number, and size of the diffusers should be designed in such a way that the total bubble surface area is maximized.

In a multichamber unit, the contaminated groundwater is treated in stages within a single unit (Figure 11.4). As a result, the water flows in a plug flow manner through the unit. The more stages provided in the system, the closer the process will be to true plug flow, resulting in a better performance. Furthermore, the chambers are interconnected so as to force the flow in a sinuous route, increasing the air to fluid contact time. Due to the absence of any packing material to enhance the mass transfer efficiency, the air to water ratios in a diffused air stripper will be significantly higher compared to a packed column air stripper.

The advantages of using this type of air stripper lies in its compact size and reduced height. In addition, the filtration chambers which are attached downgradient to the diffusion chambers facilitate the removal of inorganic material such as iron and manganese precipitates, reducing the cost of operation and maintenance. Easily removable filtration material such as filter clothing can be placed in the filtration chamber to remove the precipitates.

11.3.2.3   Low Profile Sieve Tray Air Stripper

In this configuration, contaminated groundwater flows across the surface of a series of perforated trays. Airflows upward through the tray orifices, creating a turbulent froth where the mass transfer process takes place (Figure 11.5). The water becomes progressively cleaner as it flows across a tray and down from one tray to the next in a sinuous path. Since each

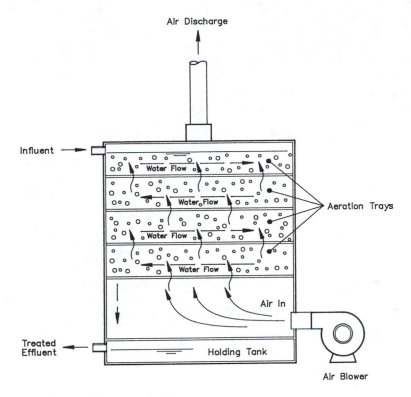

Figure 11.5 Schematic of sieve tray air stripper.

tray is relatively shallow, 9 to 15 in., the tray stripper has a low profile when compared to packed columns.

The turbulence that occurs on the tray has been reported to reduce the frequency of fouling of the stripper itself. However, if fouling occurs the trays are configured in such a way as to facilitate quick and easy clean-outs.

As with diffused aeration systems, the required air to water ratio in this type of stripper will be high due to the absence of any packing media. Removal efficiencies can be enhanced by the number of trays available, and the increased air/fluid contact time (Figure 11.6). More trays will simulate the plug flow reactor.

11.3.2.4 Significance of Water Chemistry

One important factor that must be considered during design of air strippers is water chemistry. The presence of significant amounts of naturally occurring inorganic compounds such as iron, manganese, or carbonates adversely affects an air stripper's removal efficiency. The inorganic compounds that are soluble under normal conditions become oxidized during the air stripping process and become less soluble. In addition, due to the presence of oxygen in the air and indigenous bacteria in the extracted groundwater, biological fouling can take place resulting from the degradation of biodegradable organic compounds. Furthermore, if the groundwater extraction wells are not designed properly, significant amount of suspended solids will be deposited in the air strippers. In summary, air stripper fouling can be divided into four different types:

- iron and manganese precipitation
- biological film growth

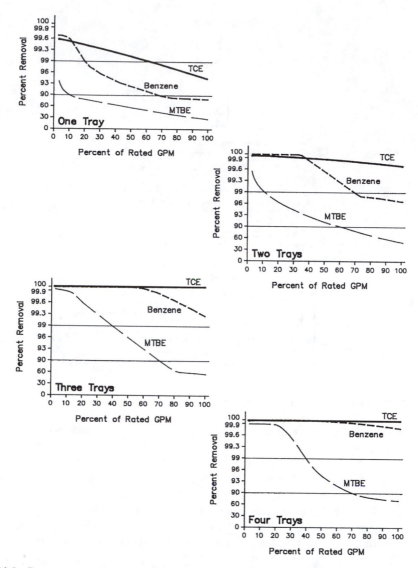

Figure 11.6 Percent removal of various contaminants in a multitray stripper.

- scaling due to carbonate deposition
- solids sedimentation

The oxidation of iron and manganese, the growth of biofilms, and the deposition of calcium carbonate are all by-products of the aeration process, regardless of the aeration mechanism used. In the presence of dissolved minerals or microbes, any aeration device will eventually have to contend with precipitate buildup. The level of these foulants, other water quality parameters, the combinations they occur in, the type of packing media, the airflow rate, the liquid loading rate, and even the materials of construction will influence the degree and frequency of fouling. At moderate levels of inorganic foulants, routine maintenance is favored and clean-outs may be required every 18 months. At higher inorganic levels, clean-outs may be required every 6 months, and at very high inorganic levels the system can fail within days of operation. As a result of potential for fouling, significant safety factors should be provided in system design for constantly meeting the treatment efficiencies.

As a rule of thumb, for iron and manganese concentrations of less than 5 mg/l, routine maintenance is sufficient. For concentrations greater than 10 mg/l, preaeration and filtration of precipitates should be provided. Adjustment of pH to acidic levels or addition of iron sequestering agents can also help to mitigate the problem.

For hardness concentrations of less than 150 mg/l (as $CaCO_3$), routine maintenance would be sufficient; filtration may be necessary. For hardness concentrations greater than 300 mg/l (as $CaCO_3$), more frequent cleanup, pH control, or filtration may be required.

Both Langelier and Ryznar indices are techniques of predicting whether water will tend to dissolve or precipitate calcium carbonate. To determine both Langelier's and Ryznar's indices, the Langelier's saturation pH (pHs) must be determined. This value is a function of the relationship between calcium hardness, total alkalinity, total solids concentration, and water temperature. Using this value, Langelier's and Ryznar's indices are calculated as follows:

$$\text{Langelier's Index} = \text{pH (actual)} - \text{pH (saturation)}$$

$$\text{Ryznar's Index} = 2\,\text{pH (saturation)} - \text{pH (actual)}.$$

Positive Langelier's index values indicate scaling, zero indicates neither scaling nor corrosion, and negative values indicate corrosion. Ryznar index is based on empirical observations, and generally, a value of 6.5 is the neutral point where no scaling or corrosion occurs. A value lower than 6.5 indicates scaling; higher than 6.5 indicates corrosion.

For total biodegradable organics concentration of more than 10 mg/l, microbial film growth around aeration devices, pipes, walls, and the packing media will cause operational problems. Periodic cleanup will solve the problem.

In the presence of iron in the extracted groundwater, a group of microorganisms called "iron bacteria" can cause significant fouling in the stripping towers. The two most important physiological properties that make "iron bacteria" a problem are (1) ferric hydroxide deposition outside the cell, and (2) extracellular polysaccharide excretion (capsule).

The capsule is a protective coat that these bacteria excrete to stabilize their own microenvironment. The presence of biofilm grown by other heterotrophic bacteria present in the system and the activity of the "iron bacteria" can result in a red slime formation, which can clog up the entire system rapidly. If the problem is very severe, periodic cleanup will be very costly and the use of a bactericide may be necessary to overcome biological fouling.

11.3.2.5   Effluent Air Treatment

When the level of volatile organics being discharged from an air stripper exceeds the guidelines established by federal, state, or local authorities, it is necessary to provide a control technology to treat the effluent air discharge.

There are three basic types of technologies commonly applied for the treatment of air discharges: vapor-phase activated carbon, thermal oxidation, and biofiltration. Selection of any of the above technologies will depend on the airflow rate, the type of contaminants, and the mass loading. Use of vapor-phase activated carbon will require the removal of a substantial amount of moisture present in the effluent air from the air stripper. Detailed descriptions of these technologies are provided in Chapter 3.

11.3.2.6   Steam Stripping

Steam stripping for groundwater treatment is essentially a distillation process where the heavy product is water and the light product is a mixture of volatile organics. The process of steam stripping takes place at high temperatures compared to air stripping, usually very close to the boiling point of water. This process is more suitable for compounds that are very

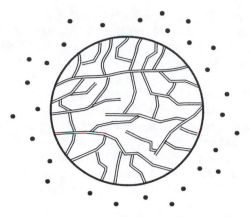

Greatly magnified cross–section of a granule of activated
carbon surrounded by contaminant molecules.

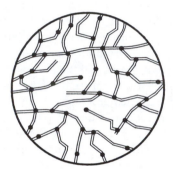

Contaminants are adsorbed and held by the carbon granule.

Figure 11.7   Illustration of contaminant adsorption within a carbon granule. A. Greatly magnified cross
section of a granule of activated carbon surrounded by contaminant molecules. B. Contaminants are adsorbed and held by the carbon granule.

volatile and have a low Henry's law constant due to their high solubility. Such compounds
include acetone, methyl ethyl ketone (MEK), methyl tertiary butyl ether (MTBE), and alcohols. Since the volatility of the organics is a very strong function of temperature, the high
stripping temperatures inherent in steam stripping allow for the removal of more soluble
organics that are not strippable by air.

Another very important feature of steam stripping is the fact that no off-gas treatment
is needed and the only waste stream generated is a small amount of very concentrated organics.

During steam stripping, steam is injected at the bottom of a packed or trayed tower to
provide heat and vapor flow. Contaminated water is fed at the top of the tower and clean
water leaves the bottom of the tower. Steam leaves at the top heavily laden with organic
materials. The steam/organics combination is condensed and processed further.

11.3.3   Carbon Adsorption

Adsorption occurs when an organic molecule is brought to the activated carbon surface
by diffusion and held there by physical and/or chemical forces, (Figure 11.7). The quantity
of a compound or group of compounds that can be adsorbed by activated carbon is determined

by balance between the forces that keep the compound in solution and those that attract the compound to the carbon surface. Factors affecting this balance include

- adsorptivity, which increases with lower solubility, higher molecular weight, low polarity, and low volatility
- the pH of the water; for example, organic acids adsorb better under acidic conditions, while amino compounds favor alkaline conditions
- the class of chemical compounds; aromatic and halogenated compounds adsorb better than aliphatic ones
- temperature; adsorption capacity decreases with increasing temperature, although the rate of adsorption may increase
- the character of the adsorbent surface, which has a major effect on adsorption capacity and rate, the raw materials and process used to activate the carbon determine its adsorption capacity.

Activated carbon is manufactured from coal, lignite, coconut shells, and other sources of carbonaceous material. The characteristics of the activated carbon will vary with the source of the raw material. The name "activated carbon" was derived from the activation process necessary for the manufacture of high-quality carbons which is separate from the initial carbonization of the raw material. As a separate process, activation can be manipulated in such a way that many kinds of carbons with different properties can be produced. Furthermore, the same kind of controlled processing has proven useful to reactivate exhausted carbons.

Liquid-phase activated carbon is most often applied in a granular (GAC) or powdered (PAC) form and has filled a critical niche in water treatment since the early 1960s. The primary characteristic of PAC that differentiates it from GAC is its particle size, and it is generally less efficient than GAC in terms of removal efficiencies. PAC has been applied in drinking water treatment systems to reduce taste and odor. More recently, PAC has also been applied in wastewater treatment in the powdered carbon activated sludge (PACT) process. In this process, the carbon is utilized to concentrate organic molecules, which are then biodegraded by the activated sludge process.

GAC is a granular media, approximately the size of medium fine sand. The specifications and properties provide means of comparing various manufacturers' products. The most critical characteristics are

- surface area—defines the interstitial surface area of the activated carbon
- sieve size—provides critical information related to the head loss and grain size characteristics of the media
- iodine molasses numbers—describe the adsorptive character of the carbon for low and high molecular weight compounds, respectively.

The key to the ability of activated carbon to remove dissolved organic compounds from water is the carbon's high interstitial surface area, 800 to 1100 m²/g, and the surface activation that allows organic molecules to adsorb to the interstitial surface. The surface area is spread throughout the structure of each carbon granule in a complex pore structure commonly described as consisting of macropores (greater than 10,000 Å) and micropores (less than 100 Å) within the macropore structure.

11.3.3.1   Carbon Adsorption System Design

Because of the significant capital and operation and maintenance costs required to construct a full-scale GAC system, predicting the performance of GAC becomes very critical. Various techniques can be used to predict the GAC performance, as described below.

Adsorption isotherms. The adsorption isotherm is the relationship, at a given temperature, between the amount of compound adsorbed and its concentration in the surrounding solution. A reading taken at any point on the isotherm curve gives the amount of contaminant adsorbed per unit weight of carbon, which is the carbon adsorptive capacity at a particular contaminant concentration and water temperature.

Adsorption isotherms have been used extensively to describe adsorption behavior and to estimate the adsorption capacity of activated carbon. Isotherms can be used to (1) select the best carbon among the alternatives, (2) estimate the life of carbon in an adsorber, and (3) test the remaining adsorption capacity of a carbon adsorber. Isotherm capacities can be used directly or by incorporating them into a kinetic model to predict performance. Isotherm estimates and real-life performance will correlate well if the contaminated water has a single contaminant. When multiple contaminants are present, adsorptive capacity estimates may be clouded due to the competitive effects between the various contaminants present.

Bench-scale columns. Advantages of a bench-scale study include (1) assessment of both the adsorption capacity and kinetics, (2) low cost, (3) elimination of the need to use numerical models, and (4) ability to generate actual performance data of multicontaminant waste stream at a specific site.

Pilot columns. The most accurate technique for predicting the performance of a full-scale GAC system is to use pilot columns. Pilot columns use the same flow rate, type of carbon, and influent water as the full-scale system to be designed. The only significant difference between a pilot column and the full-scale system is that the pilot column has a smaller diameter, and thus uses less carbon. As a result, wall effects may cause an invalid prediction. The major disadvantage of pilot columns is the expense of the test and the time required to complete the test.

Design information that can be obtained from pilot tests include

- contact time
- bed depth
- pretreatment requirements
- breakthrough characteristics
- head loss characteristics, and
- carbon dosage (in pounds of contaminants removed per pound of carbon).

Mass transfer models. Mass transfer models may be used to predict process performance and reduce the amount of bench- and pilot-scale data required for final design. Models that use surface diffusion, pore diffusion, and liquid film mass transfer characteristics are more appropriate for predicting multicontaminant systems and background organic carbon effects. These models rely on empirically derived physical parameters that describe the behavior of organic molecules and the activated carbon. As with all models, their usefulness and applicability are directly related to the assumptions and inputs utilized in applying them.

Data gathering is critical in GAC system design. It provides the information necessary to size the treatment process and to determine the potential difficulties, interferences and constraints that may exist. For example, the presence of high levels of iron, manganese, dissolved solids, oil and grease, or turbidity will interfere with the efficient operation of a GAC column unless prefiltration is provided. Detailed discussions with GAC suppliers will benefit the designer, as they can provide accumulated empirical knowledge and experience. Relevant data gathering should include

- contaminant levels
- inorganic water chemistry
- range of influent flow over the life of the project
- estimated life of the project
- effluent discharge requirements
- site conditions, and
- regulatory restraints.

Generally, GAC systems consist of a contact vessel, pressure or gravity, with a fixed bed of activated carbon. Options for carbon contacting systems include the following:

- downflow of contaminated water through the carbon bed
- upflow of contaminated water through the carbon bed
- series or parallel operation (single or multistage)
- pressure or gravity operation in downflow contractors
- packed or expanded bed in upflow contactors, and
- materials of construction and configuration of carbon vessel.

Upflow beds have an advantage over downflow beds in the efficiency of carbon usage because of the continuous countercurrent contact. For most groundwater pump and treat applications, downflow beds are preferred. The principal reason for using a downflow bed is twofold: adsorption of organics and filtration of suspended solids. The principal advantage to the dual use of granular carbon is some reduction in capital cost. The economic gain is offset, however, by loss of efficiency in both filtration and adsorption, and higher operating costs. When the suspended solids content is high, prefiltration may be required. Provision must be made to backwash downflow beds periodically and thoroughly to relieve the pressure drop associated with accumulated suspended solids.

Downflow beds may operate in parallel or in series. Valves and piping are provided in series installations to permit the beds to be operated in any position in the series sequence, thus giving a pseudo-countercurrent operation (Figure 11.8). The adsorption process that takes place on the macroscale can best be described through the use of a mass transfer zone which moves through the carbon bed in the direction of water flow (Figure 11.9). As greater volumes of water are treated and organics adsorb onto the GAC, the transfer zone passes out of the carbon bed, and measurable levels of the organics appear in the filter effluent. This condition is termed the "breakthrough." Eventually the carbon will become loaded with organics to the point at which the level of contaminants leaving the bed equals the level entering. At this point, equilibrium has been reached. Sometime between breakthrough and equilibrium the activated carbon must be regenerated or replaced with fresh carbon. Figure 11.10 illustrates the necessity to have carbon beds in series to ensure the effluent water quality on a continuous basis. The second bed will always act as a safety factor in meeting effluent discharge limits.

The effects of compound concentration on system design are (1) higher concentrations will result in stronger diffusion gradients, and thus foster enhanced mass transfer and higher carbon capacity, and (2) higher concentrations will increase the overall amount of carbon utilized within the bed. The type of organic compounds adsorbed will influence the size of the mass transfer zones (MTZ) (Figure 11.9); easily adsorbed compounds will exhibit shorter MTZs than more difficult to adsorb compounds. As a result, the bed depth and empty bed contact time are strongly influenced by the compounds being treated.

A topic that is usually taken for granted during GAC system design is that after the carbon is spent, it has to be disposed or regenerated. It should be noted that carbon reactivation or replacement of the spent carbon may represent up to 50% of the cost of a GAC system.

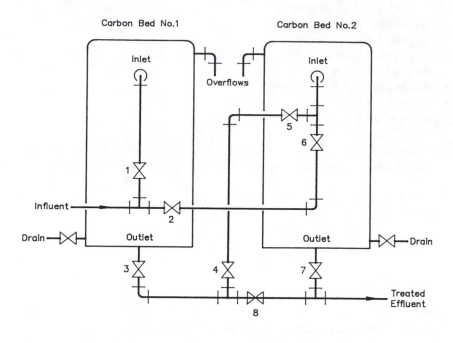

Carbon Bed No.1   Carbon Bed No.2

Figure 11.8 Two carbon beds in series with flexibility to be used interchangeably.

Operation	Valve Positions
Series	1–O/2–C/3–O/4–O/5–O/6–C/7–O/8–C
Carbon Bed No.1 Only	1–O/2–C/3–O/4–C/5–C/6–C/7–C/8–O
Carbon Bed No.2 Only	1–C/2–O/3–C/4–C/5–C/6–O/7–O/8–C

C = Closed
O = Open

11.3.4 Chemical Oxidation

During the oxidative degradation of organic compounds, an organic compound is converted, by means of an oxidizing agent, into endproducts typically having either a higher oxygen or lower hydrogen content than the original compound. The concept of using strong oxidizing agents to break down organic chemicals has been used in groundwater pump and treat applications. In general, oxidative processes that appear applicable use ozone and hydrogen peroxide (individually or together) in conjunction with ultraviolet (UV) light to destroy organic contaminants present in groundwater. This technique has been more popular for chlorinated organic compounds that are difficult to be treated by biodegradation.

Advanced oxidation processes using UV light to form free radicals were developed to enhance the utilization efficiencies of ozone (O_3) and/or hydrogen peroxide (H_2O_2). Since advanced oxidation processes are based on hydroxyl free radical chemistry, chemical interactions are highly nonspecific and nonselective. UV light significantly enhances O_3 or H_2O_2 reactivity by transformation to highly reactive hydroxyl radical. The sequence of radical formation (initiation), successive radical formation (propagation), and ultimate return to stable species (termination) is called a chain reaction (Figure 11.11).[7] Rates of destruction vary with such factors as the nature of contaminant mixture, pH, concentrations of contaminants, presence of scavengers, and inorganic foulants.

Oxidation processes do not work well in the presence of free radical scavengers, such as bicarbonate and carbonate ions. The scavengers consume the ozone and hydrogen peroxide and inhibit the effect of the UV radiation. The presence of such scavengers requires higher doses of oxidizers and larger UV fluxes.

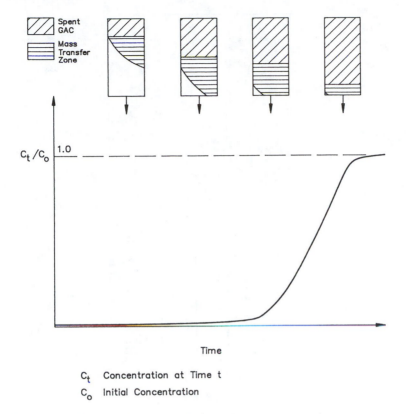

Figure 11.9   Adsorption dynamics in a carbon bed.

Another important factor is penetration of UV light through the influent water stream in the reactor. Light penetration is attenuated by high particle concentrations. Consequently, high suspended or colloidal solids content in the influent groundwater may enhance the scattering of UV light and reduce the system efficiency. Optical fouling of the quartz tubes containing the light source is another operational problem and may require the incorporation of some form of tube cleaning.

Some of the important design considerations for the UV O_3/H_2O_2 system are listed in Table 11.6. The Ultrox™ process, which utilizes UV O_3/H_2O_2, is illustrated in Figure 11.12.[8]

Information is available in the literature regarding the successful degradation of the following compounds by advanced oxidation processes.[8]

- Benzene
- Toluene
- Xylene
- Tetrachloroethylene
- Pentachlorophenol
- Methylene chloride
- Trichloroethylene
- Methyl isobutyl ketone
- Dichloroethylene
- 1,1,1-Trichloroethane
- Vinyl chloride
- Tetrahydrofuran
- Cyanide
- Ammonia
- Sulfide

11.3.5   Biodegradation

Liquid-phase biodegradation is the application of surface bioreactors for the treatment of water contaminated with organic compounds. The bioreactors support the growth and retention of desired microorganisms under optimized process conditions. Reactor design

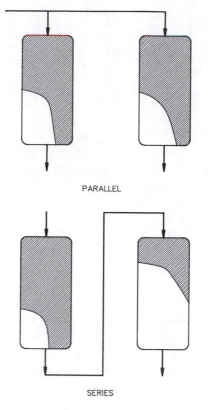

PARALLEL

SERIES

Figure 11.10 Adsorption dynamics in parallel and series beds.

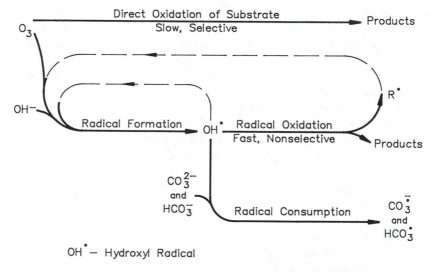

$OH^\bullet$ – Hydroxyl Radical

Figure 11.11 Reaction pathways of ozone.

requires the integration of biological concepts with reaction kinetics, mass transfer, and the flow characteristics of the contacting unit.

Biological processes have long been used for municipal and industrial wastewater treatment, and over the years several basic types have evolved, such as the activated sludge process, trickling filter, rotating biological contactor (RBC), fluidized bed reactors, anaerobic digesters, and aeration lagoons. Several versions of each of these basic processes are in current use,

Table 11.6 Design Considerations for an Advanced Oxidation Process (UV O₃/H₂O₂ Process)

Design considerations	Posttreatment requirements
• System design depends on residence time requirements and level of oxidation. • Performance depends on UV lamp geometry, wavelength of UV, and optical path length. • Efficiency of ozone generation. • Pretreatment of inorganic foulants. • Energy consumption for UV generation. • Storage of H₂O₂ and ozone generator.	• Residual ozone in off-gas must be destroyed. • Effluent may require pH adjustment, solids removal. • Effluent may require further treatment.

and most of these systems can be classified into two basic modes of biological growth: suspended growth systems and attached growth systems.

Biological treatment of hazardous wastes is a destruction technique in which the microorganisms break down and degrade the compounds present to innocuous end products such as CO_2 and H_2O. As such, there is an increase in demand for biological processes to be incorporated into remediation systems, especially for the treatment of highly soluble, aerobically biodegradable compounds such as organic acids, alcohols, and ketones. However, it is necessary to develop a system that would cater to the demands of a specific groundwater pump and treat application. The selected system should require minimum operator attention and be able to handle unexpected fluctuations in contaminant loading.

Suspended growth systems such as the activated sludge process require a significant amount of operator attention due to the need for keeping the biomass under control. As a result, attached growth systems have grown in popularity and demand for groundwater pump and treat applications.

One system that is frequently used in groundwater treatment is a submerged fixed film bioreactor. The reactor consists of a plastic media and is submerged in water. Air is bubbled at the base of the reactor to provide oxygen and mixing. The contaminated water flows through the reactor on a continuous basis (Figure 11.13). Bacterial nutrients are added to the influent upstream of the reactor. Similarly, reagents to control the pH within the optimum range for biodegradation (6.5 to 8.0) can be added if required. Bacteria will grow attached to the media, at the expense of the contaminants in the influent stream.

Bacteria has the ability to stick not only to each other, but also to surfaces through the use of biopolymers. This results in a film of bacteria growing on any surface in contact with the groundwater. Proper selection and design of media for fixed film biological reactors allow high concentrations of bacteria to be grown. As the microbial population grows, the thickness of the biofilm increases. The oxygen and substrate that diffuse across the biofilm are consumed, as they cannot penetrate the full depth of the film. As a result of having no external organic source available for cell growth, the microorganisms near the media face enter into an endogenous decay phase and lose their ability to cling to the media surface. At this point the passing liquid sloughs off the biofilm due to the hydraulic shear forces (Figure 11.14). The biofilm, reduced by sloughing, starts to grow again until it reaches dynamic equilibrium again. At steady state, the biomass concentration per unit volume of the reactor will remain constant. The solids (dead biomass) coming out with the reactor effluent can be easily removed by filtration.

The only operator attention required is to assure that the aeration system is working and that adequate nutrients are being supplied to the system. It is also necessary to control the pH within the optimum range for biodegradation.

Basic principles of biodegradation of organic contaminants are described in detail in Chapter 5. Bioreactor design criteria for the degradation of hazardous organic contaminants are few and frequently site-specific. Unlike classical wastewater treatment design where the goal is overall reduction in total organic compounds (BOD or TOC), the goal here is degradation of specific target compounds. Coupled with this is the sensitivity of treatment response to site-specific properties, physical–chemical, and microbial interactions. The com-

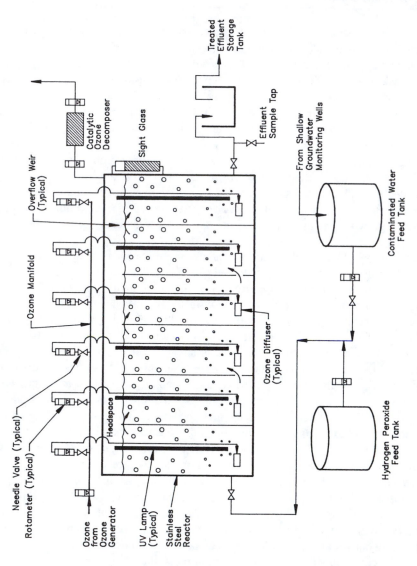

Figure 11.12 UV O₃/H₂O₂ treatment system. Source: ULTROX marketing material.

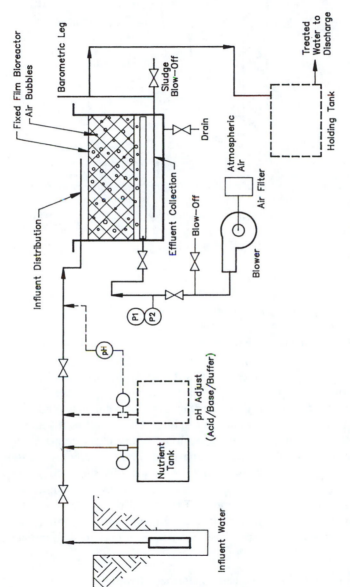

Figure 11.13 Schematic description of an above-ground submerged fixed film bioreactor.

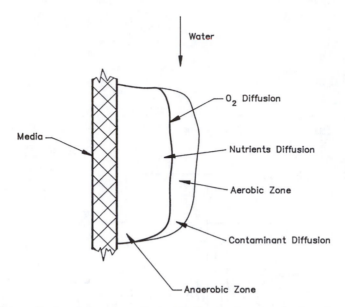

Figure 11.14 Schematic description of an attached biofilm growing on a fixed media.

plexity of these variables usually requires treatability studies for the development of design criteria. Treatability studies are extremely important for the design and optimization of the process due to unique site-specific differences in

- mixture of organic contaminants
- water and soil chemistry
- predominant species of microbes
- predominant mode of microbial metabolism, and
- influence of enhancing or inhibiting chemicals.

When dealing with easily biodegradable compounds such as BTEX compounds or acetone, the traditional approach for bioreactor design can be applied. The traditional approach synthesizes years of experience into permissible loading rates and adequate retention time coupled with a safety factor to obtain a sufficient degree of degradation efficiency.

Typical start-up procedures for a bioreactor are to inoculate the main reactor by developing growth in a batch culture reactor. The batch reactor's biomass is transferred as a suspension to the main reactor. After the introduction of seed microorganisms, primary substrate, at concentrations 2 to 10 times greater than the expected influent concentration, should be added with the required nutrients. The fixed film bioreactor should be filled with the influent groundwater and operated in a closed loop with the batch reactor. Once the biomass has developed to a sufficient level, the treatment cycle can be initiated.

Fluidized bed bioreactor. As noted earlier, attached growth bioreactors have evolved as the system of choice for groundwater pump and treat applications. However, the available biomass per unit reactor volume is modest compared to suspended growth systems. This limitation results in reduced mass loadings and bigger reactor volumes. A hybrid solution between attached growth and suspended growth systems is the fluidized bed bioreactor. Significantly increased levels of biomass per unit reactor volume can be achieved in fluidized bed reactors.

The fluidized bed reactor is operated in an upflow condition to expand the support media holding the biofilm. Drag forces caused by the fluid flow against the support media provide

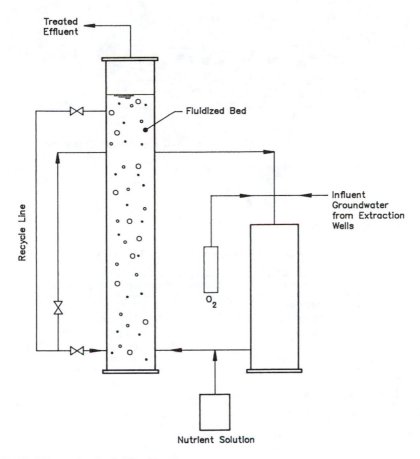

Figure 11.15 Schematic of a fluidized bed bioreactor.

bed expansion. As the biomass increases in thickness on the fluidized bed media, significant differences in the effective particle diameter and settling velocity can occur. Reactor design must distribute and control the influent flow so these density changes in bed media can be accounted for. By carefully controlling the flow velocity and/or by using expanded cross-sectional areas at the top of the bed, the biomass is retained in the reactor.

Granular activated carbon (GAC) has grown in popularity as the fluidized bed media for the treatment of hazardous organic compounds. The use of GAC creates a biophysical–chemical reactor environment where the contaminants are first adsorbed to the GAC particles and then degraded by the growth of biofilm surrounding the GAC particle. The schematic of a fluidized bed bioreactor is shown in Figure 11.15.

11.3.6 Membrane Filtration

Membrane-based water treatment processes have gained increased popularity and acceptance in the last 20 years. Often explained as a filtration technique, membrane treatment involves forcing water, under pressure, through a synthetic membrane (Figure 11.16). Membrane porosities vary in size depending upon the type of contaminant to be removed from the influent water.

In its simplest form of application, a membrane typically serves as a sieve, separating contaminants from the liquid that is forced through it. Modern membrane technology usually is categorized as microfiltration, ultrafiltration, nanofiltration, reverse osmosis, or hyperfiltration. The technology has been employed individually and in combination, in parallel and

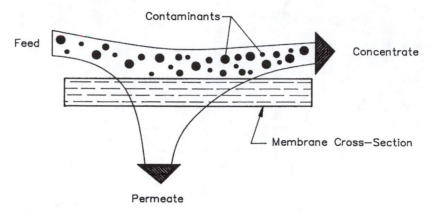

Figure 11.16 Crossflow membrane filtration.

Table 11.7 Performance of Different Membrane Filtration Technologies

Filtration technology	Size limit of contaminants, solids
Reverse osmosis	< 0.001 μm
Nanofiltration	0.001 – 0.01 μm
Ultrafiltration	0.01 – 0.1 μm
Microfiltration	0.08 – 2.0 μm
Conventional filtration	1.0 – 2000 μm

Table 11.8 Selected Characteristics of Membrane Types

Type	Membrane area/unit volume	Fouling resistance	Typical material
Spiral-wound	Large	Good	Composite polyanide
Capillary	Medium to large	Poor	Polysulfone
Fine hollow fiber	Large	Poor	Polyanide
Tubular	Small	Excellent	Polysulfone
Flat-sheet	Medium	Good	Polyolefin

Adapted from Dietrich, J. A., Membrane technology comes of age, *Pollut. Eng.*, July 1995.

in sequence, as a replacement to conventional treatment technologies and as pretreatment or polishing steps for conventional treatment systems.

The four common membrane technologies are distinguished primarily by their pore size, which determines the size and molecular weight of solids and contaminants that do not pass through the membranes (Table 11.7). With each filtration medium, however, there are several types of configurations, each with distinct advantages and capabilities, and a range of construction materials (Table 11.8).[9]

One drawback to the success of membrane filtration is the membrane's tendency to foul with the contaminants being removed. Colloidal solids, microbiological growth, and insoluble precipitates can collect on the membrane during operation. Sufficient accumulation of these deposits requires the shutdown of the system so the membrane can be cleaned. In some cases if cleaning is not performed early enough, the membrane may be damaged beyond repair. Also, excessive presence of fouling material may preclude this treatment completely under certain site-specific conditions.

Many contaminants can be removed by membrane filtration: heavy metals, organic compounds, total dissolved solids, oil emulsions, and suspended solids. In most cases, alternative approaches may be cost-effective in comparison to membrane filtration. Beyond treatment goals, other factors critical to selection of membrane technology include the following:

- influent characteristics, including constituents, flow rate, and temperature
- potential uses for recovered materials
- desired effluent characteristics
- concentrate characteristics and disposal options, and
- space constraints.

11.3.7   Ion Exchange

Ion exchange is a process in which ions of a given species are displaced from an insoluble exchange material by ions of a different species in solution. If an ion exchanger $M^- A^+$ carrying cations A^+ as the exchanger ions is placed in an aqueous solution phase containing B^+ cations, an ion exchange reaction takes place which may be represented by the following equation:

$$M^- A^+ + B^+ \quad \Leftrightarrow \quad M^- B^+ + A^+ \tag{11.1}$$

Solid   Solution   Solid   Solution

The equilibrium represented by the above equation is an example of cation exchange, where M^- is the insoluble fixed anion. In much the same way, anions can be exchanged provided that an anion-receptive medium is provided. An anion exchange reaction may be written as

$$M^+ A^- + B^- \quad \Leftrightarrow \quad M^+ B^- + A^- \tag{11.2}$$

Solid   Solution   Solid   Solution

The most important features characterizing an ideal exchanger are

- a hydrophilic structure of regular and reproducible form
- controlled and effective ion exchange capacity
- rapid rate of exchange
- chemical stability
- physical stability in terms of mechanical strength and resistance to attrition, and
- consistent particle size and effective surface area compatible with the hydraulic design requirements.

Modern ion exchange resins are primarily synthetic organic materials containing ionic functional groups. As noted earlier, the exchange reaction is reversible and concentration-dependent, and it is possible to regenerate exchange resins for reuse. Ion exchange can be used to remove the following contaminants:

- metallic elements when present as soluble species, either anionic or cationic
- inorganic anions such as halides, sulfate, nitrate, cyanide, etc
- inorganic cations such as NH_4^+
- organic acids such as carboxylics, sulfonics, and some phenols, at a pH sufficiently alkaline to produce the ions
- sorptive resins can remove a wide range of polar and nonpolar organics.

Ion exchange systems are commercially available from a number of vendors. The units are relatively compact and are not energy intensive. Start-up or shutdown can be accomplished easily and quickly.

Consideration must be given to proper disposal of contaminated ion exchange regeneration solution. In addition to proper disposal, another important consideration is the selection of regeneration chemicals. Caution must be exercised in making this selection to ensure the compatibility of the regenerating chemical with the contaminants being treated.

Ion exchange fouling can take place in various forms:

- organic fouling, which usually affects anion exchange resins
- iron fouling, which affects both cation and anion exchange resins
- oil fouling
- mud or polyelectrolyte fouling
- bacterial fouling, and
- calcium or barium fouling.

Each of the above conditions reduces a resin's ability to produce the specified effluent water quality.

11.3.8 Metals Precipitation

Chemical precipitation involves transforming a soluble metallic ion into an insoluble precipitate through the addition of chemicals, such that a supersaturated environment exists (i.e., the solubility product is exceeded). Chemical precipitation techniques are rarely used to precipitate organic compounds from solution, although organics can adsorb onto precipitate forms, such as hydrous metal oxides.

Chemical precipitation is the most common technique used for treatment of metal-containing waters. Precipitation can be broadly divided into two categories: chemical precipitation and coprecipitation/adsorption. Only chemical precipitation will be discussed in this chapter.

Chemical precipitation is a complex phenomenon resulting from the induction of supersaturation conditions. Precipitation proceeds through three stages: nucleation, crystal growth, and flocculation. Metal salt solubility can be predicted (at equilibrium) from thermodynamic calculations. These thermodynamic calculations cannot assess the kinetic rate, the influence of precipitate induction parameters, or the degree of supersaturation required to induce nucleation. It is important to note that the stability constants reported in the literature can vary by several orders of magnitude.

11.3.8.1   Hydroxide Precipitation

In hydroxide precipitation, heavy metals are removed by adding an alkali, such as caustic or lime, to adjust the groundwater pH to the point where the metal exhibits minimum solubility. In general, the solubilities of metal hydroxides in solution decrease with increasing pH to a minimum value beyond which (the isoelectric point) the metals become more soluble because of their amphoteric nature. The simplified reaction describing the metal hydroxide precipitation is:

$$Me^{++} + 2OH^- \Leftrightarrow Me\,(OH)_2 \downarrow. \qquad (11.3)$$

The advantages of hydroxide precipitation are[10]

- well proven and accepted in industry
- ease of automatic pH control
- relatively simple operation
- low cost of precipitant (lime).

The limitations of hydroxide precipitation are described below.[10]

- Hydroxide precipitates tend to resolubilize if the solution pH is changed.
- The removal of metals by hydroxide precipitation for mixed metal wastes may not be efficient because the minimum solubilities for different metals occur at different pH conditions.
- The presence of complexing agents has an adverse effect on metal removal.
- Cyanide interferes with hydroxide precipitation.
- Hydroxide sludge quantities can be substantial and are generally difficult to dewater due to their amorphous particle structure.
- Little metal hydroxide precipitation occurs at pH less than 6.
- Process is not stable for large flow and concentration variations in the influent.

11.3.8.2   Sulfide Precipitation

Sulfide precipitation is an effective alternative to hydroxide precipitation for removal of heavy metals from groundwater. The sulfide ion can be introduced in the form of soluble sodium sulfide (Na_2S) or sodium hydrosulfide (NaHS). The addition of the solution may be monitored by periodic analyses of metal contents, or it may be controlled by means of a feedback control loop employing ion-specific electrodes. Avoiding sulfide reagent overdose prevents formation of the odor-causing and harmful H_2S.

Sulfide can be also introduced in the form of slightly soluble ferrous sulfide (FeS) slurry. Since most of the heavy metals are less soluble than FeS, they will precipitate as metallic sulfides. Since the FeS has a very low solubility as a sulfide (0.02 $\mu g/l$), emission of H_2S is minimized. The simplified reaction describing sulfide precipitation is

$$Me^{++} + S^{--} \Leftrightarrow Me\, S \downarrow . \qquad (11.4)$$

The advantages of sulfide precipitation are[10]

- attachment of a high degree of metal removal even at low pH
- low retention time requirements in the reactor due to high reaction rates of sulfide
- feasibility of selective metal removal
- metal sulfide sludge exhibits better thickening and dewatering characteristics than the corresponding metal hydroxide sludge
- metal sulfide precipitation is less influenced by the presence of complexes and chelating agents than is the corresponding metal hydroxide precipitation
- metal sulfide sludge is less susceptible to leaching than metal hydroxide sludge.

The limitations of sulfide precipitation are[10]

- potential for H_2S gas evolution
- possibility of sulfide toxicity
- process is relatively complex and expensive as compared with hydroxide precipitation.

11.3.8.3   Carbonate Precipitation

Carbonate precipitation of heavy metals can be achieved by using soda ash (sodium carbonate). Carbonate precipitation has the following advantages over conventional hydroxide precipitation:[10]

- Optimum treatment occurs at lower pH conditions.
- Metal precipitates are reported to be denser than the liquid, facilitating solids separation.
- Sludges have better dewatering characteristics.

The simplified reaction describing carbonate precipitation is

$$Me^{++} + CO_3^{2-} \rightarrow Me\,CO_3 \downarrow.$$ (11.5)

In any of the above metal precipitation processes, the metallic precipitate must be separated from the process stream. The effectiveness of the solid/liquid separation is heavily dependent on the physical properties (size, density) of the metallic precipitates. Addition of coagulating/flocculant aids such as aluminum sulfate, ferric chloride, or polymers may be necessary to enhance the solids separation efficiency. However, addition of such reagents increases the weight of the sludge and, as a result, the cost of sludge disposal will also increase.

11.4 TREATMENT SYSTEM ENGINEERING

The engineering of a groundwater pump and treat system can be thought of as having two phases: (1) a process engineering phase, and (2) a mechanical and electrical engineering phase.

11.4.1 Process Engineering

There are as many process engineering issues as their are types of systems, and for that reason, the process engineering phase is discussed only in very general terms. The process engineering phase starts with the definition of the problem and the screening of alternatives, as outlined above. Since virtually all of the treatment options discussed above are available from suppliers in packaged units, the end-point of the process engineering is normally a specification for a process unit or the selection of a specific process from a specific supplier. The value added by the designer is in assuring that all of the technical issues have been addressed in the specifications.

11.4.2 Mechanical and Electrical Engineering

Groundwater treatment systems exhibit a broad range of sizes and scope, and the associated mechanical and electrical engineering approach ranges from a very rigorous design for large, long-term projects, to very simple designs for smaller, short-term projects. Accordingly, the first step in the mechanical and electrical engineering phase is the establishment of a design philosophy in light of the site-specific operating requirements and constraints. The following questions represent issues that should be addressed in developing this design philosophy.

- What is the expected duration of the pump and treat project? Long duration projects can justify the investment in more durable (and expensive) equipment and facilities, resulting in lower operation and management (O & M) costs over the life of the project.
- Is there a required or desired service factor for the system? The need for a high service factor will influence the approach taken to spare parts and in-line equipment spares.
- What is the availability of O & M personnel? The availability of O & M personnel will influence the approach taken to instrumentation and control. In a case where no

O & M personnel are available, the instrumentation and control (I & C) system must be designed to operate the system with little or no operator attention. If high service factor is an issue, the I & C system may include the capability of remote monitoring and operation, which would suggest PLC control. Conversely, if O & M personnel are available, and service factor is not a critical issue, the I & C system might be a conventional system based on electrical relays and interlocks.

- What is the climate at the site? If cold weather conditions are expected, the design must address freeze protection, particularly during shutdown conditions. Freeze protection might include a heated building or an unheated building with heat tracing of process piping. In addition, frost heave must be considered in the design of any concrete foundations. Pipe penetrations through any concrete slabs should also be addressed.

- Is the system located at an existing operating facility? Existing operating facilities often have very specific engineering standards that must be followed. In some cases they may also have architectural standards that would apply to any buildings. There also may be issues associated with consistency of mechanical and electrical equipment. For example, the facility may already use and maintain spare parts for a certain brand of pumps or valves. In this case, it would be inappropriate to use another brand.

- What are the system performance data collection and reporting requirements? A requirement for extensive performance monitoring would influence the design of the I & C system, suggesting PLC control with data storage in an associated PC. These requirements will also determine the number and type of flow, temperature, and pressure measurement devices. In some cases, on-line analyzers must be incorporated into the I & C system.

- What local building permits and inspections will be required? An understanding of local construction issues, procedures, and codes is necessary to properly design the system, and to avoid unnecessary delays during construction of the system. As with process engineering, the number of mechanical and electrical engineering issues are extensive, and a discussion of them is outside the scope of this chapter. However, the actual execution of the detailed engineering will proceed much more efficiently given the forethought outlined above.

11.5 PERMITTING

Groundwater treatment systems will almost always involve a permit associated with the discharge of treated water, and they will often involve a permit associated with air emissions.

11.5.1 Treated Water Discharge Permit

The following is a brief summary of the typical discharge options for treated groundwater and the associated permitting issues.

- Discharge into "Waters of the United States" is regulated and permitted under the National Pollutant Discharge Elimination System (NPDES) under the Clean Water Act (CWA). Waters of the United States is very broadly defined, and should be thought of as virtually any surface water. The permit application process is well structured (EPA Forms 3510-1 and 3510-2D, or the state equivalent) but time-consuming. EPA guidance suggests submittal of the application 180 days prior to start-up of the treatment system. The permit will specify the treatment system to be employed, the allowable discharge flows and contaminant concen-

trations, other water quality parameters (pH, TSS, etc.), and the monitoring requirements.
- Reinjection of treated water into the subsurface is regulated and permitted through the Underground Injection Control program under the Safe Drinking Water Act. The permitting procedure is similar to the NPDES procedure (use EPA Forms 3510-1 and 3510-4).
- Discharge of treated groundwater to a POTW does not require an NPDES permit. However, the treatment system must meet all of the pretreatment standards specified under the CWA, along with any other standards specified by the POTW. The specifics of the application process with the POTW vary, and should be addressed on a case-by-case basis.
- Discharge of treated groundwater to a privately owned treatment works (i.e., a plant wastewater treatment system, if the site is an operating facility with such a system) does not require an NPDES permit. However, such a discharge of treated water might necessitate a modification of the NPDES permit under which the privately owned treatment works is operating.

Sometimes it is necessary to obtain the air and/or water discharge permit prior to construction, and often the lead time for obtaining these permits is considerable. If a project is on a fast track, it is wise to start the permitting process as early as possible. One last issue for the designer to consider is the fact that an approach that is acceptable from a strict interpretation of the regulations may not be politically acceptable. For example, discharging to surface waters under a valid NPDES permit is often politically unattractive, and facility owners often choose to send treatment system effluent to a POTW, even though the costs are higher.

11.5.2 Air Discharge Permit

In general, air emissions from groundwater treatment systems are not an issue, with air strippers as the most common exception. The need for an air permit is dependent on the nature and quantity of the air emissions and the specific regulatory requirements of the permitting authority. Federal permitting requirements are specified in 40 CFR 70. These regulations are primarily geared toward larger industrial sources. A strict interpretation of the federal regulations suggests that an air permit is not required if the system is not a "major source," defined as having the potential to emit greater than 10 t per year of any one hazardous air pollutant, or more than 25 t per year of any combination of hazardous pollutants. However, the federal regulations also provide for delegation of permitting authority to state and local agencies, and this delegation has taken place for virtually all of the states. The state and local regulations and permitting procedures must be at least as stringent as the federal regulation, and can be significantly more stringent. In practice, the need for an air permit can vary widely depending on the location of the system, and must be worked on a case-by-case basis with the appropriate regulatory authority. A complete listing of regulatory authorities to whom permitting authority has been granted is listed at 40 CFR 60.4(b).

REFERENCES

1. National Research Council, *Alternatives for Groundwater Cleanup*, National Academy Press, Washington, DC, 1994.
2. Kelly, M. M., Applying Innovative Technologies to Site Contamination: Historical Trends and Future Demand, presented at Haz Mat South '94, Orlando, FL, February 1994.
3. Cheremisinoff, P. N., Oil/water separation, *Natl. Environ. J.*, May/June 1993.

4. Kavanaugh, M. C. and Trussell, R. R., Design of aeration towers to strip volatile contaminants from drinking water, *J. AWWA*, 684, 1980.
5. Leva, M., Tower Packings and Packed Tower Design, The United States Stone Ware Company, Akron, OH, 1951.
6. Perry, R. H., Chilton, C. H., and Kirkpatrick, S. D., *Chemical Engineer's Handbook*, 4th ed., McGraw-Hill, New York, 1963, 18–25.
7. Lowry, T. H. and Richardson K. S., *Mechanism and Theory in Organic Chemistry*, 2nd ed., Harper and Row, New York, 1981.
8. Zeff, J. D. and Barich, J. T., A Review of ULTROX™ Ultraviolet Oxidation Technology as Applied to Industrial, Groundwater, Wastewater and Superfund Sites, presented at the Summer *Natl. Mtg., Am. Inst. Chem. Eng.*, Philadelphia, PA, August 1989.
9. Dietrich, J. A., Membrane technology comes of age, *Pollut. Eng.*, July 1995.
10. Anderson, W. C., *Innovative Site Remediation Technology: Chemical Treatment*, Vol. 2, American Academy of Environmental Engineering, Washington, D.C., 1994.

12

STABILIZATION AND SOLIDIFICATION

12.1 INTRODUCTION

Solidification or stabilization, also referred to as waste fixation, are relatively simple processes. The waste is mixed with a binder or mixture of binders. The mass is then cured to form a solid matrix that will contain the contaminants.

Stabilization generally refers to processes that reduce the risk posed by a waste by converting the contaminants into a less soluble, immobile, and less toxic form. This is achieved by a purposeful chemical reaction that is carried out to make waste constituents less leachable by chemically immobilizing them or reducing their solubility. The physical nature of the waste is not necessarily changed.

Solidification refers to processes that encapsulate the waste in a monolithic solid of high structural integrity. The encapsulation may be that of fine waste particles (microencapsulation) or of a large block or container of wastes (macroencapsulation). Occasionally, solidification may refer to the process that results in a soil-like material rather than a monolithic structure. Solidification does not necessarily involve a chemical interaction between the waste and the solidifying reagents, but may mechanically bind the waste in the monolith. Contaminant migration is restricted by vastly decreasing the surface area exposed to leaching and/or by isolating the waste within an impervious capsule.[1]

Most of the processes used in the application of stabilization/solidification (S/S) are modifications of proven processes and are directed at encapsulating or immobilizing the hazardous constituents and involve excavation and processing or *in situ* mixing.[1]

12.1.1 Sorption and Surfactant Processes

Sorption processes are based on a contaminant being attracted and retained on a sorbent. For example, inorganic heavy metals can be adsorbed onto ion exchange media such as clay, humic material, and fly ashes. In general, hydrophobic organic compounds are not compatible with inorganic material such as cement. Therefore, added organic sorbents can combine with organic wastes before being solidified in cement.

Surfactants are manufactured to have different compatibilities on each end of the molecule, allowing waste material to be adsorbed on one end, while the other end is compatible with inorganic cement. Surfactants can be used also to disperse organic wastes in an aqueous phase and then combine with cement for solidification.[1]

12.1.2 Emulsified Asphalt

Asphalt emulsions are very fine droplets of asphalt dispersed in water that are stabilized by chemical emulsifying agents. The process involves adding emulsified asphalts having the appropriate charge to hydrophilic liquid or semiliquid wastes at ambient temperature. After mixing, the emulsion breaks, the water in the waste is released, and the organic phase forms a continuous matrix of hydrophobic asphalt around the waste solids. In some cases, additional neutralizing agents, such as lime or gypsum, may be required. After given a sufficient time to set and cure, the resulting solid asphalt has the waste uniformly distributed throughout and is impermeable to water.[1]

12.1.3 Bituminization

The process of bituminization combines heated bitumen and a slurry of the waste material in a heated extruder that mixes the bitumen and waste. Water is evaporated from the mixture to about 0.5% moisture. The final product is a homogeneous mixture of waste solids embedded in bitumen and encapsulated when the bitumen cools.[1]

12.1.4 Vitrification

Vitrification processes are solidification processes that employ heat to melt and convert waste materials into glass or other glass and crystalline products. The high temperature employed, 1200°C or higher, causes the glass formers (such as silicates) to form a glass matrix incorporating waste materials such as heavy metals and radionuclides.[1] The high temperatures also destroy any organic contaminants with very few by-products, which can be treated with an off-gas treatment system that generally accompanies vitrification processes. There are three types of vitrification processes:

- *Electrical processes* can be applied *in situ* by applying electrical energy via inserted graphite electrodes.
- *Thermal processes* require an external heat source and a typical reactor is a refractory-lined rotary kiln.
- *Plasma processes* can achieve temperatures up to 10,000°F via electrical discharges.

12.1.5 Modified Sulfur Cement

Modified sulfur cement is a thermoplastic material that can be easily melted, combined with waste components in a homogeneous mixture, and cooled to form a solid, monolithic waste form. It is easily melted, and a variety of common mixing devices, such as paddle mixers and pug mills, can be used.[1] The final end products (e.g., sulfur concrete) are resistant to attack by most acids and salts.

12.1.6 Inorganic Cementitious Processes

Cementitious stabilization is applicable to a wide range of industrial wastes and results in very stable products. S/S techniques that utilize Portland cement, fly ash, cement kiln dust, quick lime, and slags in various combinations have been used for a number of years all over the world. Setting and curing reactions vary among processes. Most of the commercial, cementitious S/S systems, however, solidify in highly similar reactions.

Portland cement is the most widely used of all S/S binding reagents today.[2] It was originally used for nuclear waste solidification processes in the 1950s. Portland cement is not only used alone, but is also used in many other formulations combined with fly ash, lime, soluble silicates, clays, and other materials.

Portland cement is a type of hydraulic cement, a family of materials that upon addition of water produce a hardened paste. This paste acts as a strong "glue" to bind together aggregates and other substances to form concrete, grouts, mortars, and stabilized wastes.

The following properties are associated with the five primary cement compounds.

- Tricalcium silicate: hydrates and hardens rapidly and is largely responsible for initial set and early strength.
- Dicalcium silicate: hydrates and hardens slowly and contributes largely to strength increase at ages beyond 1 week.
- Tricalcium aluminate: liberates a large amount of heat during the first few days of hydration and hardening. It also contributes slightly to early strength development. Cements with low percentages of this compound are especially resistant to sulfates contained in the wastes or soils.
- Tetracalcium aluminoferrite: reduces the clinkering temperature, thereby assisting in the manufacture of cement.
- Calcium sulfate: also known as gypsum, added during final cement grinding slows down the hydration rate. Without gypsum, a cement would set rapidly.

Cement kiln dust consists of varying amounts of components such as raw feed, calcified limestone, alkaline compounds, chloride compounds, and others. The composition varies widely depending on cement kiln operation, type of kiln, type of fuel, and several other factors. Applications of cement kiln dust in S/S systems will vary depending on the characteristics of the dust.

When cement and water are mixed together, a series of chemical reactions begins that results in stiffening, hardening, evolution of heat, and finally development of long-term strength. The overall process is called hydration, since water-containing compounds are formed. The hydrates that form from the compounds listed above determine most of the characteristics of the hardened cement paste. It is the presence of calcium hydroxide and alkalis in solution that gives the plastic cement paste its high pH, an important aspect of cement-based S/S technology.

Slags, themselves waste products, have been used for several years in waste treatment. Slags, either alone or with cementitious materials, are mixed with waste slurries to enhance settling and compaction. Slag has probably been incorporated into a number of stabilization processes along with other reagents, especially at or near the slag producers such as steel mills. As with other waste product reagents (fly ash and kiln dusts), slag usage is often not documented in the literature or promoted specifically as a commercial S/S process.

Blast furnace slag is produced when the molten slag from an iron-producing blast furnace is cooled quickly to minimize crystallization. It is a blend of amorphous silicates and aluminosilicates of calcium and other bases. Because of the presence of ferrous ion and reduced sulfur compounds, it may act as a reducing agent for metal species, such as hexavalent chromium, that are less mobile in the reduced valence state. The combination of Portland cement and slag for stabilization has been used for stabilizing heavy metal contamination.[1]

Processes that use lime pretreated with additives that cause it to become hydrophobic have also been used in S/S systems.[1] This approach allows the improved treatment of oily and other high organic content wastes. Quick lime (CaO) is treated with a surfactant that delays the reaction between CaO and water until the CaO has first interacted with the organic matter in the waste. Subsequently, the CaO–organic mixture is converted into a dispersed $Ca(OH)_2$ solid that gradually converts to limestone ($CaCO_3$) by reaction with carbon dioxide from the air.

12.1.7 Use of Additives in S/S Systems

Although cementitious materials have proven to be an effective stabilizing agent by themselves, the use of additives often enhances and optimizes the S/S mixtures. Soluble silicates have been used for more than a century in the production of commercial products, such as special cements, coatings, molded articles, and catalysts. In these mixtures, the soluble silicate is mixed with cement, lime, slag, or other sources of multivalent metal ions that promote the gelation and precipitation of silicates. Soluble silicates reduce the leachability of toxic metal ions by formation of low-solubility metal oxide/silicates and by encapsulation of metal ions in a silicate or metal–silicate–gel matrix. This characteristic is one basis for the use of soluble silicates in S/S systems.

Soluble silicate is usually added in the form sodium silicate, also known as water glass. Soluble silicates have been used both as accelerators and as anti-inhibitors for concrete and have the same function in Portland cement-based S/S systems.

Soluble phosphates and lime can be used to stabilize wastes containing heavy metals. They are primarily effective against lead and cadmium, but may be of benefit also in controlling other toxic metals. The process involves the addition of various forms of phosphate and alkali for control of pH as well as for formation of complex metal molecules of low solubility. The process is based on the conversion of heavy metals to metallic phosphates of very low solubility. Unlike most other S/S processes, soluble phosphate processes do not convert the waste into a solid, hardened, monolithic mass. Instead, the treated waste retains its particulate nature. It remains free-flowing, and increases little in volume.

Table 12.1 lists the most common kinds of wastes to which each process, described previously, is applicable.

Table 12.1   Potential Applications of Stabilization/Solidification Processes

Stabilization/solidification process	Kind of wastes
Sorbents and surfactants	Oily wastes, industrial sludges, contaminated soils with inorganics and low levels of organics.
Emulsified asphalt	Petroleum-contaminated soils, wastes from paint removal, metal plating wastes.
Bituminization	Low-level radioactive waste solutions, metal plating sludges.
Vitrification	High-level radioactive wastes, contaminated soils and sludges.
Modified sulfur cement	Predried particulate wastes, contaminated soils, sludges, metals.
Cementitious materials	Contaminated soils, low-level radioactive wastes, sludges, oily wastes, tars, petroleum refinery sludges, metal-containing wastes.

Adapted from Anderson, W. C., *Innovative Site Remediation Technology: Stabilization/Solidification*, American Academy of Environmental Engineers, Annapolis, MD, 1994.

12.2   POTENTIAL APPLICATIONS

12.2.1   Stabilization of Metals

Two features of metal chemistry may make the stabilization process difficult: complexation and variable oxidation states. Complexation of metal ions is the formation of species that are very stable in solution under a variety of conditions. Frequently, electroplating wastes contain metal complexes (e.g., cyano-metallic complexes). These complexes must be destroyed before successful S/S treatment can be achieved, but these methods are specific for each metal. Variable oxidation states are exhibited by some metals; for example, chromium, which commonly is found in the Cr(III) and Cr(VI) states. Cr(III) forms an insoluble hydroxide on addition of basic S/S reagents such as cement, but Cr(VI) does not. Pretreatment by reduction from the Cr(VI) state to Cr(III) state is recommended before S/S treatment. Ferrous sulfate is commonly used to convert Cr(VI) to Cr(III). Steps should also be taken to assure that no oxidants remain in the solidified system, which would oxidize the chromium back to Cr(VI) and, therefore, to a more soluble and leachable form.

In general, successful solidification and stabilization of metals will involve the following steps:

- control of excess acidity by neutralization
- destruction of metal complexes if necessary
- control of oxidation state as needed
- conversion to insoluble species (stabilization), and
- formation of a solid with solidification reagents.

12.2.2   Stabilization of Wastes Containing Organics

Many organics are water insoluble and prefer to remain associated with the solid phase. A low-permeability solid matrix may physically retard leaching of some organics, and this may be sufficient for many wastes with low levels of organic compounds. At low levels many organic compounds can be stabilized, but some compounds interfere with the processes of solidification and stabilization.

Interference with Portland cement or other cementitious materials can occur in several ways. Oils and grease can simply coat the cement, preventing the reaction between water and cement. Some organics can be adsorbed on cement surfaces and severely retard cement hydration. Other organics are flocculating agents and will destroy the dispersion of cement grains, which is so vital to successful growth of the hydrated cement structure. Volatile organic compounds will be expelled from the waste matrix due to the heat generated during the hydration of cementitious materials.

Some approaches can be taken to stabilize organics, but they must be carefully selected for each waste. Sorption by selective reagents may remove interfering compounds. Both activated carbon and organically modified clays are examples of sorption reagents that have been used with cement stabilization. Some organic compounds can be converted into salts during the S/S process which are less soluble than the parent compound.

12.3   TESTING REQUIRED TO EVALUATE WASTES BEFORE AND AFTER STABILIZATION/SOLIDIFICATION

12.3.1   Physical Tests

More common physical tests used to evaluate waste S/S processes are as follows:

- index property tests, which provide data that are used to relate general physical characteristics of a material (e.g., suspended solids) to process operational parameters (e.g., pumpability of a sludge)
- density tests, which are used to determine weight to volume relationships of materials
- permeability tests, which measure the relative ease with which fluids (water) will pass through the material
- strength tests, which provide a means for judging the effectiveness of an S/S process under mechanical stresses
- durability tests, which determine how well a material withstands repeated wetting and drying or freezing and thawing cycles, and
- compaction tests, which determines the moisture content that allows maximum compaction to occur so as to achieve maximum density.

12.3.2 Chemical Tests

Leach tests can be used to compare the effectiveness of various S/S processes. Leaching of waste constituents from the stabilized matrix will be influenced by the following factors:

- chemical composition of waste, stabilized matrix, and leaching medium
- physical and engineering properties of the stabilized matrix
- hydraulic gradient across the waste
- polarity of the leaching solution and waste species
- oxidation/reduction conditions and competing reaction kinetics
- bulk chemical diffusion of the waste or reactive species within the leachate pore solution or solid matrix
- concentration of reactive species, and
- accumulation of waste species in the pore solution at the particle surface.

Numerous leaching tests have been developed to test stabilized/solidified wastes. Extraction tests refer to a leaching test that generally involves agitation of ground or pulverized waste forms in a leaching solution (may be acidic or neutral). Extraction tests may involve one-time or multiple extractions. Extraction tests are generally used to determine the maximum, or saturated, leachate concentrations under a given set of test conditions.

The leach test, another type of leaching test, involves no agitation. The leaching of monolithic (instead of crushed) waste forms is evaluated in these tests. Leaching may occur under static or dynamic conditions, depending on the frequency of the leaching solution renewal. In static leach tests, the leaching solution is not replaced by a fresh solution; therefore, leaching takes place under static hydraulic conditions (low leaching velocities and maximum leachate concentrations for monolithic waste forms). In dynamic leach tests, the leaching solution is periodically replaced with new solution; therefore, this test simulates the leaching of a monolithic waste form under nonequilibrium conditions in which maximum saturation limits are not obtained and leaching rates are high. "Static" and "dynamic," therefore, refer to the velocity, not the chemistry of the leaching solution.[3]

The column leach test is another type of laboratory leaching test. This test involves pulverized waste in a column, where it continuously contacts with a leaching solution at a specified rate. The leaching solution is generally pumped through the waste in an upflow column setup. Column tests are considered to be more representative of field leaching conditions than batch extraction tests because of the continuous flux of the leaching solution through the waste. This test is not often used, however, because of problems with the reproducibility of test results caused by channeling effects, nonuniform packing of wastes, biological growth, and clogging of the column.[3]

There are many extraction tests that have been used to characterize leaching:[3]

- Toxicity Characteristic Leaching Procedure Test (TCLP)
- Extraction Procedure Toxicity Test (EP Tox)
- California Waste Extraction Test (Cal WET)
- Multiple Extraction Procedure Test (MEP)
- Monofilled Waste Extraction Procedure (MWED)
- Equilibrium Leach Test (ELT)
- Acid Neutralization Capacity Test (ANC)
- Sequential Extraction Test (SET)
- Sequential Chemical Extraction Test (SEC)

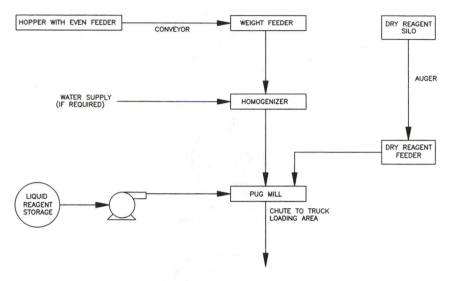

Figure 12.1 Block process flow diagram describing above-ground soil stabilization/solidification.

12.4   FIELD APPLICATIONS

12.4.1   *Ex Situ* Applications

Solidification and stabilization processes can be applied under *ex situ* and *in situ* conditions. Under *ex situ* conditions contaminated soils are excavated, screened to remove oversized material, and then homogenized to provide uniform mixing before being fed into a mixer such as a pug mill (Figure 12.1). The waste material is mixed with water if required. In the mixer the waste material is mixed with the stabilizing agent, additives, and other required chemical reagents. After it is thoroughly mixed, the treated waste is discharged from the mixer. Treated waste is a solidified mass with significant compressive strength, high stability, and a rigid texture similar to that of concrete.[4] Contaminated sludges will undergo similar treatment to that is described above.

12.4.2   *In Situ* Applications

Current methods of *in situ* solidification/stabilization utilize mechanical mixers or standard backhoe buckets. Although these methods are advantageous and cost-effective under certain conditions, tighter requirements for more homogeneous mixing of wastes combined with treatment and capture of organic vapors and dusts are increasingly required.

In situ soil mixing is a technique that has increasingly been relied upon, in the recent past, for *in situ* S/S of contaminated soils and sludges. Depending on the application, mechanical mixing augers or high pressure jets can be used to inject the reagents to modify the waste properties. A major advantage of the method is the capability to treat soils (up to 100 ft deep) without excavation, shoring, or dewatering.

The mechanical auger system utilizes a crane-mounted mixing system. The mixing head is enclosed in a bottom-opened cylinder, as depicted in Figure 12.2, to allow for closed system mixing of the waste and treatment reagents. Treatment reagents are transferred pneumatically for dry chemicals, or pumped in situations where fluid reagents would be utilized (Figure 12.3). Treatment reagents are precisely weighed (for dry systems), or volumetrically measured (for fluid systems), to allow the correct proportions to be mixed with the untreated waste soils or sludge.

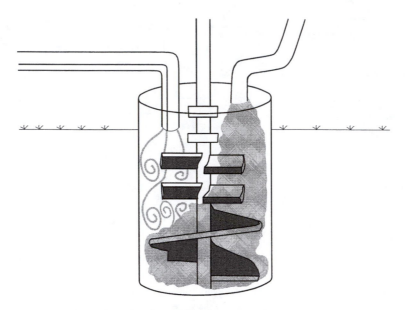

Figure 12.2 *In situ* shallow soil mixing for stabilization.

The bottom-opened cylinder described in Figure 12.2 is lowered into the waste and the mixing blades are rotated while reagents are introduced. The mixing blades mix through the total depth of the waste in an up and down motion. A negative pressure is kept on the head space of the bottom-opened cylinder to pull any vapors or dust to the vapor treatment system. At the completion of a mixed cylinder of waste, the blades are retracted inside the bottom-opened cylinder and the cylinder removed. The cylinder is then placed adjacent, and over-lapping, to the previous cylinder and the process is repeated until all the waste has been treated, as shown in Figure 12.3.

Jet mixing or jet grouting generally involves hydraulically mixing soils with a drilling fluid that carries the stabilizing reagents. A significant advantage of jet mixing over other *in situ* mixing methods is the smaller size of the equipment. The jet mixing drill can be about the size of a compact automobile and thus can maneuver inside buildings and under obstacles. The jet pump can be located some distance away from the actual location to minimize noise levels.

The first step in the method is to position a jetting pipe at the bottom of the treatment zone. The pipe can be the drill pipe itself, but in hard or rock-bearing soils, holes may need to be drilled first, and the jetting pipe placed in a separate step. Next, the jetting pipe is rotated slowly and pressurized with a grout made of cement or slag, additives, and water. The high pressure forces the grout out laterally through special jet parts located in the sides of the pipe and near its bottom. The grout exits the jet ports at very high velocity, impinges on the soil, shattering and penetrating it for several inches or feet away from the jets.

The rotating jets destroy soil formations, and intimately and uniformly mix the native soil with cement. Finally, the rotating pipe is drawn slowly upward at a carefully controlled rate so that the jets create a nearly cylindrical column of treated soil. Typical column diameters can be from 2 to 5 ft. Jet mixing conditions vary with design requirements, but typical pressures are 4000 to 6000 psi, lift rates about 1 ft/min, and cement/water ratios about 1 to 1. Column diameters and properties achieved depend greatly on the native soil.

Fluid injection and mixing creates an increase in soil volume known as swell. The amount of swell is a function of soil type, injection volume, reagent type, and operating conditions. Typical swell is about 25 to 75% of the treated volume, depending on the type of soil and amount of reagent injected. Operator skill in minimizing waste by reducing water content can reduce swell.

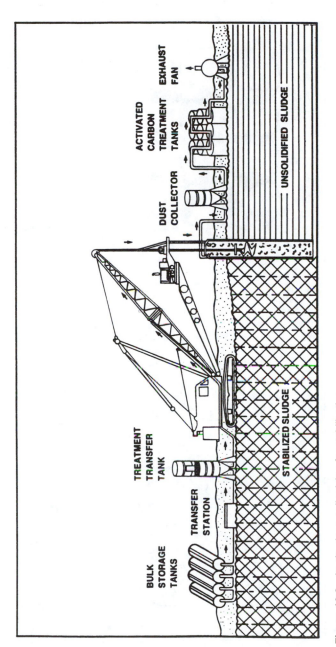

Figure 12.3 *In situ* deep soil mixing for stabilization.

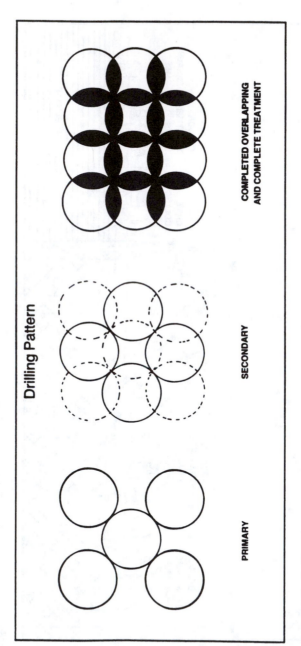

Figure 12.3 Continued.

The testing and monitoring of an *in situ* mixing project is relatively simple given adequate planning and a thorough understanding of the technology. The following items generally require documentation and verification:

- reagent materials and manufacturer's certificates
- grout mix proportioning including viscosity, density, homogeneity, etc.
- grout pressure and withdrawal rate
- reagent application and mixing ratio with soil or sludge
- column diameters, overlap and depth verification, and
- sampling of mixed soil for performance testing.

Sampling and testing can be difficult with soil–cement mixtures. Stabilized materials may be harder than soil, yet much softer than rock. Shelby tube sampling and rock coring may both fail to provide representative samples. It is recommended that samples be obtained from depth, using a special sampling device, before the mixture can set for the casting of cylinders.

REFERENCES

1. Anderson, W. C., *Innovative Site Remediation Technology: Stabilization/Solidification*, American Academy of Environmental Engineers, Annapolis, MD, 1994.
2. Portland Cement Association (PCA), *Solidification and Stabilization of Wastes Using Portland Cement*, Skokie, IL, 1991.
3. U.S. Environmental Protection Agency, Stabilization/Solidification of CERCLA and RCRA Wastes, EPA/625/6-89/022, 1989.
4. U.S. Environmental Protection Agency, Superfund Innovative Technology Evaluation Program, Technology Profiles, EPA/540/R-94/1528, 1994.
5. Geocon, Inc., marketing brochure, 1995.

Appendix A
LIST OF POTENTIAL REMEDIATION TECHNOLOGIES

List of Potential Remediation Technologies

Remediation technology	Groundwater	Soil	Vapors	*In Situ*	*Ex Situ*
Physical					
Adsorption	X		X	X	X
Reactive walls	X			X	
Resin adsorption	X		X		X
Carbon adsorption	X		X		X
Ion exchange	X				X
Air flotation	X				X
Centrifugation	X				X
Distillation	X				X
Electrokinetics		X		X	
Electroosmosis		X		X	
Evaporation	X				X
Filtration	X				X
Cartridge filters	X				X
Pressure filtration	X				X
Granular media filtration	X				X
Reverse osmosis	X				X
Microfiltration	X				X
Ultrafiltration	X				X
Vacuum membrane filtration			X		X
Flotation	X				X
Flocculation	X				X
Freeze crystallization	X				X
Gravity separation	X				X
Ground freezing		X		X	
Hot-water displacement	X	X		X	
Mechanical soil aeration		X		X	X
Metal precipitation	X	X		X	X
Reactive walls	X			X	
Reactive zones	X			X	
Sedimentation	X				X
Volatilization					
Air sparging	X			X	
Air stripping	X				X
Evaporation of water	X				X
Soil vapor extraction		X	X	X	
Washing/extraction					
Soil flushing		X		X	
Soil washing		X			X
Solvent extraction		X			X
Supercritical extraction	X	X			X
Stabilization/Containment					
Containment	X	X		X	
Drains and trenches	X	X		X	
Vertical wells	X	X		X	
Horizontal wells	X	X		X	
Secure landfilling		X			X
Slurry trench cutoff wall	X	X		X	
Sheet pile cutoff wall	X	X		X	
Soil-mixed wall	X	X		X	
Grouting	X	X		X	
Capping	X	X		X	
Stabilization					
Cement-based fixation		X		X	X
Glassification		X			X

	1	2	3	4	5
Polymerization	X	X			X
Pozzolanic-based fixation		X		X	X
Silicate-based fixation		X		X	X
Thermoplastic microencapsulation		X		X	X
Vitrification		X	X	X	

Biological

	1	2	3	4	5
Aerobic and anaerobic reactors	X				X
Activated sludge	X				X
Sequencing batch reactor	X				X
Fluidized bed aerobic	X				X
Fluidized bed anaerobic	X				X
Fixed-film bioreactor aerobic	X				X
Fixed-film bioreactor anaerobic	X				X
Aerated biofilm reactor	X				X
Membrane reactor	X				X
Trickling filter	X				X
Biotower	X				X
Anaerobic filter	X				X
Rotating bio-contactor	X				X
Bioslurry reactor		X			X
In situ biostimulation	X	X		X	X
Air sparging	X	X		X	
Bioventing		X	X	X	
Colloidal gas aphrons (O_2)	X	X		X	
Hydrogen peroxide	X			X	X
Ozone	X			X	X
Engineered soil piles		X	X		X
Bioaugmentation					
Bioseeding	X	X		X	X
Enzyme addition	X			X	X
Recirculating leachfield		X		X	X
White rot fungi		X			X
Yeast strains	X	X			
Biofiltration			X	X	X
Lagoons/ponds	X				X
Land treatment (general)		X			X
Land farming		X			X
Composting		X			X
Mycorrhizas		X			X
Phytoremediation	X	X		X	

Chemical

	1	2	3	4	5
Chlorination	X				X
Dehalogenation	X	X			X
Dechlorination	X	X			X
Hydrolysis	X	X		X	X
Ion exchange	X			X	X
Lignin adsorption		X			X
Neutralization	X	X		X	X
Oxidation (general)					
Chemical oxidation-reduction	X	X		X	X
Electrolytic oxidation	X				X
UV/ozone/H_2O_2	X				X
UV/photolysis	X				X
Polymerization		X			X
Precipitation/coagulation	X				X

Thermal

	1	2	3	4	5
Desorption		X		X	X
High temperature indirect desorber		X			X

Low temperature thermal desorbers		X			X
Radio frequency heating		X		X	
Steam/hot-air soil stripping		X		X	
Steam/vacuum extraction		X		X	
Electric reactor (1)		X			X
Fixed hearth		X			X
Incineration					
Fluidized bed incinerator		X			X
Industrial boiler		X			X
Industrial kiln		X			X
Infrared incineration		X			X
Plasma systems		X		X	X
Pure oxygen burner		X			X
Rotary kiln		X			X
Thermal/catalytic oxidation		X	X		X
Liquid injection	X	X			X
Molten glass		X			X
Molten salt		X			X
Multiple hearth		X			X
Supercritical water oxidation	X				X
Wet air oxidation	X				X

Notes: "X" indicates technology or process is applicable to the media or location identified at top of column. Blank indicates technology or process is not applicable to the media or location identified at top of column.

Appendix B
DESCRIPTION OF FLOW DEVICES

CONTROLLING AND METERING FLOW

FLOW CONTROL

In general, flow can be controlled primarily two ways: through variable speed motors and controls and, more commonly, through the use of throttling valves.

Variable speed motor drives alter the frequency of the alternating current (AC) power supply changing the rotations per minute (rpm) of the pump motor. For a particular pump impeller, since speed in rpm is proportional to flow, this is an effective means of controlling flow rate.

Throttling valves are commonly used to regulate flow rate. Throttling valves reduce flow rates by increasing the friction head (back pressure) on the pump discharge. An important concept in the selection of throttling valves is the flow characteristic curve, which compares flow rate through the valve vs. the percentage the valve is open. This curve describes, for a particular valve, what percentage of total flow will be allowed through the valve at what percentage of valve stem travel. For a throttling application, it is ideal to have a valve in which the flow increases by the same absolute percentage for every percentage increase in valve stem travel. A throttling valve which opens very quickly at the beginning of valve stem travel will make throttling difficult.

Common valve types include

- ball
- butterfly
- check
- gate
- globe
- needle
- plug

Very frequently in fluid flow applications, globe valves are employed for throttling applications. The globe valve has a circular disc or plug that moves away from the seal to open, and into the seal to close. The contour of the plug can be altered to produce various flow characteristic curves. Due to the flow direction changes that take place within a traditional globe valve, it generally produces more head loss than a variety of other valve types. This is usually acceptable due to the fact that throttling valves are properly used to apply additional head loss or backpressure to a pump discharge.

Ball valves are named as such due to the spherical closure element. Ball valves require only a quarter turn of the operating handle to move the valve from fully open to fully closed. Also, ball valves offer relatively low head loss due to the large bore and lack of severe flow direction changes. For these reasons, ball valves serve as excellent block or shutoff valves but frequently are difficult to use as flow throttling valves.

A butterfly valve has a circular disc-shaped closure element that pivots vertically one quarter turn to fully open or close. Butterfly valves are also frequently used as block valves, although they are also used as throttling valves on larger air and water lines. When used for throttling, the disc design may differ from that of a standard butterfly valve to homogenize the varying aerodynamic forces associated with disc positions. When used for throttling, the valve handle is usually replaced by a gear assembly and valve operator to provide finer control.

Another quarter turn valve is the plug valve. The plug valve operates similarly to the ball valve and uses a cylindrical or tapered cylindrical section instead of a spherical closure element.

In a gate valve, a sealing disc (or wedge) moves out of the flow stream perpendicular to the flow direction. Gate valves are typically used as isolation or block valves due to their low head loss when fully open and are generally not used for throttling applications. In addition to having poor throttling characteristics, when used for throttling, the exposed edge of the gate is subject to wear and may not seal properly when closed.

Check valves are generally used to allow flow in one direction only and are available in many types. Common types include ball, tilting disc, split disc, and spring loaded. It is important to consider the application when specifying a check valve. Some commonly desired features of check valves include low head loss, nonclogging, bubble-tight shutoff, and nonslamming.

Needle valves are usually small in size and make use of a tapered or needle-like closure element. Needle valves are typically used for precise throttling of small flows ($\frac{1}{2}$ in. pipe size and smaller).

FLOW MEASUREMENT

The measurement of flow is usually a critical part of environmental systems. Flow information is an important parameter for determining other critical information such as quantity of contaminant removed, chemical additive dosages, and impact to other system components. The throughput of environmental systems is usually metered. Frequently, regulations require the metering of extracted and discharge quantities of water and/or air. The following are some of the more common types of flowmeters:

- Positive Displacement
  - Piston
  - Nutating
  - Gear
- Turbine
- Paddlewheel
- Ultrasonic
  - Doppler
  - Transit Time
- Open Channel
- Mass
- Magnetic
- Open Channel Flow Measurement
- Variable Area Rotameters
- Pitot Tubes

Positive displacement flowmeters typically employ a mechanism within a metering chamber. For each fixed quantity of fluid which passes through the chamber, the mechanism will rotate, wobble (nutate), or be displaced a fixed quantity. The most familiar application of this type of meter is for household water metering. Positive displacement meters are generally well suited for clean nonviscous and viscous liquids and can be low in cost.

Gear-style positive displacement flowmeters employ gears (round or oval) that are moved by the fluid passing around or between them. These flowmeters are well suited to meter oils and other clean viscous fluids that can be difficult to measure with other meter types.

Within a turbine-style flowmeter (or propeller meter), a vaned rotor is exposed perpendicular to the flow. As fluid passes through the metering chamber, the turbine will rotate proportionally to the fluid flow. The turbine shaft may be mechanically or magnetically linked to the meter register or totalizer.

Frequently, the turbine is equipped with a magnet or magnets while the meter body is equipped with a magnetically actuated switch. The rotating turbine will cause the switch to actuate each time the magnet passes; in this manner, electrical "pulses" are generated. The number of pulses is proportional to the flow rate. Turbine meters are used for a wide variety of applications for accurate liquid measurement. They are typically used for viscous liquids (including water) and are moderate to high in cost.

Paddlewheel meters make use of a paddlewheel-like rotor which is turned by the flow stream to produce a pulse output proportional to flow rate in a similar manner to the turbine meter. Paddlewheel meters are versatile due to their physical installation. The paddlewheel sensor is inserted into the process pipe and the display is scaled according to the pipe size. The paddlewheel can be reused for a different flow rate range by inserting the paddlewheel into a different pipe size and recalibrating the display.

The aforementioned flow meters rely on mechanical movement of wetted parts (parts exposed to the flow stream). Flow streams that are chemically aggressive or contain particulate matter, grit, or solids pose an application problem to meters with wetted mechanisms.

Ultrasonic flowmeters rely on measuring the effect that a fluid stream has on sound waves to measure flow. A Doppler type of ultrasonic flowmeter emits a low-energy ultrasonic signal through the process pipe into the flow stream which is reflected back to a sensor by particles or gas bubbles (discontinuities) present in the fluid stream. Because the particles are moving, they will induce a frequency change in the ultrasonic sound waves reflected by them that will vary with the particles velocity. The frequency difference is directly proportional to the flow rate of liquid.

Ultrasonic flowmeters may also operate on the transit-time principle. An ultrasonic signal is emitted by one transducer and is picked up by another transducer upstream. The fluid velocity will affect the sound wave velocity; this effect is measured and converted to flow rate. A major advantage of ultrasonic flow measurement is that the meter is noninvasive. There are no wetted parts as the transducer(s) mount on the outside of process piping.

Magnetic flowmeters have the ability to measure the flow of a wide range of fluids. The operation of magnetic flowmeters is based upon Faraday's law, which states that the voltage induced across a conductor as it moves at right angles through a magnetic field is proportional to the velocity of that conductor. The meter provides the magnetic field, and the flow stream is the conductor that moves through it. For the meter to function properly, the fluid being measured must meet a minimum conductivity requirement. Deionized or distilled water are often of insufficient conductivity to be measured using a magnetic flowmeter.

Open channel flow-measurement is used by many industries. A fabricated flume or weir is installed in line and the fluid stream is directed through it. The height of liquid passing through the flume is measured and is proportional to fluid flow.

Mass-type flowmeters are used in applications where the fluid's specific volume changes with temperature or pressure. Generally the flow is measured in units of mass. There are primarily two types of mass flowmeters. Heated-tube-type meters rely on the heat transfer characteristics of the fluid. A given amount of energy (heat) is added to a fluid stream; the temperature difference measured downstream is proportional to the mass flow rate of the fluid stream.

Another type of mass flowmeter is the U-tube or Coriolis meter. Within this meter, a U-shaped tube is magnetically forced to oscillate at its natural frequency. When the fluid stream is forced through the tube, the redirection of the fluid flow coupled with the oscillations effects a torsion force in the U-shaped tube. The force of torsion is proportional to flow.

Many of the above types of flowmeters are inappropriate for measuring gas flow. However, a common type of flowmeter used to measure liquid or gas flow is the variable area rotameter. In this type of meter, a flowing liquid or gas lifts a weighted float in a tapered tube increasing the area available for passage of the fluid; the height of the float is directly proportional to the flow rate.

Pitot tubes are commonly used to determine gas velocity in a duct or conduit. A pitot tube, through its physical construction, measures static pressure and total pressure (static pressure plus velocity pressure). The pressure differential of these two is the velocity pressure which, knowing the density of a gas, can be converted to velocity. Charts are readily available to simplify the conversion. Pitot tubes have no moving parts, are generally reliable, and relatively low in cost.

PRESSURE MEASUREMENT

Gauges and Switches

Pressure gauges and switches are widely used to monitor and control environmental systems and processes. Pressure gauges provide an indication of absolute pressure and are frequently used to monitor the operation of system components. Standard pressure gauges may have either a bottom connection, back connection, or be surface- or panel-mounted. Pressure gauges are available in a variety of materials of construction, units of measurement, and ranges.

Liquid-filled pressure gauges are commonly used when an application involves pulses, shock, or vibration. The liquid, which surrounds gauge components, serves to dampen shock and vibration, lubricate gauge components, and protect components from corrosion. Common liquid fillers include glycerin and silicon. Glycerin is acceptable for temperatures of approximately 60 to 150°F. Silicon is suitable for service where temperatures range from approximately −50 to 150°F. Care should be taken to ensure the chemical compatibility of these liquids with the process stream.

Differential pressure gauges measure the difference in pressure between two points. These gauges are commonly used to monitor the status and condition of system components including pumps, blowers, filters, and compressors. For example, by monitoring the increase in differential pressure, an operator can monitor the plugging of filters.

Pressure switches can be obtained with or without indicators. Pressure switches are used to control process equipment in addition to providing an indication of pressure as discussed above. Pressure switches can be low in cost and are frequently used in environmental remediation systems. Some common uses include

- verification that sufficient vacuum or pressure is being generated by a blower
- compressor start/stop and verification of sufficient system pressure
- prevention of overpressurization of carbon or other vessels
- building pressurization applications, and
- water pump start/stop and verification of pump operation.

Pressure Transducers

A pressure transducer converts physical pressure to an electrical signal through the use of a pressure sensor and compensation network. Common transducer pressure sensors include semiconductor strain gauges and foil grid strain gauges. The electrical characteristics of these strain gauges vary with the application of an external force to them. It is this principal that allows for the conversion of physical pressure to a usable electrical signal. The electrical signal is converted into a useable pressure reading by an indicator or controller (instrumentation).

Other than directly measuring the pressure in vessels and process piping, by converting pressure to the height of a specific liquid, pressure transducers (and gauges) can be used to accurately measure liquid level. When coupled with a transducer, programmable instrumentation may be used to control pumps, control actuated throttling valves, maintain liquid levels, or record data.

LEVEL MEASUREMENT AND CONTROL

Most level controls can be sorted into two basic categories, point controls and continuous controls. Point controls are capable of actuating at a distinct level and are usually used to provide start, stop, and alarm functions. Continuous level controls provide a continuous indication of level and are capable of providing multiple actuation points or a continuous output that can be used for start, stop, alarm, or more complex functions such as throttling control. There are numerous types of level switches, indicators, and controls on the market. The requirements of each application should be considered on an individual basis to determine the most appropriate type.

Point Controllers

Common types of point actuating level controllers include

- float
- reed
- nonmagnetic, and
- conductance.

Float switches are commonly used for a variety of applications including pump on/off control. A floating head containing a switch is tethered by its cable to a fixed point. As the float rises and falls, the switch is actuated. They are usually economical and readily adjustable; in the past, some float switches used mercury to close the switch contacts, earning them the label "mercury switches." Most recently manufactured float switches do not use mercury to close the switch contacts due to environmental concerns.

Reed switches employ a float with an embedded permanent magnet. A small switch is embedded in the shaft that the float travels on. When the magnet is positioned around the switch, it is pulled closed (or open for a normally closed switch). These switches are usually durable due to the embedded, hermetically sealed switch and can be fabricated to operate in chemically aggressive environments. These switches are frequently used for tankfull probes and other tank level control applications.

Nonmagnetic float switches are used in applications where a simple, relatively low-cost level switch is acceptable and significant concentrations of magnetic inorganics (rust particles) or particulate matter are present. The switch relies on a float to rise (or fall) with the liquid level and move a rigid lever that is physically joined to the switch. Nonmagnetic float switches are available in a variety of materials of construction; mechanical joints are frequently covered with a protective boot to prevent jamming or fouling.

Conductance level controls rely on the dielectric constant of the liquid being measured. The dielectric constant reflects the liquid's ability to conduct electricity. When liquid is present between two contacts (electrodes) electric current is allowed to pass between them. The transmission of this current triggers a change in the position of a control relay. As many probes may be hung in the liquid as are needed for control points. Further, when the application allows, a conductive tank wall or conductive well casing can be substituted for the current transmitting probe. Typically, salt solutions or drinking water have a sufficient dielectric constant; however, nonaqueous materials such as oils or other hydrocarbons are not sufficiently conductive to actuate conductance probes.

Continuous Control

Continuous level sensors are usually coupled with instrumentation to display liquid level. Depending upon the choice of instrumentation, continuous level sensors can be used for point control, multipoint control, or to provide a continuous output suitable for throttling control or other functions. Common types of continuous level controllers include:

- capacitance
- ultrasonic
- radio frequency, and
- tuning fork.

Capacitance probes, as with conductance probes, rely on the dielectric constant of the liquid. A capacitor is formed between two probes or, more commonly, between a probe and a conductive vessel wall. As the material rises and "connects" the vessel wall and sensing probe, the capacitance between the two changes. This change is processed and converted into a useable electrical signal. The signal can be used to continuously display liquid level, actuate a number of setpoints, or produce a constant output.

Ultrasonic level controls are readily available as either point or continuous controllers. Point ultrasonic level controls have no moving parts and are therefore very durable and easily maintained. They rely on the process liquid filling a small (approximately $\frac{1}{2}$ in.) gap, allowing the ultrasonic signal to cross the gap and actuate the switch.

Continuous ultrasonic level sensors are noncontact and operate on a principal similar to sonar. The sensor emits directional, ultrasonic sound waves which are reflected by the liquid surface and detected by the sensor. The signals are electrically conditioned, processed and converted to be proportional to the distance between the liquid surface and the sensor. Through the use of instrumentation, this value can be converted to the height of liquid in the tank or other desirable values. Instrumentation can also produce a continuous output, or be programmed to actuate at a setpoint or multiple setpoints for control purposes.

Because they are noncontact, continuous ultrasonic level sensors are well suited for aggressive environments or for measuring dry or granular materials incompatible with many other level controls. When used with reconfigurable instrumentation, continuous ultrasonic level controls can often be reused for subsequent applications.

Radio-frequency-type (RF) level sensors rely on the presence of liquid or material to effect a phase shift in the radio wave. This phase shift is used to signal switch actuation. Typical RF level sensors are suitable for a wide variety of applications for liquids and solids, although solids must meet certain criteria including moisture content. RF is generally unsuitable for use with highly conductive solids.

Tuning fork controls rely on the principal of resonance. The tuning fork is forced to resonate at its natural frequency. When the resonance is interrupted by the presence of liquid or a solid, the interruption causes the switch to actuate. These switches are highly reliable and rarely need calibration. Due to the forced vibrations they are also somewhat self-cleaning.

ELECTRICAL CONTROL INSTRUMENTATION

Many of the controls discussed in the previous sections rely on instrumentation for the conversion of raw electrical signals, transmission of the signals, and processing of the signals to control system components. Transducers have three main types of control outputs that must be compatible with the instrumentation used. Common output types include

- millivolt (mV) output,
- volt (V) output, and
- current (milliampere, mA) output.

Millivolt output transducers are generally low in cost and small in size. Due to the low energy level, the control signal is prone to interference from other instrumentation and from higher voltage electrical lines. The distance from transducer to instrumentation is generally limited to approximately 200 ft. Millivolt output controls are well suited to nonindustrial applications where the transducer is located in close proximity to instrumentation.

Transducers with a voltage output operate in a similar manner to millivolt units; however, due to the amplified voltage output, they tend to be higher in cost. The higher voltage output affords the signal more resistance to electrical interference and a greater maximum transmission distance. Voltage transducers are frequently used in light industrial settings.

Current output transducers, frequently referred to as transmitters, are also higher in cost than millivolt output devices due to the additional signal conditioning required. Current output transducers most commonly operate on a 4 to 20 mA direct current power supply. This signal may be transmitted much farther than millivolt or voltage signals and is resistant to electrical interference. Great care still needs to be taken to properly shield control signals from other signals and higher voltage lines. Current output transducers are the most commonly used for industrial applications.

Without regard to the type of electrical equipment used, all equipment must be rated for use in the environment it will operate in. The National Electric Code (NEC) governs electrical installations and establishes guidelines to ensure safety. All electrical components should be installed in accordance with the NEC. The National Electrical Manufacturers Association (NEMA) provides a rating system for electrical equipment.

Due to the nature of environmental remediation, portions of systems are frequently required to be classified as "hazardous locations" in accordance with the NEC. Depending on which class and division is applicable, the required electrical rating will change (thus affecting price).

Appendix C
PHYSICAL PROPERTIES OF SOME
COMMON ENVIRONMENTAL CONTAMINANTS

Physical Properties of Some Common Environmental Contaminants

Compound	Molecular weight	Henry's law constant (atm·m³/mol)	Vapor pressure (mmHg)	Solubility (mg/L)	log K_{oc}
A					
Acenaphthene	154.21	7.9×10^{-5} (25°C)	0.00155 (25°C)	3.47 (25°C)	1.25
Acenaphthylene	152.20	2.8×10^{-4}	0.0290 (20°C)	3.93 (25°C)	3.68
Acetone	58.08	3.97×10^{-5} (25°C)	266 (25°C)	Miscible	−0.43
Acrolein	56.06	4.4×10^{-6} (25°C)	265 (25°C)	200,000 (25°C)	−0.28
Acrylonitrile	53.06	1.10×10^{-4} (25°C)	110–115 (25°C)	80,000 (25°C)	−1.13
Aldrin	364.92	4.96×10^{-4}	6×10^{-6} (25°C)	0.011 (25°C)	2.61
Anthracene	178.24	6.51×10^{-5} (25°C)	1.95×10^{-4} (25°C)	0.075 (25°C)	4.41
Acetaldehyde	44.05	6.61×10^{-5} (25°C)	760 (20.2°C)	Miscible	Unavailable
Acetic acid	60.05	1.23×10^{-3} (25°C)	11.4 (20°C)	Miscible	Unavailable
Acetic anhydride	102.09	3.92×10^{-6} (25°C)	5 (25°C)	12% by wt. (20°C)	Unavailable
Acetonitrile	41.05	3.46×10^{-6} (25°C)	73 (20°C)	Miscible	0.34
2-Acetylaminofluorene	223.27	—	—	—	3.20
Acrylamide	71.08	3.03×10^{-3} (20°C)	7×10^{-3} (20°C)	2.155 g/l (30°C)	Unavailable
Allyl alcohol	58.08	5.00×10^{-6} (25°C)	20 (20°C)	Miscible	0.51
Allyl chloride	76.53	1.08×10^{-2} (25°C)	360 (25°C)	—	1.68
Allyl glycidyl ether	114.14	3.83×10^{-6} (20°C)	3.6 (20°C)	141 g/l	Unavailable
4-Aminobiphenyl	169.23	3.89×10^{-10} (25°C)	6×10^{-5} (20–30°C)	842 (20–30°C)	2.03
2-Aminopyridine	94.12	—	Low	100 wt. % at (20°C)	Unavailable
Ammonia	17.04	2.91×10^{-4} (20°C)	10 atm (25.7°C)	531 g/l (20°C)	0.49
n-Amyl acetate	130.19	3.88×10^{-4} (25°C)	4.1 (25°C)	1.8 g/l (20°C)	Unavailable
sec-Amyl acetate	130.19	4.87×10^{-4} (20°C)	10 (35.2°C)	0.2 wt. % (20°C)	Unavailable
Aniline	93.13	0.136 (25°C)	0.6 (20°C)	1.3 wt. % (20°C)	1.41
o-Anisidine	123.15	1.25×10^{-6} (25°C)	0.1 (30°C)	1.3 wt. % (room temp)	Unavailable
p-Anisidine	123.15	—	—	3.3 (room temp)	Unavailable
Antu	202.27	—	≈ 0 (20°C)	600 (20°C)	Unavailable
B					
Benzene	78.11	0.00548 (25°C)	95.2 (25°C)	1800 (25°C)	1.92
Benzidine	184.24	3.88×10^{-11} (25°C)	0.83 (20°C)	500 (25°C)	1.60
Benzo[a]anthracene	228.30	8.0×10^{-6}	1.1×10^{-7} (25°C)	0.014 (25°C)	6.14
Benzo[b]fluoranthene	252.32	1.2×10^{-5} (20–25°C)	5×10^{-7} (20°C)	0.0012 (25°C)	5.74

Compound	MW				
Benzo[*k*]fluoranthene	252.32	0.00104	9.59×10^{-11} (25°C)	0.00055 (25°C)	6.64
Benzoic acid	122.12	7.02×10^{-8}	0.0045 (25°C)	3,400 (25°C)	1.48–2.70
Benzo[*ghi*]perylene	276.34	1.4×10^{-7} (25°C)	1.01×10^{-10} (25°C)	0.00026 (25°C)	6.89
Benzo[*a*]pyrene	252.32	$< 2.4 \times 10^{-6}$	5.6×10^{-9} (25°C)	0.0038 (25°C)	5.60–6.29
Benzyl alcohol	108.14	Insufficient vapor pressure data for calculation at 25°C	1 (58°C)	42,900 (25°C)	1.98
Benzyl butyl phthalate	312.37	3.78×10^{-7}	8.6×10^{-6} (20°C)	42.2 (25°C)	1.83–2.54
α-Bhc	290.83	1.3×10^{-6} (25°C)	2.5×10^{-5} (20°C)	2.0 (25°C)	3.279
β-Bhc	290.83	5.3×10^{-6} (20°C)	2.8×10^{-7} (20°C)	0.24 (25°C)	3.553
δ-Bhc	290.83	2.3×10^{-7} (20°C)	1.7×10^{-5} (20°C)	31.4 (25°C)	3.279
Bis(2-chloroethoxy)methane	173.04	2.5×10^{-7} (20–25°C)	1 (53°C)	81,000 (25°C)	2.06
Bis(2-chloroethyl)ether	143.01	1.3×10^{-5}	1.55 (25°C)	10,200 (25°C)	1.15
Bis(2-chloroisopropyl)ether	171.07	1.1×10^{-4}	0.85 (20°C)	1,700 (20°C)	1.79
Bis(2-ethylhexyl)phthalate	390.57	1.1×10^{-5} (25°C)	6.2×10^{-8} (25°C)	0.4 (25°C)	5.0
Bromodichloromethane	163.83	2.12×10^{-4}	50 (20°C)	4,500 (0°C)	1.79
Bromoform	252.73	5.6×10^{-4}	5.6 (25°C)	3.130 (25°C)	2.45
4-Bromophenyl phenyl ether	249.20	1.0×10^{-4}	0.0015 (20°C)	No data found	4.94
2-Butanone	72.11	4.66×10^{-5} (25°C)	100 (25.0°C)	25.57 wt. % (25°C)	0.09
Benzo[*e*]pyrene	252.32	4.84×10^{-7} (25°C)	5.54×10^{-9} (25°C)	0.0038 (25°C)	5.6
Benzyl chloride	126.59	3.04×10^{-4} (20°C)	1 (22.0°C)	493 (20°C)	2.28
Biphenyl	154.21	4.15×10^{-4} (25°C)	10^{-2} (25°C)	7.5 (25°C)	3.71
Bromobenzene	157.01	2.4×10^{-3} (25°C)	4.14 (25°C)	409 (25°C)	2.33
Bromochloromethane	129.39	1.44×10^{-3} (24–25°C)	141.07 (24.05°C)	0.129 *M* (25.0°C)	1.43
Bromotrifluoromethane	148.91	5.00×10^{-1} (25°C)	149 (20°C)	0.03 wt. % (20°C)	2.44
1,3-Butadiene	54.09	6.3×10^{-2} (25°C)	2,105 (25°C)	735 (20°C)	2.08
n-Butane	58.12	9.30×10^{-1} (25°C)	1,820 (25°C)	61 (20°C)	Unavailable
1-Butene	56.11	2.5×10^{-1} (25°C)	2,230 (25°C)	222 (25°C)	Unavailable
Butoxyethanol	118.18	2.36×10^{-6}	0.76 (20°C)	Miscible	Unavailable
n-Butyl acetate	116.16	3.3×10^{-4} (25°C)	15 (25°C)	5,000 (25°C)	Unavailable
sec-Butyl acetate	116.16	1.91×10^{-4} (20°C)	10 (20°C)	0.8 wt. % (20°C)	Unavailable
tert-Butyl acetate	116.16	—	—	—	Unavailable
n-Butyl alcohol	74.12	8.81×10^{-6} (25°C)	7.0 (25°C)	74,700 (25°C)	Unavailable
sec-Butyl alcohol	74.12	1.02×10^{-5} (25°C)	13 (20°C)	201,000 (20°C)	Unavailable
tert-Butyl alcohol	74.12	1.20×10^{-5} (25°C)	42 (25°C)	Miscible	Unavailable
n-Butylbenzene	134.22	1.25×10^{-2} (25°C)	1.03 (25°C)	1.26 (25.0°C)	3.40
sec-Butylbenzene	134.22	1.14×10^{-2} (25°C)	1.81 (25°C)	309 (25.0°C)	2.95
tert-Butylbenzene	134.22	1.17×10^{-2} (25°C)	2.14 (25°C)	34 (25.0°C)	2.83

n-Butyl mercaptan	90.18	7.04×10^{-3} (20–22°C)	55.5 (25°C)	590 (22°C)	Unavailable

C

Carbon dissulfide	76.13	0.0133	360 (25°C)	2,300 (22°C)	2.38–2.55
Carbon tetrachloride	153.82	0.024 (20°C)	113 (25°C)	1.160 (25°C)	2.35
Chlordane	409.78	4.8×10^{-5}	1×10^{-5} (25°C)	1.85 (25°C)	5.57
cis-Chlordane	409.78	Insufficient vapor pressure data for calculation at 25°C	No data found	0.051 (20–25°C)	6.0
trans-Chlordane	409.78	Insufficient vapor pressure data for calculation at 25°C	No data found	No data found	6.0
4-Chloroaniline	127.57	1.07×10^{-5} (25°C)	0.025 (25°C)	3.9 g/l (20–25°C)	2.42
Chlorobenzene	112.56	0.00445 (25°C)	11.8 (25°C)	502 (25°C)	1.68
p-Chloro-*m*-cresol	142.59	1.78×10^{-6}	No data found	3,850 (25°C)	2.89
Chloroethane	64.52	0.0085 (25°C)	1,064 (20°C)	5,740 (20°C)	0.51
2-Chloroethyl vinyl ether	106.55	2.5×10^{-4}	26.75 (20°C)	15,000 (20°C)	0.82
Chloroform	119.38	0.0032 (25°C)	198 (25°C)	9,300 (25°C)	1.64
2-Chloronaphthalene	162.62	6.12×10^{-4}	0.017 (25°C)	6.74 (25°C)	3.93
2-Chlorophenol	128.56	5.6×10^{-7} (25°C)	1.42 (25°C)	28,000 (25°C)	2.56
4-Chlorophenyl phenyl ether	204.66	2.2×10^{-4}	0.0027 (25°C)	3.3 (25°C)	3.6
Chrysene	228.30	7.26×10^{-20}	6.3×10^{-9} (25°C)	0.006 (25°C)	5.39
Camphor	152.24	3.00×10^{-5} (20°C)	0.18 (20°C)	0.12% (20°C)	Unavailable
Carbaryl	201.22	1.27×10^{-5} (20°C)	6.578×10^{-6} (25°C)	0.4105 (25°C)	2.42
Carbofuran	221.26	3.88×10^{-8} (30–33°C)	2×10^{-5} (33°C)	700 (25°C)	2.2
Chloroacetaldehyde	78.50	—	100 (20°C)	about 50 wt. %, forms a hemihydrate	Unavailable
α-Chloroacetophenone	154.60	—	0.012 (20°C)	Miscible	Unavailable
o-Chlorobenzylidenemalonitrile	188.61	Not applicable reacts with water	3.4×10^{-5} (20°C)	Not applicable reacts with water	Not applicable reacts with water
p-Chloronitrobenzene	157.56	$< 6.91 \times 10^{-3}$ (20°C)	< 1 (20°C)	0.003 wt. % (20°C)	2.68
1-Chloro-1-nitropropane	123.54	1.57×10^{-1} (20–25°C)	5.8 (25°C)	< 0.8 wt. % (20°C)	3.34
Chloropicrin	164.38	8.4×10^{-2}	23.8 (25°C)	1.621 g/l (25°C)	0.82
Chloroprene	88.54	3.20×10^{-2}	200 (20°C)	—	—
Chlorpyrifos	350.59	4.16×10^{-6} (25°C)	1.87×10^{-5} (25°C)	2 (25°C)	3.86
Crotonaldehyde	70.09	1.96×10^{-5}	30 (20°C)	18.1 wt. % (20°C)	Unavailable
Cycloheptane	98.19	—	—	30 (25°C)	Unavailable
Cyclohexane	84.16	1.94×10^{-1} (25°C)	95 (20°C)	58.4 (25°C)	Unavailable
Cyclohexanol	100.16	5.74×10^{-6} (25°C)	1 (20°C)	36,000 (20°C)	Unavailable

Compound	MW				
Cyclohexanone	98.14	1.2×10^{-5} (25°C)	4 (20°C)	23,000 (20°C)	Unavailable
Cyclohexene	82.15	4.6×10^{-2} (25°C)	67 (20°C)	213 (25°C)	Unavailable
Cyclopentadiene	66.10	—	—	0.0103 mol/l at room temperature	Unavailable
Cyclopentane	70.13	1.86×10^{-1} (25°C)	400 (31.0°C)	164 (25°C)	Unavailable
Cyclopentene	68.12	6.3×10^{-2} (25°C)	—	535 (25°C)	Unavailable
D					
p, p' DDD	320.05	2.16×10^{-5}	1.02×10^{-6} (30°C)	0.160 (24°C)	4.64
p, p' DDE	319.03	2.34×10^{-5}	6.49×10^{-6} (30°C)	0.0013 (25°C)	6
p, p' DDT	354.49	5.2×10^{-5}	1.9×10^{-7} (25°C)	0.0004 (25°C)	6.26
Dibenz[a,h]anthracene	278.36	7.33×10^{-9}	$\approx 10^{-10}$ (20°C)	0.00249 (25°C)	6.22
Dibenzofuran	168.20	Insufficient vapor pressure data for calculation at (25°C)	No data found	10 (25°C)	3.91–4.10
Dibromochloromethane	208.28	9.9×10^{-4}	76 (20°C)	4,000 (20°C)	1.92
Di-n-butyl phthalate	278.35	6.3×10^{-5}	1.4×10^{-5} (25°C)	400 (25°C)	3.14
1,2-Dichlorobenzene	147.00	0.0024 (25°C)	1.5 (25°C)	145 (25°C)	3.23
1,3-Dichlorobenzene	147.00	0.0047 (25°C)	2.3 (25°C)	143 (25°C)	3.23
1,4-Dichlorobenzene	147.00	0.00445 (25°C)	0.4 (25°C)	74 (25°C)	2.2
3,3'-Dichlorobenzidine	253.13	4.5×10^{-8} (25°C)	1×10^{-5} m/l (22°C)	3.11 (25°C)	3.3
Dichlorodifluoromethane	120.91	0.425 (25°C)	4.887 (25°C)	280 (25°C)	2.56
1,1-Dichloroethane	98.96	0.00587 (25°C)	234 (25°C)	5,060 (25°C)	1.48
1,2-Dichloroethane	98.96	9.8×10^{-4} (25°C)	87 (25°C)	8,300 (25°C)	1.15
1,1-Dichloroethylene	96.94	0.021	591 (25°C)	5,000 (25°C)	1.81
trans-1,2-Dichloroethylene	96.94	0.00674 (25°C)	410 (30°C)	6,300 (25°C)	1.77
2,4-Dichlorophenol	163.00	6.66×10^{-6}	0.089 (25°C)	4,500 (25°C)	2.94
1,2-Dichlorophenol	112.99	0.00294 (25°C)	50 (25°C)	2,800 (25°C)	1.71
cis-1,3-Dichloropropylene	110.97	0.00355	43 (25°C)	2,700 (25°C)	1.68
trans-1,3-Dichloropropylene	110.97	0.00355	34 (25°C)	2,800 (25°C)	1.68
Dieldrin	380.91	2×10^{-7}	1.8×10^{-7} (25°C)	0.20 (25°C)	4.55
Diethyl phthalate	222.24	8.46×10^{-7}	0.22 (±0.7) Pa (25°C)	1,000 (25°C)	1.84
2,4-Dimethylphenol	122.17	6.55×10^{-6} (25°C)	0.098 (25°C)	7,868 (25°C)	2.07
Dimethyl phthalate	194.19	4.2×10^{-6}	0.22 ± 0.7 Pa (25°C)	4,320 (25°C)	1.63
4,6-Dinitro-o-cresol	198.14	1.4×10^{-6}	5.2×10^{-5} (25°C)	250 (25°C)	2.64
2,4-Dinitrophenol	184.11	1.57×10^{-8} (18–20°C)	0.00039 (20°C)	6,000 (25°C)	1.25
2,4-Dinitrotoluene	182.14	8.67×10^{-7}	1.1×10^{-4} (20°C)	270 (22°C)	1.79
2,6-Dinitrotoluene	182.14	2.17×10^{-7}	3.5×10^{-4} (20°C)	≈ 300	1.79

Compound	Molecular weight				
Di-n-octyl phthalate	390.57	1.41 × 10⁻¹² (25°C)	0.0014 mm (25°C)	3 (25°C)	8.99
1,2-Diphenylhydrazine	184.24	4.11 × 10⁻¹¹ (25°C)	2.6 × 10⁻⁵ (25°C)	221 (25°C)	2.82
2,4-D	221.04	1.95 × 10⁻² (20°C)	0.0047 (20°C)	890 ppm (25°C)	1.68
Decahydronaphthalene	138.25	39.2 (25°C)	1 (22.5°C)	0.889 ppm (25°C)	Unavailable
n-Decane	142.28	1.87 × 10⁻¹ (25°C)	1.35 (25°C)	0.022 (25°C)	Unavailable
Diacetone alcohol	116.16	—	1 (22.0°C)	Miscible	Unavailable
1,4-Dibromobenzene	235.91	5.0 × 10⁻⁴ (25°C)	0.161 (25°C)	16.5 (25°C)	3.2
1,2-Dibromo-3-chloropropane	236.36	2.49 × 10⁻⁴ (20°C)	0.8 (21°C)	1,000 at room temperature	2.11
Dibromodifluoromethane	209.82	—	688 (20°C)	—	—
1-3-Dichloro-5,5 Dimethylhydantoin	197.03	Not applicable reacts with water	—	0.21 wt. % (25°C)	Not applicable reacts with water
Dichlorofluoromethane	120.91	≈ 2.42 × 10⁻² (20–30°C)	760 (8.9°C)	1 wt. % (20°C)	1.57
sym-Dichloromethyl ether	114.96	Not applicable reacts with water	Decomposes	Not applicable reacts with water	Not applicable reacts with water
Dichlorvos	220.98	5.0 × 10⁻³	0.0527 (25°C)	—	9.57
Diethylamine	73.14	2.56 × 10⁻⁵ (25°C)	195 (20°C)	815,000 (14°C)	Unavailable
2-Diethylaminoethanol	117.19	—	1 (20°C)	Miscible	Unavailable
1,1-Difluorotetrachloroethane	203.83	1.07 × 10⁻¹ (20°C)	40 (19.8°C)	—	—
1,2-Difluorotetrachloroethane	203.83	6.36 × 10⁻⁴ (20°C)	40 (19.8°C)	0.01 wt. % (20°C)	2.78
Diisobutyl ketone	142.24	—	1.7 (20°C)	0.05 wt. % (20°C)	Unavailable
Diisopropylamine	101.19	—	60 (20°C)	Miscible	Unavailable
N,N-Dimethylacetamide	115.18	1.77 × 10⁻⁵ (25°C)	1.3 (25°C)	Miscible	Unavailable
Dimethylamine	45.08	4.98 × 10⁻⁶ (20°C)	1,520 (10°C)	Miscible	Unavailable
p-Dimethylaminoazobenzene	225.30	1.943 (25°C)	—	13.6 (20–30°C)	3
Dimethylaniline	121.18	1.18 (25°C)	1 (29.5°C)	1,105.2 (25°C)	Unavailable
2,2-Dimethylbutane	86.18	3.54 × 10⁻¹ (25°C)	319.1 (25°C)	21.2 (25°C)	Unavailable
2,3-Dimethylbutane	86.18	8.70 × 10⁻¹ (25°C)	234.6 (25°C)	19.1 (25°C)	Unavailable
cis-1,2-Dimethylcyclohexane	112.22	—	14.5 (25°C)	6.0 (25°C)	Unavailable
trans-1,4-Dimethylcyclohexane	112.22	—	22.65 (25°C)	3.84 ppm (25°C)	Unavailable
Dimethylformamide	73.09	2.45 × 10⁻⁹ (25°C)	3.7 mm (25°C)	Miscible	Unavailable
1,1-Dimethylhydrazine	60.10	—	157 (25°C)	Miscible	-0.7
2,3-Dimethylpentane	100.20	1.73 (25°C)	100 (33.3°C)	5.25 (25°C)	Unavailable
2,4 Dimethylpentane	100.20	3.152 (25°C)	98.4 (25°C)	5.50 (25°C)	Unavailable
3,3-Dimethylpentane	100.20	1.84 (25°C)	82.8 (25°C)	5.94 (25°C)	Unavailable
2,2-Dimethylpropane	72.15	2.18 (25°C)	1,287 (25°C)	33.2 (25°C)	Unavailable
2,7-Dimethylquinoline	157.22	—	—	1.795 (25°C)	Unavailable

Dimethyl sulfate	126.13	2.96×10^{-6} (20°C)	0.5 (20°C)	2.8 wt. % (20°C)	0.61
1,2-Dinitrobenzene	168.11	$< 1.47 \times 10^{-3}$ (20°C)	<1 (20°C)	0.015 wt. % (20°C)	Unavailable
1,3-Dinitrobenzene	168.11	2.75×10^{-7} (35°C)	8.15×10^{-4} (35°C)	0.05 wt. % (20°C)	2.18
1,4-Dinitrobenzene	168.11	4.79×10^{-7} (35°C)	2.25×10^{-4} (35°C)	0.01 wt. % (20°C)	Unavailable
Dioxane	88.11	4.88×10^{-6} (25°C)	37 (25°C)	Miscible	0.54
Diuron	233.11	1.46×10^{-9} (25–30°C)	2×10^{-7} (30°C)	42 (25°C)	2.51
n-Dodecane	174.34	24.2 (25°C)	0.057 (25°C)	0.008 (25°C)	Unavailable
E					
α-Endosulfan	406.92	1.01×10^{-4} (25°C)	10^{-5} (25°C)	0.530 (25°C)	3.31
β-Endosulfan	406.92	1.91×10^{-5} (25°C)	10^{-5} (25°C)	0.280 (25°C)	3.37
Endosulfan sulfate	422.92	Insufficient vapor pressure data for calculation	No data found	0.117	3.37
Endrin	380.92	5.0×10^{-7} (25°C)	7×10^{-7} (25°C)	0.26 (25°C)	3.92
Endrin aldehyde	380.92	3.86×10^{-7} (25°C)	2×10^{-7} (25°C)	0.26 (25°C)	4.43
Epichlorohydrin	92.53	$2.38–2.54 \times 10^{-5}$ (20°C)	13 (20°C)	60,000 (20°C)	1
EPN	323.31	—	0.0003 (100°C)	—	3.12
Ethanolamine	61.08	—	<1 (25.9°C)	Miscible	Unavailable
Ethylbenzene	106.17	0.00868 (25°C)	10 (25.9°C)	152 (25°C)	1.98
2-Ethoxyethanol	90.12	—	4 (20°C)	Miscible	Unavailable
2-Ethoxyethyl acetate	132.18	9.07×10^{-7} (20°C)	2 (20°C)	23 wt. % (20°C)	Unavailable
Ethyl acetate	88.11	1.34×10^{-4} (25°C)	94.5 (25°C)	100 ml/l (25°C)	Unavailable
Ethyl acrylate	100.12	$1.94–2.59 \times 10^{-3}$ (20°C)	29.5 (20°C)	1.5 wt. % (20°C)	Unavailable
Ethylamine	45.08	1.07×10^{-5} (25°C)	400 (2.0°C)	Miscible	Unavailable
Ethyl bromide	108.97	7.56×10^{-3} (25°C)	386 (20°C)	0.9 wt. % (20°C)	2.67
Ethylcyclopentane	98.19	2.10×10^{-2} (25°C)	40 (25.0°C)	245 (25°C)	Unavailable
Ethylene chlorohydrin	80.51	—	8 (25°C)	Miscible	Unavailable
Ethylenediamine	60.10	1.73×10^{-9} (25°C)	10 (21.5°C)	Miscible	Unavailable
Ethylene dibromide	187.86	7.06×10^{-4} (25°C)	11 (25°C)	3,370	1.64
Ethylenimine	43.07	1.33×10^{-3} (25°C)	250 (30°C)	Miscible	0.11
Ethyl ether	74.12	1.28×10^{-3} (25°C)	442 (20°C)	6.05 wt. % (25°C)	Unavailable
Ethyl formate	74.08	2.23×10^{-4} (25°C)	194 (20°C)	118,000 (25°C)	Unavailable
Ethyl mercaptan	62.13	2.74×10^{-3} (25°C)	527.2 (25°C)	1.3 wt. % (20°C)	Unavailable
4-Ethylmorpholine	115.18	—	6.1 (20°C)	Miscible	Unavailable
2-Ethylthiophene	112.19	—	60.9 (60.3°C)	292 (25°C)	Unavailable

F

Fluoranthene	202.26	0.0169 (25°C)	5.0×10^{-6} (25°C)	0.265 (25°C)	4.62
Fluorene	166.22	2.1×10^{-4}	10 (146°C)	1.98 (25°C)	3.7
Formaldehyde	30.03	3.27×10^{-7}	400 (−33°C)	Miscible	0.56
Formic acid	46.03	1.67×10^{-7} at pH 4	35 (20°C)	Miscible	Unavailable
Furfural	96.09	1.52–3.05×10^{-6} (20°C)	2 (20°C)	8.3 wt. % (20°C)	Unavailable
Furfuryl alcohol	98.10	—	0.4 (20°C)	Miscible	Unavailable

G

Glycidol	74.08	—	0.9 (25°C)	Miscible	Unavailable

H

Heptachlor	373.32	0.0023	4×10^{-4} (25°C)	180 ppb (25°C)	4.34
Heptachlor epoxide	389.32	3.2×10^{-5}	2.6×10^{-6} (20°C)	0.350 (25°C)	4.32
Hexachlorobenzene	284.78	0.0017	1.089×10^{-5} (20°C)	0.006 (25°C)	3.59
Hexachlorobutadiene	260.76	0.026	0.15 (20°C)	3.23 (25°C)	3.67
Hexachlorocyclopentadiene	272.77	0.016	0.081 (25°C)	1.8 (25°C)	3.63
Hexachloroethane	236.74	0.0025	0.8 (30°C)	27.2 (25°C)	3.34
2-Hexanone	100.16	0.00175 (25°C)	3.8 (25°C)	35,000 (25°C)	2.13
n-Heptane	100.20	2.035 (25°C)	45.85 (25°C)	2.24 (25°C)	Unavailable
2-Heptanone	114.19	1.44×10^{-4} (25°C)	2.6 (20°C)	0.43 wt. % (25°C)	Unavailable
3-Heptanone	114.19	4.20×10^{-5} (20°C)	1.4 (25°C)	14,300 (20°C)	Unavailable
cis-2-Heptene	98.19	4.13×10^{-1} (20°C)	48 (25°C)	15 (25°C)	Unavailable
trans-2-Heptene	98.19	4.22×10^{-1} (25°C)	49 (25°C)	15 (25°C)	Unavailable
n-Hexane	86.18	1.184 (25°C)	151.5 (25°C)	9.47 (25°C)	Unavailable
1-Hexene	84.16	4.35×10^{-1} (25°C)	186.0 (25°C)	50 (25°C)	Unavailable
sec-Hexyl acetate	144.21	4.38–5.84×10^{-3} (20°C)	4 (20°C)	0.013 wt. % (20°C)	Unavailable
Hydroquinone	110.11	$< 2.07 \times 10^{-9}$ (20–25°C)	1 (132.4)	70,000 (25°C)	0.98

I

Indeno[1,2,3-cd]pyrene	276.34	2.96×10^{-20} (25°C)	10^{-10} (25°C)	0.062	7.49
Isophorone	138.21	5.8×10^{-6}	0.38 (20°C)	12,000 (25°C)	1.49
Indan	118.18	—	—	88.9 (25°C)	2.48
Indole	117.15	—	—	3,558 (25°C)	1.69
Indoline	117.15	—	—	10,800 (25°C)	1.42

Compound					
1-Iodopropane	169.99	9.09×10^{-3}	43.1 (25°C)	0.1065 wt. % (23.5°C)	2.16
Isoamyl acetate	130.19	5.87×10^{-2} (25°C)	4 (20°C)	0.2 wt. % (20°C)	1.95
Isoamyl alcohol	88.15	8.89×10^{-6} (20°C)	2.3 (20°C)	26,720 (22°C)	Unavailable
Isobutyl acetate	116.16	4.85×10^{-4} (25°C)	20 (25°C)	6,300 (25°C)	Unavailable
Isobutyl alcohol	74.12	9.25×10^{-6} (20°C)	10.0 (20°C)	8.7 wt. % (20°C)	3.9
Isobutyl benzene	134.22	1.09×10^{-2} (25°C)	2.06 (25°C)	33.71 (25°C)	Unavailable
Isopropyl acetate	102.13	2.81×10^{-4} (25°C)	73 (25°C)	18,000 (20°C)	Unavailable
Isopropylamine	59.11	—	478 (20°C)	Miscible	3.45
Isopropylbenzene	120.19	1.47×10^{-2} (25°C)	4.6 (25°C)	48.3 (25°C)	
Isopropyl ether	102.18	9.97×10^{-3} (25°C)	150 (25°C)	0.65 wt. % (25°C)	Unavailable
K					
Kepone	490.68	3.11×10^{-2} (25°C)	2.25 (25°C)	2.7 (20–25°C)	4.74
L					
Lindane	290.83	4.8×10^{-7}	6.7×10^{-5} (25°C)	7.52 (25°C)	3.03
M					
Malathion	330.36	4.89×10^{-9} (25°C)	7.95×10^{-6} (25°C)	330 (30°C)	2.46
Maleic anhydride	98.06	Not applicable reacts with water	5×10^{-5} (20°C)	—	Not applicable reacts with water
Methoxychlor	345.66	Insufficient vapor pressure data for calculation at (25°C)	No data found	0.1 (25°C)	4.9
Methyl bromide	94.94	0.2	1,633 (25°C)	13,000 (25°C)	1.92
Methyl chloride	50.48	0.010 (25°C)	3,789 (20°C)	7,400 (25°C)	1.4
Methylene chloride	84.93	0.00269 (25°C)	455 (25°C)	13,000 (25°C)	0.94
2-Methylnaphthalene	142.20	Insufficient vapor pressure data for calculation	No data found	25.4 (25°C)	3.93
4-Methyl-2-pentanone	100.16	1.49×10^{-5} (25°C)	15 (20°C)	1.91 wt. % (25°C)	0.79
2-Methylphenol	108.14	1.23×10^{-6} (25°C)	0.24 (25°C)	25,000 (25°C)	1.34
4-Methylphenol	108.14	7.92×10^{-7} (25°C)	0.108 (25°C)	23,000 (25°C)	1.69
Mesityl oxide	98.14	4.01×10^{-6} (20°C)	8.7 (20°C)	3 wt. % (20°C)	Unavailable
Methyl acetate	74.08	9.09×10^{-5} (25°C)	235 (25°C)	240,000 (20°C)	Unavailable
Methyl acrylate	86.09	$1.23–1.44 \times 10^{-4}$ (20°C)	70 (20°C)	52,000	Unavailable
Methylal	76.10	1.73×10^{-4} (25°C)	400 (25°C)	33 wt. % (20°C)	Unavailable
Methyl alcohol	32.04	4.66×10^{-6} (25°C)	127.2 (25°C)	Miscible	Unavailable

Methylamine	31.06		3.1 atm (20°C)	9.590 (25°C)	Unavailable
Methylaniline	107.16	1.81×10^{-2} (25°C)	< 1.0 (20°C)	5.624 g/l (25°C)	Unavailable
2-Methylanthracene	192.96	1.19×10^{-5} (25°C)	—	0.039 (25°C)	5.12
2-Methyl-1,3-butadiene	68.12		550.1 (25°C)	642 (25°C)	Unavailable
2-Methylbutane	72.15	7.7×10^{-2} (25°C)	687.4 (25°C)	49.6 (25°C)	Unavailable
3-Methyl-1-butene	70.13	1.35 (25°C)	902.1 (25°C)	130 (25°C)	Unavailable
Methyl cellosolve	76.10	5.35×10^{-1} (25°C)	6 (20°C)	Miscible	Unavailable
Methyl cellosolve acetate	118.13		7 (20°C)	Miscible	Unavailable
Methylcyclohexane	98.19		46.3 (25°C)	16.0 (25°C)	Unavailable
o-Methylcyclohexanone	112.17	4.35×10^{-1} (25°C)	≈ 1 (20°C)	—	—
1-Methylcyclohexene	96.17		—	52 (25°C)	Unavailable
Methylcyclopentane	84.16		137.5 (25°C)	41.8 (25°C)	Unavailable
Methyl formate	60.05	3.62×10^{-1} (25°C)	625 (25°C)	30 wt. % (25°C)	Unavailable
3-Methylheptane	114.23	2.23×10^{-4} (25°C)	19.5 (25°C)	0.792 (25°C)	Unavailable
5-Methyl-3-heptanone	128.21	3.70 (25°C)	2 (25°C)	0.26 wt. % (20°C)	Unavailable
2-Methylhexane	100.20	1.30×10^{-4} (20°C)	65.9 (25°C)	2.54 (25°C)	Unavailable
3-Methylhexane	100.20	3.42 (25°C)	61.6 (25°C)	4.95 (25°C)	Unavailable
Methylhydrazine	46.07	1.55–1.64 (25°C)	49.6 (25°C)	Miscible	Unavailable
Methyl iodide	141.94		405 (25°C)	2 wt. % (20°C)	1.36
Methyl isocyanate	57.05	5.87×10^{-3} (25°C)	348 (25°C)	6.7 wt. % (20°C)	Unavailable
Methyl mercaptan	48.10	3.89×10^{-4} (20°C)	1,516 (25°C)	23.30 g/l (20°C)	Unavailable
Methyl methacrylate	100.12	3.01×10^{-3} (25°C)	40 (26°C)	1.5 wt. % (20°C)	Unavailable
4-Methyloctane	128.26	2.46×10^{-4} (20°C)	7 (25°C)	0.115 (25°C)	Unavailable
2-Methylpentane	86.18	10.27 (25°C)	211.8 (25°C)	13.8 (25°C)	Unavailable
3-Methylpentane	86.18	1.732 (25°C)	189.8 (25°C)	17.9 (25°C)	Unavailable
2-Methyl-1-pentene	84.16	1.693 (25°C)	195.4 (25°C)	78 (25°C)	Unavailable
4-Methyl-1-pentene	84.16	2.77×10^{-1} (25°C)	270.8 (25°C)	48 (25°C)	Unavailable
1-Methylphenanthrene	192.26	6.15×10^{-1} (25°C)	—	269 ppb (25°C)	4.56
2-Methylpropane	58.12		10 atm (66.8°C)	48.9 (25°C)	Unavailable
2-Methylpropene	56.11	1.171 (25°C)	2.270 (25°C)	263 (25°C)	Unavailable
α-Methylstyrene	118.18	2.1×10^{-1} (25°C)	1.9 (20°C)	—	—
Mevinphos	224.16		0.003 (20°C)	Miscible	Unavailable
Morpholine	87.12		13.4 (25°C)	Miscible	Unavailable

N

Naphthalene	128.18	4.6×10^{-4}	0.23 (25°C)	30 (25°C)	2.74
2-Nitroaniline	138.13	9.72×10^{-5} (25°C)	8.1 (25°C)	1,260 (25°C)	1.23–1.62

3-Nitroaniline	138.13	Insufficient vapor pressure data for calculation	1 (119.3°C)	890 (25°C)	1.26
4-Nitroaniline	138.13	1.14×10^{-8} (25°C)	0.0015 (20°C)	800 (18.5°C)	1.08
Nitrobenzene	123.11	2.45×10^{-5}	0.28 (25°C)	2,000 (25°C)	2.36
2-Nitrophenol	139.11	3.5×10^{-6}	0.20 (25°C)	2,000 (25°C)	1.57
4-Nitrophenol	139.11	3.0×10^{-5} (20°C)	10^{-4} (20°C)	16,000 (25°C)	2.33
N-Nitrosodimethylamine	74.09	0.143 (25°C)	8.1 (25°C)	Miscible	1.41
N-Nitrosodiphenylamine	198.22	2.33×10^{-8} (25°C)	No data found	35.1 (25°C)	2.76
N-Nitrosodi-r-propylamine	130.19	Insufficient vapor pressure data to calculate	No data found	9,900 (25°C)	1.01
Naled	380.79	—	2×10^{-4} (20°C)	—	Not applicable reacts with water
1-Naphthylamine	143.19	1.27×10^{-10} (25°C)	6.5×10^{-5} (20–30°C)	1,700	3.51
2-Naphthylamine	143.19	2.01×10^{-9} (25°C)	2.56×10^{-4} (20–30°C)	586 (20–30°C)	2.11
Nitrapyrin	230.90	2.13×10^{-3}	0.0028 (20°C)	40	2.64
4-Nitrobiphenyl	199.21	—	—	—	—
Nitroethane	75.07	4.66×10^{-5} (25°C)	15.6 (20°C)	45 ml/l (20°C)	Unavailable
Nitromethane	61.04	2.86×10^{-5}	27.8 (20°C)	22 ml/l (20°C)	Unavailable
1-Nitropropane	89.09	8.68×10^{-5} (25°C)	7.5 (20°C)	1.4 wt. % (25°C)	Unavailable
2-Nitropropane	89.09	1.23×10^{-4} (25°C)	12.9 (20°C)	1.7 wt. % (20°C)	Unavailable
2-Nitrotoluene	137.14	4.51×10^{-5} (20°C)	0.15 (20°C)	0.06 wt. % (20°C)	Unavailable
3-Nitrotoluene	137.14	5.41×10^{-5} (20°C)	0.25 (25°C)	0.05 wt. % (20°C)	Unavailable
4-Nitrotoluene	137.14	5.0×10^{-5} (25°C)	5.484 (26.0°C)	0.005 wt. % (20°C)	Unavailable
n-Nonane	128.26	5.95 (25°C)	4.3 (25°C)	0.122 (25°C)	Unavailable

O

Octachloronaphthalene	403.73	—	< 1 (20°C)	—	Unavailable
n-Octane	114.23	3.225 (25°C)	14.14 (25°C)	0.431 (25°C)	Unavailable
1-Octene	112.22	9.52×10^{-1} (25°C)	17.4 (25°C)	2.7 (25°C)	0.89
Oxalic acid	90.04	1.43×10^{-10} (pH 4)	< 0.001 (20°C)	9.81 wt. % (25°C)	—

P

PCB-1016	257.90	750	4×10^{-4} (25°C)	0.22–0.25	4.7
PCB-1221	192.00	3.24×10^{-4}	0.0067 (25°C)	1.5 (25°C)	2.44
PCB-1232	221.00	8.64×10^{-4}	0.0046 (25°C)	1.45 (25°C)	2.83
PCB-1242	154–358 with an average value of 261	5.6×10^{-4}	4.06×10^{-4} (25°C)	0.24 (25°C)	3.71

Compound	MW				log K_{ow}
PCB-1248	222–358 with an average value of 288	0.0035	4.94×10^{-4} (25°C)	0.054	5.64
PCB-1254	327 (average)	0.0027	7.71×10^{-5} (25°C)	0.012 (25°C)	5.61
PCB-1260	324–460 370 (average)	0.0071	4.05×10^{-5} (25°C)	0.080 (24°C)	6.42
Parathion	291.27	8.56×10^{-8} (25°C)	9.8×10^{-6} (25°C)	24 (25°C)	3.68
Pentachlorophenol	266.34	3.4×10^{-6}	1.7×10^{-4} (20°C)	20–25 (25°C)	2.96
Phenanthrene	178.24	2.56×10^{-5} (25°C)	6.80×10^{-4} (25°C)	1.18 (25°C)	3.72
Phenol	94.11	3.97×10^{-7} (25°C)	0.34 (25°C)	93,000 (25°C)	1.43
Pentachlorobenzene	250.34	0.0071 (20°C)	6.0×10^{-3} (20–30°C)	2.24×10^{-6} M (25°C)	6.3
Pentachloroethane	202.28	2.45×10^{-3} (25°C)	4.5 (25°C)	7.69 and 500 were reported at 25°C and 20°C	3.28
1,4-Pentadiene	68.12	1.20×10^{-1} (25°C)	734.6 (25°C)	558 (25°C)	Unavailable
n-Pentane	72.15	1.255 (25°C)	512.8 (25°C)	39.5 (25°C)	Unavailable
2-Pentanone	86.13	6.44×10^{-5} (25°C)	16 (25°C)	5.51 wt. % (25°C)	Unavailable
1-Pentene	70.13	4.06×10^{-1} (25°C)	637.7 (25°C)	148 (25°C)	Unavailable
cis-2-Pentene	70.13	2.25×10^{-1} (25°C)	494.6 (25°C)	203 (25°C)	Unavailable
trans-2-Pentene	70.13	2.34×10^{-1} (25°C)	505.5 (25°C)	203 (25°C)	Unavailable
Pentycyclopentane	140.28	—	—	0.115 (25°C)	Unavailable
p-Phenylenediamine	108.14	—	—	38,000 (24°C)	Unavailable
Phenyl ether	170.21	2.13×10^{-4} (20°C)	0.12 (30°C)	21 (25°C)	Unavailable
Phenylhydrazine	108.14	—	< 0.1 (20°C)	—	Unavailable
Phthalic anhydride	148.12	6.29×10^{-9} (20°C)	2×10^{-4} (20°C)	0.62 wt. % (20°C)	1.9
Picric acid	229.11	$< 2.15 \times 10^{-5}$ (20°C)	< 1 (20°C)	1.4 wt. % (20°C)	—
Pindone	230.25	—	—	18 (25°C)	2.95
Propane	44.10	7.06×10^{-1} (25°C)	8.6 atm (20°C)	62.4 (25°C)	Unavailable
β-Propiolactone	72.06	7.63×10^{-7} (25°C)	3.4 (25°C)	37 vol. % (25°C)	Unavailable
n-Propyl acetate	102.12	1.99×10^{-4} (25°C)	35 (25°C)	18,900 (20°C)	Unavailable
n-Propyl alcohol	60.10	6.74×10^{-6} (25°C)	20.8 (25°C)	Miscible	Unavailable
n-Propylbenzene	120.19	1.0×10^{-2} (25°C)	3.43 (25°C)	55 (25°C)	2.87
Propylcyclopentane	112.22	8.90×10^{-1} (25°C)	12.3 (25°C)	2.04 (25°C)	Unavailable
Propylene oxide	58.08	8.34×10^{-5} (20°C)	445 (20°C)	41 wt. % (20°C)	Not applicable reacts with water
n-Propyl nitrate	105.09	—	18 (20°C)	3,640 (20°C)	Unavailable
Propyne	40.06	1.1×10^{-1} (25°C)	4,310 (25°C)		Unavailable
Pyrene	202.26	1.87×10^{-5}	6.85×10^{-7} (25°C)	0.148 (25°C)	4.66
Pyridine	79.10	8.88×10^{-6} (25°C)	20 (25°C)	Miscible	Unavailable

Q

p-Quinone	108.10	9.48×10^{-7} (20°C)	0.1 (20°C)	1.5 wt. % (20°C)	Unavailable

R

Ronnel	321.57	8.46×10^{-6} (25°C)	8×10^{-4} (25°C)	40 (25°C)	2.76

S

Styrene	104.15	0.00261	6.45 (25°C)	0.031 wt. % (25°C)	2.87
Strychnine	334.42	—	—	0.02 wt. % (20°C)	2.45
Sulfotepp	322.30	2.88×10^{-6} (20°C)	0.00017 (20°C)	25	2.87

T

TCDD	321.98	5.40×10^{-23} (18–22°C)	7.2×10^{-10} (25°C)	0.0193 ppb (22°C)	6.66
1,1,2,2-Tetrachloroethane	167.85	4.56×10^{-4} (25°C)	6 (25°C)	2,970 (25°C)	2.07
Tetrachloroethylene	165.83	0.0153	20 (25°C)	150 (25°C)	2.42
Toluene	92.14	0.00674 (25°C)	22 (20°C)	490 (25°C)	2.06
Toxaphene	413.82	0.063	0.2–0.4 (25°C)	0.2–0.4 (25°C)	3.18
1,2,4-Trichlorobenzene	181.45	0.00232	0.29 (25°C)	31.3 (25°C)	2.7
1,1,1-Trichloroethane	133.40	0.0162 (25°C)	124 (25°C)	950 (25°C)	2.18
1,1,2-Trichloroethane	133.40	9.09×10^{-4} (25°C)	19 (20°C)	4,500 (20°C)	1.75
Trichloroethylene	131.39	0.0091	72.6 (25°C)	1,100 (25°C)	1.81
Trichlorofluoromethane	137.37	1.73 (25°C)	792 (25°C)	1,240 (25°C)	2.2
2,4,5-Trichlorophenol	197.45	1.76×10^{-7} (25°C)	0.022 (25°C)	1.2 g/l (25°C)	2.85
2,4,6-Trichlorophenol	197.45	9.07×10^{-8} (25°C)	0.017 (25°C)	800 (25°C)	3.03
2,4,5-T	255.48	4.87×10^{-8} (25°C)	6.46×10^{-6} (25°C)	278 (25°C)	1.72
1,2,4,5-Tetrabromobenzene	393.70	—	—	0.040	4.82
1,1,2,2-Tetrabromoethane	345.65	6.40×10^{-5} (20°C)	0.1 (20°C)	0.07 wt. % (20°C)	2.45
1,2,3,4-Tetrachlorobenzene	215.89	6.9×10^{-3} (20°C)	2.6×10^{-2} (25°C)	5.92 (25°C)	5.4 average value
1,2,3,5-Tetrachlorobenzene	215.89	1.58×10^{-3} (25°C)	1 (58.2°C)	5.19 (25°C)	6.0 average
1,2,4,5-Tetrachlorobenzene	215.89	1.0×10^{-2} (20°C)	< 0.1 (25°C)	0.465 (25°C)	6.1 average
Tetraethyl pyrophosphate	290.20	—	1.55×10^{-4} (20°C)	Miscible	Not applicable reacts with water
Tetrahydrofuran	72.11	7.06×10^{-5} (25°C)	145 (20°C)	Miscible	Unavailable
1,2,4,5-Tetramethylbenzene	134.22	2.49×10^{-2} (25°C)	0.49 (25°C)	3.48 (25°C)	3.79
Tetranitromethane	196.03	—	13 (25°C)	—	—

Compound					
Tetryl	287.15	< 1.89 × 10⁻³ (20°C)	< 1 (20°C)	0.02 wt. % (20°C)	2.37
Thiophene	84.14	2.93 × 10⁻³ (25°C)	79.7 (25°C)	3,015 (25°C)	1.73
Thiram	269.35	—	—	30	—
2,4-Toluene disocyanate	174.15	—	0.01 (20°C)	Not applicable reacts with water	Not applicable reacts with water
o-Toluidine	107.16	1.88 × 10⁻⁶ (25°C)	0.1 (20°C)	15,000 (25°C)	2.61
1,3,5-Tribromobenzene	314.80	—	—	2.51 × 10⁻⁶ (25°C)	4.05
Tributyl phosphate	266.32	—	—	0.1 wt. % (20°C)	2.29
1,2,3-Trichlorobenzene	181.45	8.9 × 10⁻³ (20°C)	1 (40°C)	18.0 (25°C)	3.87
1,3,5-Trichlorobenzene	181.45	1.9 × 10⁻³ (20°C)	0.58 (25°C)	6.01 (25°C)	5.7 (average)
1,2,3-Trichloropropane	147.43	3.18 × 10⁻⁴ (25°C)	3.4 (20°C)	—	2.59
1,1,2-Trichlorotrifluoroethane	187.38	3.33 × 10⁻¹ (20°C)	270 (20°C)	0.02 wt. % (20°C)	3.37
Tri-o-cresyl phosphate	368.37	—	—	3.1 (25°C)	—
Triethylamine	101.19	4.79 × 10⁻⁴ (20°C)	54 (20°C)	15,000 (20°C)	Unavailable
Trifluralin	335.29	4.84 × 10⁻⁵ (23°C)	1.1 × 10⁻⁴ (25°C)	240	3.73
1,2,3-Trimethylbenzene	120.19	3.18 × 10⁻³ (25°C)	1.51 (25°C)	75.2 (25°C)	3.34
1,2,4-Trimethylbenzene	120.19	5.7 × 10⁻³ (25°C)	2.03 (25°C)	51.9 (25°C)	3.57
1,3,5-Trimethylbenzene	120.19	3.93 × 10⁻³ (25°C)	2.42 (25°C)	48.2 (25°C)	3.21
1,1,3-Trimethylcyclohexane	126.24	—	—	1.77 (25°C)	Unavailable
1,1,3-Trimethylcyclopentane	112.22	1.57 (25°C)	39.7 (25°C)	3.73 (25°C)	Unavailable
2,2,5-Trimethylhexane	128.26	2.42 (25°C)	16.5 (25°C)	1.15 (25°C)	Unavailable
2,2,4-Trimethylpentane	114.23	3.01 (25°C)	49.3 (25°C)	2.05 (25°C)	Unavailable
2,3,4-Trimethylpentane	114.23	2.98 (25°C)	27.0 (25°C)	1.36 (25°C)	Unavailable
2,4,6-Trinitrotoluene	227.13	—	4.26 × 10⁻³ (54.8°C)	0.013 wt. % (20°C)	2.48
Triphenyl phosphate	326.29	5.88 × 10⁻² (20–25°C)	< 0.1 (20°C)	0.001 wt. % (20°C)	3.72
V					
Vinyl acetate	86.09	4.81 × 10⁻⁴	115 (25°C)	25,000 (25°C)	0.45
Vinyl chloride	62.50	2.78	2,660 (25°C)	1,100 (25°C)	0.39
W					
Warfarin	308.33	—	—	17 (20°C)	2.96
X					
o-Xylene	106.17	0.00535 (25°C)	6.6 (25°C)	213 (25°C)	2.11
m-Xylene	106.17	0.0063 (25°C)	8.287 (25°C)	173 (25°C)	3.2
p-Xylene	106.17	0.0063 (25°C)	8.763 (25°C)	200 (25°C)	2.31

Sources: Montgomery, J. H. and Welkom, L. M., *Groundwater Chemicals Desk Reference*, Lewis Publishers, Chelsea, MI, 1990; Montgomery, J. H., *Groundwater Chemicals Desk Reference*, Vol. 2, Lewis Publishers, Chelsea, MI, 1991.

Appendix D
ENVIRONMENTAL DEGRADATION RATES FOR SELECTED ORGANIC COMPOUNDS

Environmental Degradation Rates

A

	Soil high	Soil low	Surface water high	Surface water low	Groundwater high	Groundwater low
Acetone	168 h Ae	24 h Ae	168 h Ae	24 h Ae	336 h Ae	48 h Ae
2-Acetylaminofluorene	4320 h Ae	672 h Ae	4320 h Ae	672 h Ae	8640 h Ae	1344 h Ae
4-Aminoazobenzene	672 h Ae	168 h Ae	672 h Ae	62.4 h Ph	1344 h Ae	336 h Ae
Acetamide	168 h Ae	24 h Ae	168 h Ae	24 h Ae	336 h Ae	48 h Ae
Amitrole	4320 h Ae	672 h Ae	4320 h Ae	672 h Ae	8640 h Ae	1344 h Ae
Acetonitrile	672 h Ae	168 h Ae	672 h Ae	168 h Ae	8640 h Ae	336 h Ae
Acrylic acid	168 h Ae	24 h Ae	168 h Ae	24 h Ae	4320 h An	48 h Ae
1-Amino-2-methylanthraquinone	672 h Ae	268 h Ae	672 h Ae	62.4 h Ae	1344 h Ae	336 h Ae
Acenaphthene	2448 h Ae	295 h Ae	300 h Ph	3 h Ph	4896 h Ae	590 h Ae
O-Anisidine	4320 h Ae	672 h Ae	4320 h Ae	62.4 h Ph	8640 h Ae	1344 h Ae
4-Aminobiphenyl	168 h Ae	24 h Ae	168 h Ae	24 h Ae	336 h Ae	48 h Ae
P-Anisidine	672 h Ae	168 h Ae	672 h Ae	62.4 h Ph	1344 h Ae	336 h Ae
Acrolein	672 h Ae	168 h Ae	672 h Ae	168 h Ae	1344 h Ae	336 h Ae
Acrylonitrile	552 h Ae	30 h Ae	552 h Ae	30 h Ae	1140 h Ae	60 h Ae
Allyl alcohol	168 h Ae	24 h Ae	168 h Ae	24 h Ae	336 h Ae	48 h Ae
Azaserine	1344 h Ae	196 h Hy	1344 h Ae	196 h Hy	1344 h Ae	196 h Hy
Aldicarb	8644 h Ae	480 h Ae	8664 h Ae	480 h Ae	15,240 h An	990 h Ae
2-Aminoanthraquinone	672 h Ae	168 h Ae	672 h Ae	62.4 h Ph	1344 h Ae	336 h Ae
Aziridine	672 h Ae	168 h Ae	672 h Ae	168 h Ae	8640 h Ae	336 h Ae
Anthracene	11,040 h Ae	1200 h Ae	1–7 Ph	0.58 h Ph	22,080 h Ae	2400 h Ae
Acenaphthylene	1440 h Ae	1020 h Ae	1440 h Ae	1020 h Ae	2880 h Ae	2040 h Ae
Aldrin	14,200 h Ae	504 h Ae	14,200 h Ae	504 h Ae	28,400 h Ae	24 h An
Auramine	4320 h Ae	672 h Ae	4320 h Ae	672 h Ae	8640 h Ae	1344 h Ae
Aflatoxin B1	672 h Ae	168 h Ae	672 h Ae	168 h Ae	1344 h Ae	336 h Ae

B

	Soil high	Soil low	Surface water high	Surface water low	Groundwater high	Groundwater low
Benzo[a]pyrene	12,720 h Ae	1368 h Ae	1.1 h Ph	0.37 h Ph	25,440 h Ae	2736 h Ae
Benzamide	360 h Ae	48 h Ae	360 h Ae	48 h Ae	720 h Ae	96 h Ae
Benz[a]anthracene	16,320 h Ae	2448 h Ae	3 h Ph	1 h Ph	32,640 h Ae	4896 h Ae
1-Butanol	168 h Ae	24 h Ae	168 h Ae	24 h Ae	1296 h An	48 h An
Benzene	384 h Ae	120 h Ae	384 h Ae	120 h Ae	17,280 h An	240 h Ae

Compound					
Bromoform	4320 h Ae	672 h Ae	4320 h Ae	672 h Ae	8640 h Ae	1344 h Ae
tert-Butyl alcohol	4800 h	360 h	4320 h Ae	672 h Ae	8640 h Ae	1344 h Ae
sec-Butyl alcohol	168 h Ae	24 h Ae	168 h Ae	24 h Ae	336 h Ae	48 h Ae
Butyl benzyl phthalate	168 h Ae	24 h Ae	168 h Ae	24 h Ae	4320 h	48 h
Butylglycolyl butyl phthalate	168 h Ae	24 h Ae	168 h Ae	24 h Ae	336 h Ae	48 h Ae
Biphenyl	168 h Ae	36 h Ae	168 h Ae	36 h Ae	336 h Ae	72 h Ae
Benzidine	192 h Ae	48 h Ae	192 h Ae	31.2 h ph	384 h Ae	96 h Ae
Benzoyl peroxide	48 h	4 h	168 h Ae	24 h Ae	336 h Ae	48 h Ae
Benzotrichloride	3 min Hy	11 sec Hy	3 min Hy	11 sec Hy	3 min Hy	11 sec Hy
Benzoylchloride	150 secs Hy	17 sec Hy	150 sec Hy	17 sec Hy	150 sec Hy	17 sec Hy
Benzylchloride	290 h Hy	15 h Hy	290 h Hy	15 h Hy	290 h Hy	15 h Hy
Bis(4-dimethylamino-phenyl)methane	4320 h Ae	672 h Ae	2626 h	26.3 h	8640 h Ae	1344 h Ae
Bis(2-ethylhexyl) adipate	672 h Ae	19.2 h Hy	672 h Hy	19.2 h Hy	1344 h Ae	19.2 h Hy
1,4-Benzoquinone	120 h Ae	1 h Ae	120 h Ae	1 h Ae	240 h Ae	2 h Ae
1,2-Butylene oxide	310 h Hy	168 h Hy	310 h Hy	168 h Hy	30 h Hy	168 h Hy
1,3-Butadiene	672 h Ae	168 h Ae	672 h Ae	168 h Ae	1344 h Ae	336 h Ae
1,3-Benzenediamine	672 h Ae	168 h Ae	672 h Ae	31 h Ph	1344 h Ae	336 h Ae
Bis-(2-chloroisopropyl) ether	4320 h Ae	432 h Ae	4320 h Ae	432 h Ae	8640 h Ae	864 h Ae
Bis-(2-chloroethyl) ether	4320 h Ae	672 h Ae	4320 h Ae	672 h Ae	8640 h Ae	1344 h Ae
Bis-(2-ethylhexyl) phthalate	550 h Ae	120 h Ae	550 h Ae	120 h Ae	9336 h An	240 h Ae
Butyraldehyde	168 h Ae	24 h Ae	168 h Ae	24 h Ae	336 h Ae	48 h Ae
Butyl acrylate	168 h Ae	24 h Ae	168 h Ae	24 h Ae	336 h Ae	48 h Ae
Benzo[ghi]perylene	15,600 h	14,160 h	15,600 h	14,160 h	31,200 h	28,320 h
Benzo[b]fluoranthene	14,640 h	8640 h	720 h Ph	8.7 h Ph	29,280 h Ae	17,280 h Ae
Benzo[k]fluoranthene	51,360 h Ae	21,840 h Ae	499 h Ph	3.8 h Ph	102,720 h Ae	42,680 h Ae
Benz[c]acridine	8760 h Ae	4320 h Ae	8760 h Ae	4320 h Ae	17,520 h Ae	8640 h Ae
Bis(chloromethyl) ether	0.106 h Hy	0.0106 h Hy	0.0106 h Hy	0.0106 h Hy	0.0106 h Hy	0.0106 h Hy
Bromoethylene	4320 h Ae	672 h Ae	4320 h Ae	672 h Ae	69,000 h An	1344 h Ae
Bromooxynil octanoate	528 h Ae	0.667 h Hy	528 h Ae	0.667 h Hy	1056 h Ae	0.0667 h Hy
Benefin	2880 h Ae	504 h Ae	864 h Ph	288 h Ph	5760 h Ae	144 h An

C

Compound					
Cyclophosphamide	672 h Ae	168 h Ae	672 h Ae	168 h Ae	9976 h Hy	336 h Ae
Carbon tetrachloride	8640 h Ae	4320 h Ae	8640 h Ae	4032 h Ae	8640 h	168 h
Chlordane	33264 h	5712 h	33,264 h	5712 h	66,528 h Ae	11,424 h Ae
Carbaryl	720 h	3.2 h	200 h	200 h	1440 h	3.2 h
Chloroform	4320 h Ae	672 h Ae	4320 h Ae	672 h Ae	4320 h Ae	1344 h Ae

Compound					
Chloroethane	672 h Ae	168 h Ae	168 h Ae	1344 h Ae	336 h Ae
Chloroacetic acid	168 h Ae	24 h Ae	24 h Ae	336 h Ae	48 h Ae
Cumene hydroperoxide	672 h Ae	168 h Ae	168 h Ae	1344 h Ae	336 h Ae
o-Cresol	168 h Ae	24 h Ae	24 h Ae	336 h An	48 h Ae
C.I. solvent yellow 3	672 h Ae	168 h Ae	62.4 h Ph	1344 h Ae	336 h Ae
Cumene	192 h Ae	48 h Ae	48 h Ae	384 h Ae	96 h Ae
P-Cresol	16 h Ae	1 h Ae	1 h Ae	672 h An	2 h Ae
3-Chloropropene	335 h Hy	166 h Hy	166 h Hy	335 h Hy	166 h Hy
Chloromethyl methyl ether	0.033 h Hy	0.0108 h	0.0108 h	0.033 h Hy	0.0108 h.
m-Cresol	696 h Ae	48 h Ae	48 h Ae	1176 h An	96 h Ae
Chlorobenzene	3600 h Ae	1632 h Ae	1632 h Ae	7200 h Ae	3264 h Ae
Cyclohexane	4320 h Ae	672 h Ae	672 h Ae	8640 h Ae	1344 h Ae
p-Cresidine	4320 h Ae	672 h Ae	672 h Ae	8640 h Ae	1344 h Ae
Catechol	168 h Ae	24 h Ae	24 h Ae	336 h Ae	48 h Ae
Crotonaldehyde (*trans*)	168 h Ae	24 h Ae	24 h Ae	336 h Ae	48 h Ae
Chloroprene	4320 h Ae	672 h Ae	672 h Ae	8640 h Ae	1344 h Ae
C.I. vat yellow 4	4320 h Ae	672 h Ae	672 h Ae	88,640 h Ae	1344 h Ae
Captan	1440 h Ae	48 h	10.5 min Hy	10.3 h Hy	10.5 min Hy
Cupferron	4320 h Ae	672 h Ae	672 h	8640 h Ae	1344 h Ae
Cyanamide, calcium salt	672 h Ae	168 h Ae	168 h Ae	1344 h Ae	336 h Ae
Chrysene	24,000 h Ae	8904 h Ae	13 h Ph	48,000 h Ae	17,808 h Ae
Chlorobenzilate	840 h	168 h	168 h Ae	1680 h Ae	3360 h Ae
2-Chloroacetophenone	672 h Ae	168 h Ae	168 h Ae	1344 h Ae	336 h Ae
C.I. basic green	4320 h Ae	672 h Ae	672 h Ae	8640 h Ae	1344 h Ae
C.I. solvent yellow 14	672 h Ae	168 h Ph	640 h Ph	1344 h Ae	336 h Ae
Cresol (s)	168 h Ae	24 h Ae	24 h Ae	336 h Ae	48 h Ae
D					
2-4-DB	168 h	24 h	24 h Ae	336 h Ae	48 h Ae
1,2-Dichlorobenzene	4320 h Ae	672 h Ae	672 h Ae	8640 h Ae	1344 h Ae
2,4-Diaminotoluene	4320 h Ae	1740 h Ph	31 h Ph	8640 h Ae	1344 h Ae
1,2-Dibromo-3-chloropropane	4320 h Ae	672 h Ae	672 h Ae	8640 h Ae	1344 h Ae
(Dichloromethyl)benzene	0.6 h Hy	0.1 h Hy	06.6 h Hy	0.6 h Hy	0.1 h Hy
1,3-Dinitrobenzene	4320 h Ae	720 h Ph	554 h Ph	8640 h Ae	48 h An
4,4'-Diaminodiphenyl ether	4320 h Ae	3480 h	62.4 h	8640 h Ae	1344 h Ae
2,4-Dimethylphenol	168 h Ae	24 h Ae	24 h Ae	336 h Ae	48 h Ae
p-Dichlorobenzene	4320 h Ae	672 h Ae	672 h Ae	88,640 h Ae	1344 h Ae

Compound					
1,2-Dichloroethane	4320 h Ae	2400 h Ae	4320 h Ae	8640 h Ae	2400 h Ae
Diethanolamine	168 h Ae	14.4 h Ae	168 h Ae	336 h Ae	28.8 h Ae
Diethylene glycol, monoethyl ether	672 h	168 h Ae	672 h Ae	1344 h Ae	336 h Ae
Di(n-octyl)phthalate	672 h Ae	168 h Ae	672 h Ae	8760 h An	336 h Ae
3,3'-Dimethoxybenzidine	4320 h Ae	672 h Ae	1740 h Ph	8640 h Ae	672 h An
3,3'-Dimethylbenzidine	168 h Ae	24 h Ae	168 h Ae	336 h Ae	48 h Ae
Dimethyl terephthalate	672 h Ae	168 h Ae	672 h Ae	8640 h Ae	336 h Ae
2,4-Dichlorophenol	1680 h Ae	168 h Ae	3 h Ph	1032 h An	133 h Ae
2,4-Dinitrotoluene	4320 h Ae	672 h Ae	33 h Ph	8640 h Ae	48 h An
N,N-Dimethylaniline	4320 h Ae	176 h Ph	1925 h Ph	8640 h Ae	1344 h Ae
Diphenylamine	672 h Ae	672 h Ae	672 h Ae	1344 h Ae	336 h Ae
1,4-Dioxane	4320 h Ae	672 h Ae	4320 h Ae	8640 h Ae	1344 h Ae
Dibromochloromethane	336 h Ae	168 h Ae	79 h Ae	158 h Ae	4 h Ae
Dimethylamine	86 h Ae	336 h Ae	4320 h Ae	4320 h	336 h
Dimethyl phthalate	4320 h Ae	672 h Ae	168 h Ae	336 h Ae	48 h Ae
Dibenzofuran	168 h Ae	24 h Ae	672 h Ae	835 h Ae	205 h Ae
DDT	672 h Ae	168 h Ae	8400 h Ph	—	—
2,4-Dinitrophenol	1.4×10^5 h Ae	17,520 h Ae	3840 h	77 h	68 h An
Dibenz[a,h]anthracene	6312 h Ae	1622 h Ae	782 h Ph	12,624 h Ae	17,328 h Ae
Diethylstilbestrol	22,560 h Ae	8664 h Ae	3840 h Ph	45,120 h Ae	1344 h Ae
1,1-Dimethyl hydrazine	4320 h Ae	672 h Ae	528 h	8640 h Ae	384 h Ae
7,12-Dimethylbenz(a)anthracene	528 h Ae	192 h Ae	672 h Ae	1056 h Ae	960 h Ae
Dimethylaminoazobenzene	672 h Ae	480 h Ae	672 h Ae	1344 h Ae	336 h Ae
Dimethoate	672 h Ae	168 h Ae	1344 h Ae	1344 h Ae	528 h Ae
Dieldrin	888 h	264 h	25,920 h Ae	2688 h Ae	24 h An
Diethyl sulfate	25,920 h Ae	4200 h Ae	12 h Hy	51,840 h	1.7 h Hy
DDD	12 h Hy	1.7 h Hy	1.4×10^5 h Ae	12 h Hy	1680 h
DDE	1.4×10^5 h Ae	17,520 h Ae	146 h Ph	2.7×10^5 h Ae	384 h
Dichloromethane	672 h Ae	168 h Ae	672 h Ae	1344 h Ae	336 h Ae
1,1-Dichloroethane	3696 h	768 h	3696 h	8640 h Ae	1344 h Ae
1,1-Dichloroethylene	4320 h Ae	672 h Ae	4320 h Ae	3168 h	1344 h
Dichlorodifluoromethane	4320 h Ae	672 h Ae	4320 h Ae	8640 h Ae	1344 h Ae
Dalapon	1440 h Ae	336 h Ae	1440 h Ae	2880 h Ae	672 h Ae
Dimethyl sulfate	12 h Hy	12 h Hy	12 h Hy	12 h Hy	1.2 h Hy
1,2-Dichloropropane	30,936 h Ae	4008 h Ae	30,936 h Ae	61,872 h Ae	8016 h Ae
Dimethylcarbamyl chloride	195 sec Hy	14 sec Hy	195 sec Hy	105 sec Hy	14 sec Ae
Diethyl phthalate	1344 h Ae	72 h Ae	1344 h	2688 h Ae	144 h Ae

Compound					
Di-n-butyl phthalate	552 h Ae	336 h	24 h	552 h	48 h
Dinoseb	2952 h Mi	2952 h	1032 h mi	5904 h Ae	96 h An
3,3'-Dichlorobenzidine	4320 h Ae	0.075 h	0.025 h Ph	8640 h Ae	1344 h Ae
Dihydrosafrole	672 h Ae	672 h	168 h Ae	1344 h Ae	336 h Ae
2,4-Dichlorophenoxyacetic acid	1200 h Ae	96 h Ph	48 h Ph	4320 h	480 h
1,2,7,8-Dibenzopyrene	8644 h Ae	8644 h Ae	5568 h Ae	17,328 h Ae	11,136 h Ae
Disulfoton	504 h Ae	504 h Ae	72 h Ae	1008 h Ae	144 h Ae
1,2-Dimethyl hydrazine	672 h Ae	672 h Ae	168 h Ae	8640 h Ae	336 h Ae
m-Dichlorobenzene	4320 h Ae	4320 h Ae	672 h Ae	8640 h Ae	1344 h Ae
1,3-Dichloropropene	271 h Hy	271 h Hy	133 h Hy	271 h Hy	133 h Hy
2,3-Dinitrotoluene	4320 h Ae	50 h Ph	45 h Ph	8640 h Ae	48 h An
2,6-Dinitrotoluene	4320 h Ae	17 h Ph	2 h Ph	8640 h Ae	48 h An
3,4-Dinitrotoluene	4320 h Ae	326 h Ph	319 h Ph	8640 h Ae	48 h An
2,5-Dinitrotoluene	4320 h Ae	2.78 h Ph	1.85 h Ph	8640 h Ae	48 h An
Dibutylnitrosamine	504 h Ae	8 h Ph	4 h Ph	8640 h Ae	1008 h Ae
Decarbomomophenyl oxide	88,760 h Ae	8760 h Ae	4320 h Ae	17,520 h Ae	8640 h Ae
Diepoxybutane	183 h Hy	138 h Hy	92 h Hy	1138 h Hy	92 h Hy
N,N'-Diethyl hydrazine	672 h Ae	67 h Ph	168 h Ph	8640 h Ae	336 h Ae
Dimethyl tetrachloroterephthalate	2208 h Ae	2208 h Ae	432 h Ae	4416 h Ae	864 h Ae
Diallate	2160 h Ae	2160 h Ae	252 h Ae	4320 h Ae	504 h Aae
Diaminotoluenes'	4320 h Ae	1740 h Ph	31 h Ph	8640 h Ae	11,344 h Ae
2,3,7,8-TCDD (Dioxiin)	14,160 h	14,160 h Ae	10,032 h Ae	28,320 h Ae	20,064 h Ae
4,6-Dinitro-o-cresol	504 h	504 h	77 h Ph	1008 h Ae	68 h An
1,2 Dichloroethylene	4320 h Ae	4320 h Ae	672 h Ae	69,000 h An	1344 h Ae
E					
Ethyl methane sulfonate	77 h Hy	77 h Hy	46 h Hy	77 h Hy	46 h Hy
Ethanol	24 h	26 h Ae	6.5 h Ae	52 h Ae	13 h Ae
Ethylene	672 h	672 h Bio	24 h Bio	1344 h Bio	48 h Bio
Ethylene oxide	285 h Hy	285 h Hy	251 h Hy	285 h Hy	251 h Hy
Ethyl carbethoxymethyl phthalate	672 h Ae	672 h Ae	168 h Ae	8640 h Ae	336 h Ae
Ethylenethiourea	168 h Ae	672 h Ae	168 h Ae	1394 h Ae	336 h Ae
Ethylbenzene	240 h Ae	240 h Ae	72 h Ae	5472 h	144 h
Epichlorohydrin	672 h Ae	672 h Ae	168 h Ae	1344 h Ae	336 h Ae
Ethylene dibromide	4320 h Ae	4320 h Ae	672 h Ae	2880 h Ae	470 h An
Ethylene glycol	288 h Ae	288 h Ae	48 h Ae	576 h Ae	96 h Ae
2-Ethoxyethanol	672 h Ae	672 h Ae	168 h Ae	1344 h Ae	336 h Ae

Compound						
Ethylene glycol, monobutyl ether	672 h Ae	168 h Aae	672 h Ae	168 h Ae	1344 h Ae	336 h Ae
Endosulgan	218 h Hy	4.5 h Hy	218 h Hy	4.5 h Hy	218 h Hy	4.5 h Hy
Ethyl acrylate	168 h Ae	24 h Ae	168 h Ae	24 h Ae	336 h Ae	48 h Ae
Ethyl acetate	168 h Ae	24 h Ae	168 h Ae	24 h Ae	336 h Ae	48 h Ae
Ethyl chloroformate	0.89 h Hy	0.55 h Hy	0.89 h Hy	0.55 h Hy	0.89 h Hy	0.55 h Hy
Ethyl N-methyl-N-nitrosocarbamate	24 h Hy	12 h Hy	24 h Hy	12 h Hy	24 h Hy	12 h Hy
Ethyl carbamate	168 h Ae	24 h Ae	168 h Ae	24 h Ae	336 h Ae	48 h Ae
F						
Formaldehyde	168 h Ae	24 h Ae	168 h Ae	24 h Ae	336 h Ae	48 h Ae
Formic acid	24 h	2.6 h	26 h Ae	6.5 h Ae	52 h Ae	13 h Ae
Fluorene	1440 h Ae	768 h Ae	1440 h Ae	768 h Ae	2880 h Ae	1536 h Ae
Furan	672 h Ae	168 h Ae	672 h Ae	13.8 h Ph	2688 h Ae	672 h Ae
Fluoranthene	3360 h Ae	3360 h Ae	63 h Ph	21 h Ph	21,120 h Ae	6720 h Ae
Fluridone	4608 h	1056 h	864 h Ph	288 h Ph	9216 h Ae	2112 h Ae
G						
Glycidylaldehyde	672 h Ae	168 h Ae	672 h Ae	168 h Ae	8640 h Ae	336 h Ae
H						
γ-Hexachlorocyclohexane (Lindane)	5765 h Hy	330 h Hy	5765 h Hy	330 h Hy	5765 h Hy	142 h Hy
Hexachloroethane	4320 h Ae	672 h Ae	4320 h ae	672 h Ae	8640 h Ae	1344 h Ae
Hexachlorophene	782 h Ae	6000 h Ae	7872 h Ae	6000 h Ae	15,744 h Ae	12,000 h Ae
Hydrocyanic acid	4320 h ae	672 h Ae	4320 h Ae	672 h Ae	8640 h Ae	1344 h Ae
Heptachlor	129.4 h Hy	23.1 h Hy	129.4 h Hy	23.1 h Hy	129.4 h Hy	23.1 h Hy
Hexachlorocyclopentadiene	672 h Ae	168 h Ae	173 h Hy	1.0 min Ph	1344 h Ae	173 h Hy
Hexachlorobutadiene	4320 h Ae	672 h Ae	4320 h Ae	672 h Ae	8640 h Ae	1344 h Ae
Hexachlorobenzene	50,136 h Ae	23,256 h Ae	50,136 h Ae	23,256 h Ae	10,272 h Ae	46,512 h Ae
Hydrazobenzene	4320 h Ae	672 h Ae	1740 h Ph	31 h Ph	8640 h Ae	1344 h Ae
Hydroquinone	168 h Ae	24 h Ae	19.3 h Ph	0.39 h Ph	36 h Ae	48 h Ae
Hydrazine	168 h Ae	24 h Ae	168 h Ae	24 h Ae	336 h Ae	48 h Ae
α-Hexachlorocyclohexane	3240 h Ae	330 h Hy	3240 h Ae	330 h Hy	6480 h Ae	330 h Hy
β-Hexachlorocyclohexane	2976 h Ae	330 h Hy	29,976 h Ae	330 h Hy	5952 h Ae	330 h Hy
δ-Hexachlorocyclohexane	2400 h Ae	330 h Hy	2400 h Ae	330 h Hy	4800 h Ae	330 h Hy
Heptachlor epoxide	13,248 h Ae	792 h Ae	13,248 h Ae	792 h Ae	26,496 h	24 h
Hexachloronaphthalene	8760 h Ae	4320 h Ae	8760 h Ae	4320 h Ae	17,520 h Ae	8640 h Ae

Compound						
I						
Isopropanol	168 h Ae	168 h Ae	168 h Ae	24 h Ae	336 h Ae	48 h Ae
Isophorone	672 h Ae	672 h Ae	672 h Ae	168 h Ae	1344 h Ae	336 h Ae
Isoprene	672 h Ae	672 h Ae	672 h Ae	18 h Ae	1344 h Ae	336 h Ae
Isobutyl alcohol	173 h Ae	173 h Ae	173 h Ae	43 h Ae	346 h Ae	86 h Ae
Isobutylraldehyde	168 h Ae	168 h Ae	168 h Ae	24 h Ae	336 h Ae	48 h Ae
4,4'-Isopropylidenediphenol	4320 h Ae	24 h Ae	3833 h Ph	24 h Aae	8640 h Ae	48 h Ae
Isosafrole	672 h Ae	672 h Ae	672 h Ae	168 h Ae	1344 h Ae	336 h Ae
Indeno(1,2,3-*cd*)pyrene	17,520 h Ae	14,400 h Ae	6000 h Ph	3000 h Ph	35,040 h Ae	28,800 h Ae
Isopropalin	2520 h Ae	408 h Ae	864 h Ph	288 h Ph	5040 h Ae	96 h An
K						
Kepone	17,280 h Ae	7488 h Ae	17,280 h Ae	7488 h Ae	34,560 h Ae	14,976 h Ae
L						
Lasiocarpine	672 h Ae	168 h Ae	672 h Ae	168 h Ae	144 h Ae	336 h Ae
Linuron	4272 h	672 h	4272 h Ae	672 h Ae	8544 h Ae	1344 h Ae
M						
Malathion	168 h Ae	72 h Ae	1236 h Ae	100 h Ae	2472 h Ae	200 h Ae
Mecoprop	2400 h Ae	168 h Ae	240 h Ae	168 h Ae	4320 h	336 h
Melamime	4320 h Ae	672 h Ae	672 h Ae	24 h Ae	8640 h Ae	1344 h Ae
Methanol	168 h Ae	24 h Ae	168 h Ae	24 h Ae	168 h Ae	24 h Ae
Methoxychlor	8760 h Ae	4320 h Ae	5.4 h Ph	2.2 h Ph	8760 h Ae	1200 h An
2-Methoxyethanol	672 h Ae	168 h Ae	672 h Ae	168 h Ae	1344 h Ae	336 h Ae
Methyl acrylate	168 h Ae	24 h Ae	168 h Ae	24 h Ae	336 h Ae	48 h Hy
2-Methylaziridine	870 h Hy	87 h Hy	870 h Hy	87 h Hy	912 h Hy	87 h Hy
Methyl bromide	672 h Ae	168 h Ae	672 h Ae	168 h Ae	870 h Hy	336 h Ae
Methyl *t*-butyl ether	4320 h Ae	672 h Ae	4320 h Ae	672 h Ae	8640 h Ae	1344 h Ae
Methyl chloride	672 h Ae	168 h Ae	672 h Ae	168 h Ae	1344 h Ae	336 h Ae
2-Methyl-4-chlorophenoxyacetic acid	168 h	96 h	168 h Ae	96 h Ae	4320 h	192 h
3-Methylcholanthrene	33,600 h Mi	14,616 h Mi	33,600 h Mi	14,616 h Mi	67,200 h Mi	29,232 h Ae
4,4'-Methylenebis (2-chloroaniline)	4320 h Ae	672 h Ae	1740 h Ph		8640 h Ae	1344 h Ae
Methylenebis (Phenylisocyanate)			31 h Ph			
Methylene bromide	672 h Ae	168 h Ae	672 h Ae	168 h Ae	1344 h Ae	336 h Ae

Chemical						
4,4'-Methylenedianiline	168 h Ae	24 h Ae	168 h Ae	24 h Ae	336 h Ae	48 h Ae
Methyl ethyl ketone	168 h Ae	24 h Ae	168 h Ae	24 h Ae	336 h Ae	48 h Ae
Methyl ethyl ketone perioxide	672 h Ae	168 h Ae	672 h Ae	168 h Ae	1344 h Ae	336 h Ae
Methylhydrazine	576 h Ae	312 h Ae	576 h Ae	312 h Ae	1152 h Ae	624 h Ae
Methyl iodide	672 h Ae	168 h Ae	672 h Ae	168 h Ae	1344 h Ae	336 h Ae
Methyl isobutyl ketone	168 h Ae	24 h Ae	168 h Ae	24 h Ae	336 h Ae	48 h Ae
Methylisocyanate	0.326 h Hy	0.144 h Hy	0.326 h Hy	0.144 h Hy	0.326 h Hy	0.144 h Hy
Methyl methacrylate	672 h Ae	168 h Ae	672 h Ae	168 h Ae	1344 h Ae	336 h Ae
N-Methyl-N-nitrosourea	3.5 h Hy	0.013 h Hy	3.5 h Hy	0.013 h Hy	3.5 h Hy	0.013 h Hy
Methyl parathion	8640 h Ae	240 h Ae	912 h Ph	192 h Ph	1680 h	24 h
Methylthiouracil	672 h Ae	168 h Ae	672 h Ae	168 h Ae	8640 h Ae	336 h Ae
Michler's ketone	672 h Ae	168 h Ae	672 h	31.2 h Ph	1344 h Ae	336 h Ae
Mitomycin C	672 h Ae	168 h Ae	672 h Ae	168 h Ae	8640 h Ae	336 h Ae
Mustard gas	0.26 h Hy	0.065 h Hy	0.26 h Hy	0.065 h Hy	0.26 h Hy	0.065 h Hy

N

Chemical						
Naphthalene	1152 h Ae	398	480 h Ae	12 h Ae	6192 h An	24 h Ae
α-Naphthylamine	4320 h Ae	672 h Ae	3480 h	62.4 h	8640 hAe	1344 h Ae
β-Naphthylamine	4320 h Ae	672 h Ae	3480 h Ph	62 h Ph	8640 h Ae	1344 h Ae
Nitrilotriacetic acid	672 h Ae	72 h Ae	672 h Ae	168 h Ae	1344 h Ae	336 h Ae
5-Nitro-o-anisidine	672 h Ae	24 h Ae	672 h Ae	24 h Ae	1344 h Ae	48 h An
Nitrobenzene	150 sec Hy	17 sec Hy	150 sec Hy	17 sec Hy	150 sec Hy	17 sec Hy
4-Nitrobiphenyl	672 h Ae	24 h Ae	672 h Ae	24 h Ae	1344 h Ae	48 h An
Nitrogen mustard	24 h Hy	30 min Hy	24 h Hy	30 min Hy	24 h Hy	30 min Hy
Nitroglycerin	168 h Ae	48 h Ae	168 h Ae	48 h Ae	336 h Ae	96 h Ae
2-Nitrophenol	672 h	168 h	672 h Ae	168 h Ae	672 h	336 h
4-Nitrophenol	29 h Ae	17 h Ae	168 h Ae	18.2 h Ae	235 h	36.4 h
2-Nitropropane	4320 h Ae	672 h Ae	4320 h Ae	672 h Ae	8640 h Ae	1344 h Ae
N-Nitrosodiethanolamine	4320 h Ae	120 h Ae	4320 h Ae	120 h Ae	8640 h Ae	240 h Ae
N-Nitrosodiethylamine	4320 h Ae	480 h Ae	8 h Ph	4 h Ph	8640 h Ae	960 h Ae
N-Nitrosodimethylamine	4320 h Ae	504 h Ae	1 h Ph	0.5 h Ph	8640 h Ae	1008 h Ae
N-Nitrosodiphenylamine	816 h Ae	240 h Ae	816 h Ae	240 h Ae	1632 h Ae	480 h Ae
P-Nitrosodiphenylamine	4320 h Ae	672 h Ae	3480 h	62.4 h	8640 h Ae	1344 h Ae
N-Nitrosodipropylamine	4320 h Ae	504 h Ae	1 h Ph	0.17 h Ph	8640 h Ae	1008 h Ae
N-Nitroso-N-ethylurea	3.5 h Hy	0.013 h Hy	3.5 h Hy	0.013 h Hy	3.5 h Hy	0.013 h Hy
N-Nitroso-N-methyl-N-nitroguanidine	26 h Hy	0.71 h Hy	26 h Hy	0.71 h Hy	26 h Hy	0.71 h Hy
N-Nitrosomethylvinylamine	4320 h Ae	672 h Ae	4320 h Ae	672 h Ae	8640 h Ae	1344 h Ae

N-Nitrosomorpholine	4320 h Ae	672 h Ae	170 h Ph	8640 h Ae	1344 h Ae
N-Nitrosopiperidine	4320 h Ae	672 h Ae	225 h Ph	8640 h Ae	1344 h Ae
N-Nitrosopyrrolidine	4320 h Ae	672 h Ae	4320 h Ae	8640 h Ae	1344 h Ae
5-Nitro-o-toluidine	672 h Ae	24 h Ae	24 h Ae	1344 h Ae	48 h An
O					
Octachloronaphthalene	88,760 h	4320 h	8760 h	17,520 h	8460 h
P					
Pentachlorobenzene	8280 h Ae	4656 h Ae	8280 h Ae	16,560 h Ae	9312 h Ae
Pentachloronitrobenzene	16,776 h Ae	5112 h Ae	16,776 h Ae	16,776 h	216 h
Pentachlorophenol	4272 h Ae	552 h Ae	110 h Ph	36,480 h An	1104 h Ae
Peracetic acid	48 h	4 h	4 h Ph	336 h Ae	48 h
Phenacetin	672 h Ae	168 h Ae	168 h Ae	1344 hAe	336 h Ae
Phenanthrene	4800 h Ae	384 h Ae	672 h Ae	9600 h Ae	768 h Ae
Phenobarbitol	672 hAe	168 h Ae	25 h Ph	8640 h Ae	336 h Ae
Phenol	240 h Ae	24 h Ae	672 h Ae	168 h Ae	12 h Ae
Phenylalanine mustard	24 h Hy	4.62 h Hy	56.5 h	24 h Hy	4.62 h Hy
P-Phenylenediamine	672 h Ae	168 h Ae	24 h Hy	1344 h Ae	336 h Ae
2-Phenylphenol	168 h Ae	24 h Ae	672 h Ae	336 h Ae	48 h Ae
Phosgene	1 h Hy	0.05 h Hy	168 h Ae	1 h Hy	0.05 h Hy
Phthalic anhydride	27 min Hy	32 sec Hy	1 h Hy	27 min Hy	32 sec Hy
Picric acid	4320 h Ae	672 h Ae	27 min Hy	8640 h Ae	48 h An
Propane sultone	672 h Ae	8.5 h Hy	4320 h Ae	1344 h Ae	8.5 h Hy
β-Propiolactone	3.4 h Hy	0.058 h Hy	672 h Ae	3.4 h Hy	0.058 h Hy
Propionaldehyde	168 h Ae	24 h Ae	3.4 h Hy	336 h Ae	48 h Ae
Propoxur	672 h Ae	38 h Hy	168 h Ae	1344 h Ae	38 h Hy
Propylene	672 h Ae	168 h Ae	672 h Ae	1344 h Ae	336 h Ae
Propylene glycol, monoethyl ether	672 h Ae	168 h Ae	672 h Ae	1344 h Ae	336 h Ae
Propylene glycol, monomethyl ether	672 h Ae	168 h Ae	672 h Ae	1344 h Ae	336 h Ae
Pyrene	45,600 h Ae	5040 h Ae	2.04 h Ph	91,200 h Ae	10,080 h Ae
Pyridine	168 h Ae	24 h Ae	168 h Ae	336 h Ae	48 h Ae
Q					
Quinoline	240 h Ae	72 h Ae	240 h	480 h Ae	144 h Ae

S

Saccharin	168 h Ae	672 h Ae	168 h Ae	1344 h Ae	336 h Ae
Safrole	168 hAe	672 h Ae	168 h Ae	1344 h Ae	336 h Ae
Streptozotocin	168 h Ae	672 h Ae	168 h Ae	1344 h Ae	336 h Ae
Strychnine	168 h Ae	672 h Ae	168 h Ae	1344 h Ae	336 h Ae
Styrene	336 h Ae	672 h Ae	336 h Ae	5040 h Ae	672 h Ae
Styrene oxide	0.00385 h Hy	27.5 h Hy	0.00385 h Hy	27.5 h Hy	0.00385 h Hy

T

2,3,7,8-TCDD (Dioxin)	168 h Ae	672 h Ae	168 h Ae	88,640 h Ae	336 h Ae
Terephthalic acid	24 h Ae	168 h Ae	24 h Ae	336 h Ae	48 h Ae
1,2,4,5-Tetrachlorobenzene	672 h Ae	4320 h Ae	672 h Ae	8640 h Ae	1344 h Ae
1,1,1,2-Tetrachloroethane	16 h Hy	1640 h Hy	16 h Hy	1604 h Hy	16 h Hy
1,1,2,2-Tetrachloroethane	10.7 h Hy	1056 h Hy	10.7 h Hy	1056 h Hy	10.7 h Hy
Tetrachloroethylene	4320 h Ae	8640 h Ae	4320 h Ae	17,280 h Ae	8640 h Ae
2,3,4,6-Tetrachlorophenol	672 h Ae	336 h Ph	1 h Ph	8640 h Ae	1344 h Ae
Tetraethyl lead	168 h Ae	9.0 h Ph	2.3 h Ph	1344 h Ae	336 h Ae
Thioacetamide	24 h Ae	168 h Ae	24 h Ae	336 h Ae	48 h Ae
Thiodianiline	168 h Ae	672 h Ae	31.2 h Ph	1344 h Ae	336 h Ae
Thiourea	24 h Ae	168 h Ae	24 h Ae	336 h Ae	48 h Ae
Toluene	96 h Ae	528 h Ae	96 h Ae	672 h Ae	168 h Ae
Toluene-2,6-diisocyanate	12 h Hy	24 h Hy	12 h Hy	24 h Hy	12 h Hy
O-Toluidine	24 h Ae	68 h Ae	24 h Ae	336 h Ae	48 h Ae
Triaziguone	168 h Hy	744 h Hy	168 h Hy	744 h Hy	336 h Hy
1,2,4-Trichlorobenzene	672 h Ae	4320 h Ae	672 h Ae	8640 h Ae	1344 h Ae
1,1,1-Trichloroethane	3360 h Hy	6552 h Ae	3360 h Ae	13,104 h Ae	3360 h
1,1,2-Trichloroethane	3263 h Ae	8760 h Ae	3263 h Hy	17,520 h Ae	3263 h Hy
Trichloroethylene	4320 h Ae	8460 h Ae	4320 h Ae	39,672 h An	7704 h Hy
Trichlorofluoromethane	4320 h Ae	8640 h Ae	4032 h Ae	17,280 h Ae	8640 h Ae
Trichlorofon	24 h Ae	588 h Hy	22 h Hy	588 h Hy	22 h Hy
2,4,5-Trichlorophenol	552 h Ae	336 h Ph	0.5 h Ph	43,690 h An	1104 h Ae
2,4,6-Trichlorophenol	168 h Ae	96 h Ph	2 h Ph	43,690 h An	336 h Ae
2,4,5-Trichlorophenoxyacetic acid	240 h	480 h Ae	240 h Ae	4320 h	480 h
1,2,3-Trichloropropane	4320 h Ae	8640 h Ae	4320 h Ae	17,280 h Ae	8640 h Ae
1,1,2-Trichloro-1,2,2-trifluoroethane	4320 h Ae	8640 h Ae	4320 h Ae	17,280 h Ae	1440 h Ae

1,2,4-Trimethylbenzene	672 h Ae	168 h Ae	672 h Ae	168 h Ae	1344 h Ae	336 h Ae
2,4,6-Trinitrotoluene	4320 h Ae	672 h Ae	1.28 h Ph	0.160 h Ph	8640 h Ae	672 h An
Tris(2,3-dibromopropyl)phosphate	168 h Ae	24 h Ae	168 h Ae	24 h Ae	336 h Ae	48 h Ae
U						
Uracil mustard	24 h Hy	30 min Hy	24 h Hy	30 min Hy	24 h Hy	30 min Hy
V						
Vinyl chloride	4320 h Ae	672 h Ae	4320 h Ae	672 h Ae	69,000 h	1344 h
W						
Warfarin	672 h Ae	168 h Ae	672 h Ae	168 h Ae	1344 h Ae	336 h Ae
X						
M-Xylene	672 h Ae	168 h Ae	672 h Ae	168 h Ae	8640 h Ae	336 h Ae
O-Xylene	672 h Ae	168 h Ae	672 h Ae	168 h Ae	8640 h Ae	336 h Ae
P-Xylene	672 h Ae	168 h Ae	672 h Ae	168 h Ae	8640 h Ae	336 h Ae
Xylene	672 h Ae	168 h Ae	672 h Ae	168 h Ae	8640 h Ae	336 h Ae
2-Xylidine	7584 h	72 h	3480 h Ph	62.4 h Ph	8640 h Ae	1344 h Ae

Note: Ae = Aerobic half life; An = Anaerobic half life; Bio = ; Hy = Hydrolysis; Mi = Mineralization; Ph = Photolysis; Phoxy = Photooxidation.

Source: Howard, P. H., Boethling, R. S., Jarvis, W. F., Meylan, W. M., and Michalenko, E. M., *Handbook of Environmental Degradation Rates*, Lewis Publishers, Chelsea, MI, 1991. With permission.

INDEX

A

Abiotic processes, 3
  adsorption, 17–19
  chlorinated aliphatic hydrocarbons, 133
  hydrolysis, 21–22
  ion exchange, 20–21
  precipitation and dissolution, 23
  reactive walls
    chemical oxidation, 205
    dechlorination, 189, 199–202
  redox reactions, 22–23
Absorption, defined, 37, see also Adsorption
Acetone, 7, 94, 119, 189, 191
Acid-base conditions, see also pH
  and hydrolysis, 21
  and ionization, 20–21
Acidophiles, microbial, 142
Activated carbon, see Carbon absorption
Addition reactions, 6
Additives, 303–304
Adsorbed phase, 4
Adsorption, 23
  carbon systems, see Carbon absorption
  contaminant partitioning, 33
    interactions with subsurface materials, 17–19
    models of, 25
    phases of contaminants in soil matrix, 34
    soil moisture and, 39–43
    vapor-phase partitioning characteristics, 37, 39–40
  ion exchange and, 20
  pump and treat systems, 270, 271, 280–285, 293–294
    carbon adsorption, 270, 271, 280–284, 285
    ion exchange, 270, 271, 293–294
    metal precipitation, 270, 271, 294–296
  reactive walls, 189, 202–204
  reactive zones *in situ*, 225
  vapor treatment, 78, 81–84
Advection, 13, 16–17, 43
Advective-diffusive airflow, 43
Aerobic metabolism, 94, 125
  aliphatic hydrocarbon degradation, 127
  bio-buffering concept, 151
  bioventing systems, airflow rate and, 72

chlorinated aliphatic hydrocarbons, 134–135
  ozone and, 223
  reactive walls, 189
After-burners, 79
Agrobacterium tumefaciens, 262
Air, soil, see Soil gas; Soil properties
AIR 3D, 56
Air channels concept, air sparging, 96
Air compressor, air sparging equipment, 111–112
Air discharge permit, 298
Air distribution (zone of influence), air sparging, 101–104
Airflow characteristics
  air sparging, 96, 97, 99
  bioventing systems, 72–73
  models of, 56–71
  soil vapor extraction, 28–33
    mathematical evaluation, 28–30
    soil-air permeability, 30–33
Air injection test, 31
Air injection into water-saturated soils, 96–97
Air inlet wells, 42–44
Air sparging, *in situ*, 91–120, 117
  applicability of, 94–96
    examples, 94–95
    geologic considerations, 95–96
  cleanup rates, 116–118
  definition of, 91–92
  design parameters, 100–107
    air distribution (zone of influence), 101–104
    contaminant type and distribution, 106
    depth of air injection, 104
    pressure and flow rate of air injection, 105
    pulsing or continuous injection modes, 106
    well construction, 106–107
  equipment, 110–113
  fracture technique applicability with, 238
  knowledge gap, 119
  limitations, 118
  literature summary, 119–120
  mechanisms
    air stripping, 93
    biodegradation, 94
    direct volatilization, 93–94